Advance in Molecularly Imprinted Polymers

Advance in Molecularly Imprinted Polymers

Editor

Michał Cegłowski

Basel • Beijing • Wuhan • Barcelona • Belgrade • Novi Sad • Cluj • Manchester

Editor
Michał Cegłowski
Faculty of Chemistry
Adam Mickiewicz University
Poznań
Poland

Editorial Office
MDPI
St. Alban-Anlage 66
4052 Basel, Switzerland

This is a reprint of articles from the Special Issue published online in the open access journal *Polymers* (ISSN 2073-4360) (available at: www.mdpi.com/journal/polymers/special_issues/frontier_molecularly_imprinted_polymers).

For citation purposes, cite each article independently as indicated on the article page online and as indicated below:

Lastname, A.A.; Lastname, B.B. Article Title. *Journal Name* **Year**, *Volume Number*, Page Range.

ISBN 978-3-0365-8597-0 (Hbk)
ISBN 978-3-0365-8596-3 (PDF)
doi.org/10.3390/books978-3-0365-8596-3

© 2023 by the authors. Articles in this book are Open Access and distributed under the Creative Commons Attribution (CC BY) license. The book as a whole is distributed by MDPI under the terms and conditions of the Creative Commons Attribution-NonCommercial-NoDerivs (CC BY-NC-ND) license.

Contents

About the Editor vii

Michał Cegłowski
Editorial: Advance in Molecularly Imprinted Polymers
Reprinted from: *Polymers* 2023, 15, 3199, doi:10.3390/polym15153199 1

Egla Yareth Bivián-Castro, Abraham Zepeda-Navarro, Jorge Luis Guzmán-Mar, Marcos Flores-Alamo and Brenda Mata-Ortega
Ion-Imprinted Polymer Structurally Preorganized Using a Phenanthroline-Divinylbenzoate Complex with the Cu(II) Ion as Template and Some Adsorption Results
Reprinted from: *Polymers* 2023, 15, 1186, doi:10.3390/polym15051186 4

Ruslan Kondaurov, Yevgeniy Melnikov and Laura Agibayeva
Application of Molecular Imprinting for Creation of Highly Selective Sorbents for Extraction and Separation of Rare-Earth Elements
Reprinted from: *Polymers* 2023, 15, 846, doi:10.3390/polym15040846 19

Anamaria Zaharia, Ana-Mihaela Gavrila, Iuliana Caras, Bogdan Trica, Anita-Laura Chiriac and Catalina Ioana Gifu et al.
Molecularly Imprinted Ligand-Free Nanogels for Recognizing Bee Venom-Originated Phospholipase A2 Enzyme
Reprinted from: *Polymers* 2022, 14, 4200, doi:10.3390/polym14194200 40

Sarah H. Megahed, Mohammad Abdel-Halim, Amr Hefnawy, Heba Handoussa, Boris Mizaikoff and Nesrine A. El Gohary
Molecularly Imprinted Solid Phase Extraction Strategy for Quinic Acid
Reprinted from: *Polymers* 2022, 14, 3339, doi:10.3390/polym14163339 55

Monika Sobiech, Dorota Maciejewska and Piotr Luliński
N-(2-Arylethyl)-2-methylprop-2-enamides as Versatile Reagents for Synthesis of Molecularly Imprinted Polymers
Reprinted from: *Polymers* 2022, 14, 2738, doi:10.3390/polym14132738 79

Stanislav S. Piletsky, Alvaro Garcia Cruz, Elena Piletska, Sergey A. Piletsky, Eric O. Aboagye and Alan C. Spivey
Iodo Silanes as Superior Substrates for the Solid Phase Synthesis of Molecularly Imprinted Polymer Nanoparticles
Reprinted from: *Polymers* 2022, 14, 1595, doi:10.3390/polym14081595 91

Michał Cegłowski, Joanna Kurczewska, Aleksandra Lusina, Tomasz Nazim and Piotr Ruszkowski
EGDMA- and TRIM-Based Microparticles Imprinted with 5-Fluorouracil for Prolonged Drug Delivery
Reprinted from: *Polymers* 2022, 14, 1027, doi:10.3390/polym14051027 99

Talkybek Jumadilov, Ruslan Kondaurov and Aldan Imangazy
Application of the Remote Interaction Effect and Molecular Imprinting in Sorption of Target Ions of Rare Earth Metals
Reprinted from: *Polymers* 2022, 14, 321, doi:10.3390/polym14020321 113

Hsiu-Wen Chien, Chien-Hsin Yang, Yan-Tai Shih and Tzong-Liu Wang
Upconversion Nanoparticles Encapsulated with Molecularly Imprinted Amphiphilic Copolymer as a Fluorescent Probe for Specific Biorecognition
Reprinted from: *Polymers* **2021**, *13*, 3522, doi:10.3390/polym13203522 . . . . **134**

Ut Dong Thach, Hong Hanh Nguyen Thi, Tuan Dung Pham, Hong Dao Mai and Tran-Thi Nhu-Trang
Synergetic Effect of Dual Functional Monomers in Molecularly Imprinted Polymer Preparation for Selective Solid Phase Extraction of Ciprofloxacin
Reprinted from: *Polymers* **2021**, *13*, 2788, doi:10.3390/polym13162788 . . . . **145**

Aleksandra Lusina and Michał Cegłowski
Molecularly Imprinted Polymers as State-of-the-Art Drug Carriers in Hydrogel Transdermal Drug Delivery Applications
Reprinted from: *Polymers* **2022**, *14*, 640, doi:10.3390/polym14030640 . . . . **159**

Zhimin Liu, Zhigang Xu, Dan Wang, Yuming Yang, Yunli Duan and Liping Ma et al.
A Review on Molecularly Imprinted Polymers Preparation by Computational Simulation-Aided Methods
Reprinted from: *Polymers* **2021**, *13*, 2657, doi:10.3390/polym13162657 . . . . **180**

About the Editor

Michał Cegłowski

Michał Cegłowski graduated from the Faculty of Chemistry at the Adam Mickiewicz University, Poland, in 2011 and obtained his Ph.D. in 2015 from the same university. His PostDoc was conducted at Ghent University in the laboratory directed by Prof. Richard Hoogenboom. He is currently an Associate Professor of Chemistry in the Faculty of Chemistry at the Adam Mickiewicz University, Poland. His research group focuses on preparing functional and selective adsorbents that can be applied in solid-phase extraction. Much of the emphasis is focused on the preparation of adsorbents that, in combination with other analytical techniques, can be used to quantify persistent organic pollutants. Materials of interest range from molecularly imprinted polymers, which can be programmed to recognize particular individuals, to hybrid materials, which connect the benefits of organic and inorganic materials.

Editorial

Editorial: Advance in Molecularly Imprinted Polymers

Michał Cegłowski

Faculty of Chemistry, Adam Mickiewicz University, Uniwersytetu Poznańskiego 8, 61-614 Poznan, Poland; michal.ceglowski@amu.edu.pl

Molecularly imprinted polymers (MIPs), due to their unique recognition properties, have found various applications, mainly in extraction and separation techniques; however, their implementation in other research areas, such as sensor construction and drug delivery, has also been substantial. These advances could not be achieved without developing new polymers and monomers that can be successfully applied in MIP synthesis and improve the range of their applications. Although much progress has been made in MIP development, more investigation must be conducted to obtain materials that can fully deserve the name of artificial antibodies.

The "Advance in Molecularly Imprinted Polymers" Special Issue connects original research papers and reviews presenting recent advances in the design, synthesis, and broad applications of molecularly imprinted polymers. The Special Issue content covers various topics related to MIP chemistry, which include their molecular modeling, application of new monomers and synthesis techniques for MIP preparation, synthesis, and application of ion-imprinted materials, selective extraction of organic molecules, sensor preparation, and MIP applications in medicine. The presented collection of scientific papers shows that research in MIP chemistry is diversified, ultimately improving the desired properties and creating new potential applications for these materials.

Bivián-Castro et al. [1] obtained a Cu(II)-imprinted polymer using a phenanthroline–divinylbenzoate complex as a functional monomer and ethylene glycol dimethacrylate as a cross-linker. The Cu(II)-loaded MIP appeared to be a mesoporous material with a pore diameter of around 4 nm. The Cu(II)-unloaded ion-imprinted polymer was considered a microporous material, as the pore diameters were less than 2 nm. The maximum adsorption capacity of the ion imprinted polymer was 287.45 mg g^{-1}. The findings of this study indicate the potential of the obtained ion-imprinted polymers for the selective extraction of heavy metal ions from polluted waters.

Ion-imprinted polymers were studied by Kondaurov et al. [2]. They obtained molecularly imprinted polymers based on methacrylic acid and 4-vinylpyridine that were selective towards samarium and gadolinium ions. The obtained data showed that the sorption degree of samarium and gadolinium ions by the obtained MIPs reached almost 90%. The additional benefit of the MIPs was the fact that simultaneous sorption of accompanying metal ions was not observed. The results suggested that these MIPs could be highly effective alternatives to the existing sorption technologies.

Zaharia et al. [3] obtained ligand-free nanogels that were molecularly imprinted with bee venom-derived phospholipase A2 (PLA2) enzyme. This enzyme acts synergistically with the polyvalent cations in the venom, creating an increased hemolytic effect and quick access of toxins into the bloodstream. The research aim was to obtain nanogel materials capable of binding PLA2, which would could remove PLA2 proteins from the bite zone, reducing the venom's toxicity. The rebinding experiments proved the specificity of the molecularly imprinted ligand-free nanogels for PLA2, with rebinding capacities up to 8-fold higher than the reference non-imprinted nanogel. The in vitro cell viability experiments showed the potential of the obtained ligand-free nanogels and suggested that they can be used for future therapies against bee envenomation.

Citation: Cegłowski, M. Editorial: Advance in Molecularly Imprinted Polymers. *Polymers* **2023**, *15*, 3199. https://doi.org/10.3390/polym15153199

Received: 12 July 2023
Accepted: 25 July 2023
Published: 27 July 2023

Copyright: © 2023 by the author. Licensee MDPI, Basel, Switzerland. This article is an open access article distributed under the terms and conditions of the Creative Commons Attribution (CC BY) license (https://creativecommons.org/licenses/by/4.0/).

Megahed et al. [4] obtained MIPs that can be used for the selective extraction of quinic acid from coffee bean extract. The authors used computational modeling to optimize the process of MIP preparation. It was used to calculate the optimum ratio for allylamine, methacrylic acid, and 4-vinylpyridine, which were used as functional monomers. The application of these MIP materials to extract quinic acid from aqueous coffee extract showed a high recovery, reaching 82%. The developed procedure is promising for the selective extraction of quinic acid from different complex herbal extracts and may be scaled to industrial applications.

Six aromatic N-(2-arylethyl)-2-methylprop-2-enamides with various substituents in the benzene ring were obtained by Sobiech et al. [5] using 2-arylethylamines and methacryloyl chloride as the primary reagents. The obtained compounds were used for covalent imprinting in the synthesis of molecularly imprinted polymers, followed by the hydrolysis of an amide linkage. The adsorption studies proved a high affinity towards tyramine and L-norepinephrine, which were used as target biomolecules. This method proves the applicability of covalent imprinting with the use of tailor-made monomers.

Piletsky et al. [6] presented the application of iodo silanes and compared it to commonly used amino silanes for the solid-phase synthesis of MIP nanoparticles. The silanes were used for peptide immobilization, and the obtained MIPs were specific toward an epitope of the epidermal growth factor receptor. The results showed that both silanes produced MIPs with excellent affinity; nevertheless, for the iodo silanes, fewer experimental steps were needed and the procedure required less expensive reagents.

The synthesis of 5-fluorouracil-imprinted microparticles and their application in prolonged drug delivery was reported by Cegłowski et al. [7]. The authors prepared MIPs using two different cross-linkers: ethylene glycol dimethacrylate and trimethylolpropane trimethacrylate. The calculated highest cumulative release was highly dependent on the cross-linker applied during the synthesis. The overall cumulative release was much higher for trimethylolpropane trimethacrylate-based MIPs. The highest cumulative release was obtained at pH 7.4 and the lowest at pH 2.2. As a result, it was demonstrated that the selection of the cross-linker should be considered during the design of materials used for drug delivery.

Jumadilov et al. [8] studied the effectiveness of the application of intergel systems and MIPs for the selective sorption and separation of neodymium and scandium ions. The intergel system method was proven cheaper and easier in this application; however, some accompanying sorption of another metal from the model solution was observed. On the other hand, the method based on MIPs was more expensive but showed higher sorption properties. The obtained results can be successfully applied to upgrade the existing sorption technologies.

Chien et al. [9] developed a fluorescent probe for specific biorecognition by a facile method in which amphiphilic random copolymers were encapsulated with hydrophobic upconversion nanoparticles. The self-folding ability of the amphiphilic copolymers allowed the formation of MIPs with template-shaped cavities that were selective towards albumin and hemoglobin. The results showed that the fluorescence was quenched when hemoglobin was adsorbed on the fluorescent probes. This effect was not observed for albumin. These fluorescent probes have the potential to be applied for specific biorecognition.

MIPs possessing dual functional monomers (methacrylic acid and 2-vinylpyridine) were synthesized by Thach et al. [10] to yield materials that can be used for selective solid phase extraction of ciprofloxacin. The batch adsorption experiments demonstrated that the obtained MIPs showed a high adsorption capacity and selectivity toward ciprofloxacin. High recovery values were obtained when aqueous solutions were used. The described analytical procedure allows the designing of new adsorbents with high adsorption capacity and good extraction performance for highly polar template molecules.

In their review article, Lusina and Cegłowski [11] explored the mechanisms that can be used for controlled drug release from MIP hydrogels for transdermal drug delivery.

They discussed applications of thermo-responsive, pH-responsive, and dual/multiple-responsive MIP hydrogels.

The other review article, prepared by Liu et al. [12], describes computational simulation modeling methods and the theoretical optimization methods of various molecular simulation calculation software for MIP preparation. The review summarizes the progress in research on and application of MIPs prepared by computational simulations and computational software in the past two decades.

This Special Issue has brought together experts that have studied and explored various aspects of MIPs. I want to thank all researchers who have contributed to the production of this Special Issue of *Polymers*. In addition, I would like to express my gratitude to the Editorial Team who helped prepare the "Advance in Molecularly Imprinted Polymers" Special Issue.

Funding: This work was supported by the National Science Centre, Poland, under Grant Number 2020/37/B/ST5/01938.

Conflicts of Interest: The author declares no conflict of interest.

References

1. Bivián-Castro, E.Y.; Zepeda-Navarro, A.; Guzmán-Mar, J.L.; Flores-Alamo, M.; Mata-Ortega, B. Ion-imprinted polymer structurally preorganized using a phenanthroline-divinylbenzoate complex with the cu(ii) ion as template and some adsorption results. *Polymers* **2023**, *15*, 1186. [CrossRef] [PubMed]
2. Kondaurov, R.; Melnikov, Y.; Agibayeva, L. Application of molecular imprinting for creation of highly selective sorbents for extraction and separation of rare-earth elements. *Polymers* **2023**, *15*, 846. [CrossRef] [PubMed]
3. Zaharia, A.; Gavrila, A.-M.; Caras, I.; Trica, B.; Chiriac, A.-L.; Gifu, C.I.; Neblea, I.E.; Stoica, E.-B.; Dolana, S.V.; Iordache, T.-V. Molecularly imprinted ligand-free nanogels for recognizing bee venom-originated phospholipase a2 enzyme. *Polymers* **2022**, *14*, 4200. [CrossRef] [PubMed]
4. Megahed, S.H.; Abdel-Halim, M.; Hefnawy, A.; Handoussa, H.; Mizaikoff, B.; El Gohary, N.A. Molecularly imprinted solid phase extraction strategy for quinic acid. *Polymers* **2022**, *14*, 3339. [CrossRef] [PubMed]
5. Sobiech, M.; Maciejewska, D.; Luliński, P. N-(2-arylethyl)-2-methylprop-2-enamides as versatile reagents for synthesis of molecularly imprinted polymers. *Polymers* **2022**, *14*, 2738. [CrossRef] [PubMed]
6. Piletsky, S.S.; Garcia Cruz, A.; Piletska, E.; Piletsky, S.A.; Aboagye, E.O.; Spivey, A.C. Iodo silanes as superior substrates for the solid phase synthesis of molecularly imprinted polymer nanoparticles. *Polymers* **2022**, *14*, 1595. [CrossRef] [PubMed]
7. Cegłowski, M.; Kurczewska, J.; Lusina, A.; Nazim, T.; Ruszkowski, P. Egdma- and trim-based microparticles imprinted with 5-fluorouracil for prolonged drug delivery. *Polymers* **2022**, *14*, 1027. [CrossRef] [PubMed]
8. Jumadilov, T.; Kondaurov, R.; Imangazy, A. Application of the remote interaction effect and molecular imprinting in sorption of target ions of rare earth metals. *Polymers* **2022**, *14*, 321. [CrossRef]
9. Chien, H.-W.; Yang, C.-H.; Shih, Y.-T.; Wang, T.-L. Upconversion nanoparticles encapsulated with molecularly imprinted amphiphilic copolymer as a fluorescent probe for specific biorecognition. *Polymers* **2021**, *13*, 3522. [CrossRef]
10. Thach, U.D.; Nguyen Thi, H.H.; Pham, T.D.; Mai, H.D.; Nhu-Trang, T.-T. Synergetic effect of dual functional monomers in molecularly imprinted polymer preparation for selective solid phase extraction of ciprofloxacin. *Polymers* **2021**, *13*, 2788. [CrossRef] [PubMed]
11. Lusina, A.; Cegłowski, M. Molecularly imprinted polymers as state-of-the-art drug carriers in hydrogel transdermal drug delivery applications. *Polymers* **2022**, *14*, 640. [CrossRef] [PubMed]
12. Liu, Z.; Xu, Z.; Wang, D.; Yang, Y.; Duan, Y.; Ma, L.; Lin, T.; Liu, H. A review on molecularly imprinted polymers preparation by computational simulation-aided methods. *Polymers* **2021**, *13*, 2657. [CrossRef] [PubMed]

Disclaimer/Publisher's Note: The statements, opinions and data contained in all publications are solely those of the individual author(s) and contributor(s) and not of MDPI and/or the editor(s). MDPI and/or the editor(s) disclaim responsibility for any injury to people or property resulting from any ideas, methods, instructions or products referred to in the content.

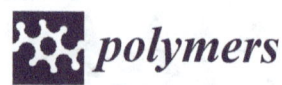

Article

Ion-Imprinted Polymer Structurally Preorganized Using a Phenanthroline-Divinylbenzoate Complex with the Cu(II) Ion as Template and Some Adsorption Results

Egla Yareth Bivián-Castro [1,*], Abraham Zepeda-Navarro [1], Jorge Luis Guzmán-Mar [2], Marcos Flores-Alamo [3] and Brenda Mata-Ortega [1]

[1] Centro Universitario de los Lagos, Universidad de Guadalajara, Av. Enrique Díaz de León 1144, Col. Paseos de la Montaña, Lagos de Moreno 47460, Jalisco, Mexico
[2] Facultad de Ciencias Químicas, Universidad Autónoma de Nuevo León (UANL), Ave. Universidad s/n, Cd Universitaria, San Nicolás de los Garza 66455, Nuevo León, Mexico
[3] Facultad de Química, Universidad Nacional Autónoma de México, Ciudad Universitaria, Ciudad de México 04510, Ciudad de México, Mexico
* Correspondence: egla.bivian@academicos.udg.mx; Tel.: +52-(474)-7424374 (ext. 66576)

Abstract: The novel [Cuphen(VBA)$_2$H$_2$O] complex (phen: phenanthroline, VBA: vinylbenzoate) was prepared and used as a functional monomer to preorganize a new ion-imprinted polymer (IIP). By leaching the Cu(II) from the molecular imprinted polymer (MIP), [Cuphen(VBA)$_2$H$_2$O-co-EGDMA]$_n$ (EGDMA: ethylene glycol dimethacrylate), the IIP was obtained. A non-ion-imprinted polymer (NIIP) was also prepared. The crystal structure of the complex and some physicochemical spectrophotometric techniques were also used for the MIP, IIP, and NIIP characterization. The results showed that the materials are nonsoluble in water and polar solvents, which are the main features of polymers. The surface area of the IIP is higher than the NIIP demonstrated by the blue methylene method. The SEM images show monoliths and particles smoothly packed together on spherical and prismatic-spherical surfaces in the morphology of MIP and IIP, respectively. Moreover, the MIP and IIP could be considered as mesoporous and microporous materials, shown by the size of the pores determined by the BET and BJH methods. Furthermore, the adsorption performance of the IIP was studied using copper(II) as a contaminant heavy metal. The maximum adsorption capacity of IIP was 287.45 mg/g at 1600 mg/L Cu^{2+} ions with 0.1 g of IIP at room temperature. The Freundlich model was found to best describe the equilibrium isotherm of the adsorption process. The competitive results indicate that the stability of the Cu-IIP complex is higher than the Ni-IIP complex with a selectivity coefficient of 1.61.

Keywords: copper(II) ion template; 4-vinylbenzoic acid; adsorption capacity; ion-imprinted polymer; heavy metals

Citation: Bivián-Castro, E.Y.; Zepeda-Navarro, A.; Guzmán-Mar, J.L.; Flores-Alamo, M.; Mata-Ortega, B. Ion-Imprinted Polymer Structurally Preorganized Using a Phenanthroline-Divinylbenzoate Complex with the Cu(II) Ion as Template and Some Adsorption Results. *Polymers* **2023**, *15*, 1186. https://doi.org/10.3390/polym15051186

Academic Editor: Michał Cegłowski

Received: 29 November 2022
Revised: 17 February 2023
Accepted: 21 February 2023
Published: 26 February 2023

Copyright: © 2023 by the authors. Licensee MDPI, Basel, Switzerland. This article is an open access article distributed under the terms and conditions of the Creative Commons Attribution (CC BY) license (https://creativecommons.org/licenses/by/4.0/).

1. Introduction

Heavy metals represent a critical contamination in bodies of water around the world; they are persistent pollutants that can never be destroyed, and they tend to bio-accumulate in living organisms. Nickel (Ni), copper (Cu), manganese (Mn), cadmium (Cd), iron (Fe), cobalt (Co), zinc (Zn), arsenic (As), chromium (Cr), mercury (Hg), and lead (Pb) are some examples of metals belonging to this classification; these metals cause an environmental impact due to their toxicity, and they have physicochemical characteristics such as high density, mass, and atomic weight above 20. The contribution of these metals to the hydrological cycle comes from various sources, one of them being lithogenic or geochemical in origin from minerals due to erosion, rain, etc. In addition, industry, economy, domestic waste, etc. are some examples of anthropogenic sources of heavy metals with an important environmental impact [1,2]. Although metals like copper are essential for all living organisms as trace dietary minerals and their deficiency alters the normal functions of the human body, causing several diseases, they can be toxic at higher or even at low levels, thus posing health risks. In nature, copper comes from minerals and it is an essential trace element for the healthy function of the living organisms, because it is part important of metalloen-zymes as their active center. Copper can coordinate with various ligands of oxygen (O), nitrogen (N), and sulfur (S) donor atoms, forming a variety of geometric structures such as a flat square, a square pyramidal, a trigonal bipyramidal, or an octahedral. Then, taking advantage of the characteristics mentioned, copper has been used as a template to preorganize several desirable structures with the objective to design materials inspired by nature with specific cavities. These new materials have been applied to selectively remove heavy metals with chemical characteristics, with radii and charge coming from the metal ion templates used [3,4]. Throughout the world, different analyses have been carried out, referring to the evaluation of the levels of heavy metals in surface waters. During the lockdown period of COVID-19 (coronavirus pandemic 2019), the levels of pollution diminished, representing an opportunity to improve the quality of natural resources [5]. Hence, several methods were applied for the extraction of contaminated metal ions. An accessible methodology to purify water is adsorptive separation by substances with a significant active increase in surface area, chemical and thermal stability, a variety of functional groups, and significant adsorption efficiency and efficacy [6,7]. A search for effective adsorbents to remove heavy metals is a popular endeavor of the scientific community around the world. Some previous reports suggested adsorbents of natural (e.g., biomass, minerals, agriculture, and animal waste) [8,9] and synthetic (e.g., inorganic polymers, chelating resins, and cross-linked polymers) [7,10] origin. Recently, ion-imprinted polymers (IIPs) have been used in water treatment for the removal and recovery of heavy metals due to their high selectivity, being optimal candidates for this type of treatment. Imprinted polymers technology is based on the elaboration of highly stable synthetic polymers called molecular imprinted polymers (MIPs). These have selective molecular recognition characteristics because there are locations within the polymer matrix that are complementary to the analyte in terms of functional group shape and position, which are able to identify the template molecule and are based on the functioning models of biological systems [11,12]. In this paper, a novel adsorbent, with the characteristics mentioned above of IIP, was prepared. This new material has specific cavities to catch heavy metals from contaminated waters. The chemical structure of the IIP was carefully built step by step. Firstly, for the first time, a copper complex of [Cuphen(VBA)$_2$H$_2$O] (phen: phenanthroline, VBA: vinylbenzoate) was prepared; its structure was completely elucidated by X-ray diffraction results using the corresponding blue crystals. The copper complex had a double purpose: 1. as the functional monomer and 2. as the ion template, the Cu(II). A radical polymerization reaction was found using EGDMA (ethylene glycol dimethacrylate) as the cross-linker agent and dimethylformamide as the porogen. The corresponding MIP of [Cuphen(VBA)$_2$H$_2$O-co-EGDMA]$_n$ was obtained as a green crystalline material. Finally, after a soft acidic leaching of the Cu(II) from the MIP, the IIP was obtained as pale yellow crystals. In a similar way, with the exception of the copper complex step, the non-ion-imprinted polymer (NIIP) was prepared. Physicochemical tests and spectrophotometric techniques were used to study the chemical structure of the materials. In addition, the surface morphology of the materials was analyzed, and some adsorption experiments were performed, suggesting the capacity, the selectivity, and the method for heavy metal adsorption by the IIP.

2. Materials and Methods

Preparation and characterization of Cu(II)-phenanthroline vinylbenzoate complex and the corresponding MIP, IIP, and NIIP.

2.1. Reagents

The 4-Vinylbenzoic acid (VBA), ethylene glycol dimethacrylate (EGDMA), azobi-isobutyronitrile (AIBN), and 1,10-phenanthroline (phen) were provided by Sigma-Aldrich (Estado de México, México). All other reagents were of AR grade.

2.2. Apparatus

A freshly methanolic solution of 0.001 M was prepared for conductivity measurements; a Conductronic PC45 electrode (Conductronic, Puebla, México) was used. A Sherwood Scientific LTD magnetic balance Magaway MSB Mk1 (Sherwood Scientific, Cambridge, United Kindom) model was used for the magnetic susceptibility determinations. Pascal's constants were used for the diamagnetism corrections [13]. FT-IR spectra were recorded in the frequency range 4000–400 cm^{-1} by the KBr pellet method using Perkin Elmer Spectrum RXI (Perkin Elmer, Waltham, MA, USA). Elemental analyses (C, H, N) were performed at ALS Environmental's Tucson Laboratory (ALS Environmental, Tucson, AZ, USA). N_2-physisorption analysis of materials was performed on previously out-gassed samples at 120 °C using a Micromeritics Instrument Corporation model, Tristar II Plus (Micromeritics Instrument Co., Norcross, GA, USA). The Brunauer-Emmett-Teller (BET) technique was used to calculate the specific surface area, and the Barret-Joyner-Halenda (BJH) method was used to calculate the pore volume. Scanning electron microscopy (SEM) was used to identify morphology and particle size, using a JEOL model, JSM-6490LV (JEOL Ltd., Tokyo, Japan), at 20 kV.

2.3. Synthesis of [Cuphen(VBA)$_2$H$_2$O] Functional Monomer

In 20 mL of deionized water, 1 mmol (0.148 g) of vinylbenzoic acid and 1 mmol of sodium hydroxide were completely mixed; then, 1 mmol of Cu(NO$_3$)$_2$·3H$_2$O previously dissolved in 5 mL of deionized water was added. Finally, the ethanolic solution of 1 mmol of phenanthroline (0.180 g) in 10 mL of ethanol was added, and the reaction was left under a stirrer for 1 h at room temperature. The resulting blue precipitate was filtered off and then dissolved in 15 mL of methanol and maintain in the refrigerator to obtain crystals available for X-ray diffraction. The results were as follows: [Cuphen(VBA)$_2$H$_2$O] Λ_M 73 cm^2/Ω mol, χ_g = 3.2 × 10^{-6} cm^3/g, M.B. = 2.0. El. Anal. Exp. C 60.31, H 4.53, N 5.82% Calc. C 57.01, H 4.78, N 6.43%. The general steps of the functional monomer preparation are represented in Scheme 1.

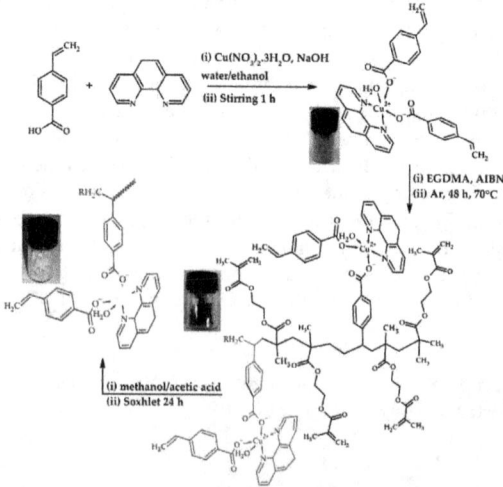

Scheme 1. General steps of the preparation of the materials: the functional monomer, the MIP, and the IIP.

2.4. Preparation of [Cuphen(VBA)$_2$H$_2$O-co-EGDMA]$_n$ (MIP), the NIIP, and [Phen(VBA)$_2$H$_2$O-co-EGDMA]$_n$ (IIP)

Ten mL of dimethylformamide (DMF) was dissolved in 0.18 mmol (0.1 g) of the copper complex, [Cuphen(VBA)$_2$H$_2$O]. Then, 3.59 mmol (678 µL) of EGDMA and 0.06 mmol (0.0099 g) of AIBN previously recrystallized in methanol were added. The reaction was left under constant stirring for 48 h at 70 °C and under inert atmospheric conditions. A dark green gel formed; and after being washed with deionized water, the MIP was obtained as a green solid with crystalline morphology. When nitromethane is the porogen instead of DMF and the amount of initiator is diminished, a brownish-green precipitate is formed (see Supplementary Material). A similar procedure was performed for the NIIP preparation: First, 0.035 mmol (0.0529 g) of VBA with 1 mmol of NaOH was dissolved in 10 mL of DMF, and 0.017 mmol (0.0324 g) of phen was added to this mixture. Then, 3.59 mmol of EGDMA and 0.06 mmol of AIBN were added. The polymerization reaction proceeded under the same conditions of the MIP. The NIIP was obtained as a light-brown crystalline solid. To obtain the ion-imprinted polymer, the copper(II) ion template was removed by an acid wash as follows: First, 0.3 g of MIP was washed in 150 mL of methanol/acetic acid (9:1, v/v) in a Soxhlet apparatus for 24 h. After a deionized water wash, a pale-yellow crystalline solid of the IIP was finally recovered. To remove the excess water, the crystalline polymers were finally washed with methanol. The general steps of the preparation of the materials, the MIP and the IIP, are represented in Scheme 1.

2.5. Crystallography of the Functional Monomer, [Cuphen(VBA)$_2$H$_2$O]

A suitable crystal of copper complex [Cuphen(VBA)$_2$H$_2$O] was mounted onto glass fiber by using perfluoropolyether oil and cooled rapidly in a stream of cold nitrogen gas. Diffraction data were collected by using the Oxford Diffraction Gemini Atlas diffractometer (Oxford Diffraction Ltd., Abingdon, United Kingdom) at 130 K, and intensity data were collected with ω scans; these processes were conducted using the CrysAlisPro and CrysAlis RED software packages (Oxford Diffraction Ltd., Abingdon, United Kingdom) [14]. All the data were corrected by Lorentz and polarization effects, and final cell constants were determined by a global refinement; collected data were corrected for absorbance by analytical numeric absorption correction [15] using a multifaceted crystal model based on expressions from the Laue symmetry using equivalent reflections. The space group was determined based on a check of the Laue symmetry and systematic absences, and it was verified by utilizing the structure solution. The molecular structure was then solved and refined with the SHELXS-2018 [16] and SHELXL-2018 [17] programs (Institute of Inorganic Chemistry, Göttingen, Germany). All non-hydrogen atoms located in successive Fourier maps were treated as a riding model on their parent C atoms, while H atoms of the water (O—H) group were located in a difference map and refined isotropically with Uiso(H) of 1.5 Ueq for H—O. Anisotropic thermal parameters were applied for all non-H atoms, and fixed isotropic parameters were employed for H atoms. Drawing of the molecular structure was performed by utilizing ORTEP [18]. Crystal data and structure refinement and selected bond lengths (Å) and bond angles (o) for [Cuphen(VBA)$_2$H$_2$O] are shown in Tables 1 and 2, respectively. Crystallographic data have been deposited at the Cambridge Crystallographic Data Center as Supplementary Material CCDC: 2223464.

Table 1. Crystal data and structure refinement for the functional monomer [Cuphen(VBA)$_2$H$_2$O].

Empirical Formula	$C_{30}H_{24}CuN_2O_5$
Formula weight	556.05
Temperature	130(2) K
Wavelength	0.71073 Å
Crystal system	Triclinic
Space group	P-1

Table 1. Cont.

Empirical Formula	$C_{30}H_{24}CuN_2O_5$
Unit cell dimensions	a = 8.0188(4) Å
	b = 10.7541(10) Å
	c = 15.7861(14) Å
	α = 101.821(7)°
	β = 94.259(6)°
	γ = 109.056(7)°
Volume	1244.68(18) Å³
Z	2
Density (calculated)	1.484 g/cm³
Absorption coefficient	0.923 mm⁻¹
F(000)	574
Crystal size	0.540 × 0.380 × 0.240 mm³
Theta range for data collection	3.751 to 26.055°
Index ranges	−9 <= h <= 9, −13 <= k <= 10, −19 <= l <= 19
Reflections collected	8643
Independent reflections	4908 [R(int) = 0.0293]
Completeness to theta = 25.242°	99.7%
Refinement method	Full-matrix least-squares on F²
Data/restraints/parameters	4908/5/350
Goodness-of-fit on F2	1.056
Final R indices [I > 2sigma(I)]	R1 = 0.0412, wR2 = 0.0950
R indices (all data)	R1 = 0.0528, wR2 = 0.1029
Largest diff. peak and hole	0.483 and −0.556 e.Å⁻³

Table 2. Selected bond lengths [Å] and angles [°] for the functional monomer [Cuphen(VBA)$_2$H$_2$O].

Bond Lengths (Å)		Bond Angles (°)	
C(1)—N(1)	1.323(3)	O(3)—Cu(1)—O(1W)	94.99(8)
C(1)—C(2)	1.402(4)	O(3)—Cu(1)—N(2)	89.33(8)
C(13)—O(1)	1.258(3)	O(1W)—Cu(1)—N(2)	165.47(8)
C(13)—O(2)	1.266(3)	O(3)—Cu(1)—N(1)	166.89(8)
C(29)—C(30)	1.301(4)	O(1W)—Cu(1)—N(1)	91.43(8)
Cu(1)—O(3)	1.9493(18)	N(2)—Cu(1)—N(1)	81.77(8)
Cu(1)—O(1W)	1.9823(19)	O(3)—Cu(1)—O(1)	99.29(7)
Cu(1)—N(2)	2.007(2)	O(1W)—Cu(1)—O(1)	91.17(7)
Cu(1)—N(1)	2.026(2)	N(2)—Cu(1)—O(1)	101.83(7)
Cu(1)—O(1)	2.2622(17)	N(1)—Cu(1)—O(1)	91.97(7)

2.6. Surface Area

The surface area of the IIP and NIIP was measured using the methylene blue absorption method [19,20]. From a stock solution of methylene blue (0.0176 g/L), a set of working standards were prepared to create a calibration curve. Each set was analyzed by UV-Vis spectrophotometry (Perkin Elmer, Waltham, MA, USA) at λ = 600 nm; 0.1 g of IIP and NIIP each were equilibrated with 25 mL of methylene blue solution, and samples of 1 mL by were analyzed by triplicate until the absorbance became constant. The amount of methylene blue adsorbed was evaluated based on its concentration before and after adsorption. The surface area of IIP and NIIP was calculated using the following equation:

$$As = \frac{GN_{AV} \Phi \times 10^{-20}}{MM_W}$$

where As is the surface area in m²/g, N_{AV} is Avogadro's number (6.02 × 10²³ mol⁻¹), M is the mass of adsorbent (g), and G, Φ, and M_W are the amount adsorbed (g), the molecular cross section (197.2 Å²), and the molecular weight (373.9 g/mol) of methylene blue, respectively.

2.7. Adsorption Capacity and Adsorption Isotherm

An amount of 0.1 g of IIP was added to a series of Cu(II) solutions of 25 mL with 400–1600 mg/L concentrations. All the solutions were left for 6 h under stirring at room temperature. Then, the polymer was filtered off, and the quantity of Cu(II) remaining was analyzed by atomic absorption spectrometry (AAS) using a SpectrAA 220FS model Varian (Agilent Technologies Co., Santa Clara, CA, USA). The amount of adsorbed copper ions per gram of IIP was evaluated as the adsorption capacity (Q), and it was calculated with the following equation:

$$Q = \frac{(C_i - C_f)V}{W}$$

where C_i and C_f are the initial and final concentration of Cu(II) (mg/L), V is the volume of the solution, and W is the mass of the sorbent, respectively.

The Freundlich adsorption isotherm was applied to understand the Cu^{2+} adsorption mechanism of the ion-imprinted polymer. The Freundlich model assumes a heterogeneous surface of the sorbent with multiple adsorption sites (bilayer) and adjacent interactions between adsorbate molecules. The linearized Freundlich equation is

$$Log Q = Log K + \frac{1}{n} log C_e$$

where Q is the metal mass adsorbed per mass unit of solid (mg/g), C_e is the final concentration of solute in the aqueous phase at equilibrium (mg/L), K and n are constants, and the calculated values could be obtained from the adsorption isotherm [21,22].

2.8. Selective Recognition

Heavy metals as Ni(II) could be chosen as a competitor to compare the selectivity of an adsorbent like the ion-imprinted polymer by Cu(II). Niquel ion usually coordinates well with the diamine and carboxylic ligands, it has the same charge and similar ionic radii as copper. Then in principle the IIP could bind Ni(II) as well as Cu(II). A 25 mL of Ni(II) solution 426 mg/L was used, and once the adsorption equilibrium was reached, the concentration of the non-adsorbed ions in the liquid phase was determined by AAS. The distribution coefficient (K_d), was calculated using the following equation:

$$K_d = \frac{(C_i - C_f)V}{C_f W}$$

where C_i, and C_f, are the initial, and final solution concentrations, V is the volume of the solution, W is the mass of the sorbent. The selectivity coefficient (k), was calculated for the ratio of binding of the metal ion in a study related with the competitor species, and using the following equation:

$$k = \frac{K_d \, [Cu(II)]}{K_d \, [Ni(II)]}$$

where k represents the selectivity coefficient, K_d[Cu(II)] and K_d[Ni(II)] represent the distribution ratios of Cu(II) and Ni(II), respectively [21,23].

3. Results and Discussion

3.1. Crystal Structure of the Functional Monomer

Here, the copper(II) complex was used as the functional monomer to prepare the imprinted structures. The functional monomer yields single crystals suitable for X-ray measurements. The discrete unit of coordination compound [Cuphen(VBA)$_2$H$_2$O] shows the coordination of the metal and atomic labeling (see Figure 1). Table 1 shows the improved cell characteristics as well as other important crystal data. In Table 2, some selected bond lengths and bond angles are shown. Copper ion has square pyramidal coordination in this molecule. The pyramid's base is created by two Cu—N bonds made by the phenanthroline ligand's two nitrogen atoms, and two Cu—O bonds formed by two oxygen atoms, one from the carboxylate group of the VBA ligand and one oxygen atom of the water molecule. The apical position is occupied by one oxygen atom from one of the carboxylate groups of the VBA ligand. At the base of the pyramid, the phenanthroline ligand is coplanar with the VBA ligand, with 1° deviation with a defined mean square plane 6.734 (4) x − 0.74 (10) y − 9.413 (9) z = 1.522 (3).

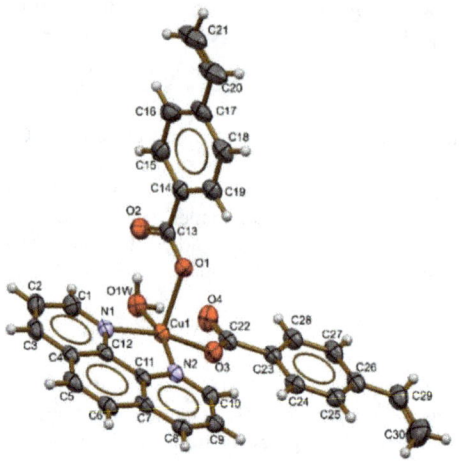

Figure 1. Ortep diagram of functional monomer [Cuphen(VBA)$_2$H$_2$O] ellipsoids at 50% of probability.

The major deviations of the atoms are 0.0213 Å, whereas the O(3) and the O(1W), which are also forming part of the pyramid base, are about 0.0877(2) and 0.0517(2) Å above the plane. At the apical position of the pyramid, a second vinylbenzoate anion is coordinated, with Cu—O(1) = 2.2623(16) Å, a distance similar to that observed in related compounds of 2.2623(16) [24].

The angle formed between the mean plane of the pyramid base and the apical ligand is 88.35(12)°, with a τ value of 0.025, very close to a square pyramid geometry [25]. The C(22)—O(3) coordinated bond length is longer than the C(22)—O(4) uncoordinated bond length, which is consistent with the creation of a bond between the anionic carboxylate oxygen atom and copper cation. However, a slight difference between C(13)—O(1) and C(13)—O(2) bond lengths is found due to the apically elongated square-pyramidal coordination. It is important to notice that in the complex, the C(20) and C(29) of the vinyl group are almost coplanar to the phenyl ring of the vinylbenzoate ligands. The vinyl group of the apical ligand presents some disorder, and the C(20)—C(21) and C(20A)—C(21A) bond length average corresponds to the well-established vinyl group of C(29)—C(30) = 1.302 Å bond length from the equatorial vinylbenzoate ligand [26].

In the discrete unit of [Cuphen(VBA)$_2$H$_2$O], the water molecule coordinated to the metal center shows two intramolecular interactions of the hydrogen bond, O(1W)—H(1WA) ... O(2) and O(1W)—H(1WB) ... O(4), forming a $S^1_1(6)$ motif; in the crystal array, there are two types of intermolecular interactions—one type is hydrogen bond C—H ... O, and the other is the π–π intermolecular contact (see Figure 2).

The intermolecular contact C(8)-H(8) ... O(4) at 2.36 Å forms the $S^1_1(8)$ motif along the b axis. Finally, the intermolecular contact of type π–π is shown between the phenyl ring C23/C27 of vinylbenzoate and N1—C1/C4—C12 of phenanthroline ligands along the plane formed by the a-c axes. All these interactions show a complex growing along the b-c plane.

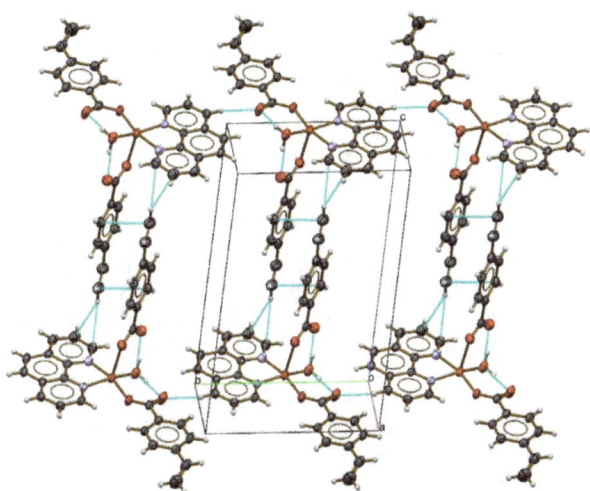

Figure 2. Crystal array of the functional monomer [Cuphen(VBA)$_2$H$_2$O], with a view along the *a* axis from the perspective of the plane formed by the *b-c* axes, emphasizing the H-bond and σ–π interactions.

3.2. Synthesis and Characterization

Considering the functionality of the copper complex to build the polymeric structures, the vinyl polymerizable groups of VBA ligands and the carboxylic coordination sites could be pending functionalities attached to the polymeric chain formed with the comonomer EGDMA; in these cases, cross-linked materials are usually obtained, some examples of which were previously reported [12,19,21,23]. A second reaction site in the copper complex could be the saturated coordination site with a water molecule that can be released by exchange reaction with ligands showing an increased binding affinity. Then, the MIP was prepared by the free radical polymerization reaction, using the copper complex as the functional monomer, EGDMA as the cross-linker comonomer, DMF as the porogen solvent, and the AIBN as the initiator. Scheme 1 shows the general steps of the preparations of the imprinted materials. The temperature of the reaction was maintained at 70 °C to guarantee the thermal decomposition of AIBN as reported by this azo initiator [27]. The MIP was obtained as a green solid with crystalline morphology. The polymerization methods varied: emulsion, mass, and coprecipitation synthesis, where the choice of method depends on the morphology, as well as the physicochemical characteristics desired in the polymer. In emulsion polymerization, a hydrophobic organic porogen (in which the polymerization mixture was found) and an organic or aqueous dispersing medium were combined, which were immiscible and formed two phases. This type of polymerization occurred in the drops of the polymerization mixture, with agitation and constant temperature. The mass polymerization method consisted of the synthesis of the polymer in the porogen; this method did not require agitation and was performed merely by adding the initiator to the pre-polymerization mixture at a constant temperature. One of the main features in these two methods was the choice of the porogen as well as its solubility in the monomer, which resulted in different ordering variants. In this experiment, we used two methods, mainly changing the porogen. The porogens used were dimethylformamide and nitromethane (see Supplementary Material) [12,28]. The IIP was obtained as a pale-yellow crystalline solid after removing the ion copper(II) template by an acid wash of the crystalline MIP, as shown in Scheme 1. The functional monomer can be dissolved in methanol, DMF, and DMSO. However, the [Cuphen(VBA)$_2$H$_2$O-*co*-EGDMA]$_n$ showed poor solubility properties in solvents such as water, ethanol, methanol, acetonitrile, DMF, DMSO, and THF. The IIP and the NIIP had the same properties as the MIP in the solvents mentioned. The characteristic low solubility in organic solvents of the MIP, IIP, and NIIP prepared here could be related to the presence of a high number of double bonds resulting in cross-linking units for the resulting polymer structure [29]. To obtain chemical purity of the materials, the unreacted monomer, ligands, and copper salt were washed with water and methanol. All the prepared imprinted and non-imprinted polymers were obtained with an adequate grade of purity because of their crystallinity. It is clear from the inset photos of Scheme 1 that the color of these materials was different. The color was characteristically blue for the functional monomer with the presence of copper ion in a square pyramid coordination complex [24]. Then, the color turned green in the case of MIP, in which the addition of EGDMA to the chain and the loss of a water coordination molecule probably induced a distortion of the geometry on the coordination sphere. The pale yellow and white colors of the ion-imprinted and non-imprinted polymers could be interpreted by the presence of only the organic ligands. However, in the case of IIP, the cross-linked structure in which ligands were preorganized by the template resulted in a difference compared with the non-organized structure of NIIP. The infrared spectra of MIP, IIP, and NIP from Figure 3 helped support the comments made above. In general, it was found that the FT-IR spectra of the polymers prepared were very similar because all the materials were synthesized with the same methodology and precursors [23,30]. However, in the case of IIP, the preorganized structure was conserved, and the corresponding imprinted cavities were available for ion adsorption application. The FT-IR spectra of the prepared materials shared some expected signals around 1100–1200 cm^{-1}, and 1721 cm^{-1} due to the C=O from the EGDMA. One of the infrared signals that verified the polymerization process was the intensity change of the C=C signal in comparison to the medium intensity signal at 1546 cm^{-1} assigned to the vinyl group from the VBA ligands that appeared in the functional monomer spectra of Figure S1 [19,30]. Additionally, the tert-butyl, a structure that is formed when starting the polymerization chain, was observed at 1250–1260 cm^{-1}. It is important to notice from Figure 3 that no additional signals from free ligands have appeared in the spectra.

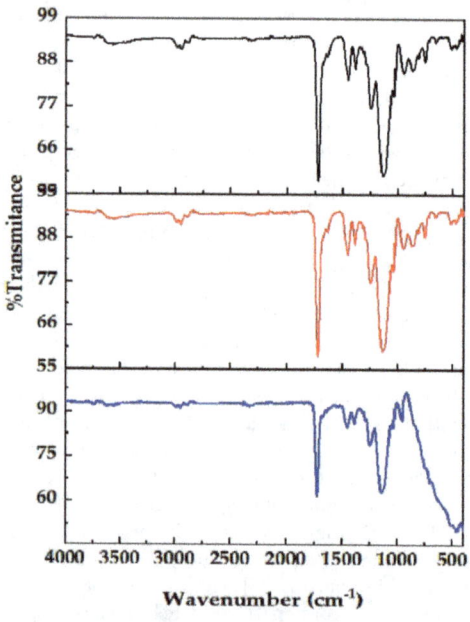

Figure 3. FT-IR spectra of the MIP, IIP, and NIIP (black, red, and blue lines, respectively).

On the other hand, the functional monomer mid-infrared spectrum (see Figure S1, Supplementary Materials) showed the typical signals of the ligands attached to the template copper ion through the N and O atoms. A strong and sharp signal at 1585 cm^{-1} was assigned to the asymmetric vibrational mode of the O—C—O group. At 1370 cm^{-1}, a strong and sharp signal corresponded to the symmetric vibrational mode of the O—C—O group. The medium-intensity C—H signals were found around 1150 cm^{-1}. Two short and sharp signals were found at 838 and 720 cm^{-1} due to the N=C of the phenanthroline ligand and the short signals around 772 cm^{-1} for the aromatic groups. The corresponding $\Delta v_{as-sym}(COO)$ for the monomer was 215 cm^{-1}. These values agree well with the monodentate coordination mode of the carboxylic groups from the VBA ligands [31]. The broad and strong signal around 3454 cm^{-1} was assigned to the O—H stretching vibrations of lattice water molecules in the functional monomer [24], which suffered a notable intensity diminished in the spectra from Figure 3, possibly attributed to the loss of water molecules during the polymerization process.

3.3. Surface Area Properties

The results obtained for the surface area properties of the imprinted polymers indicated that the Brunauer-Emmett-Teller (BET) surface area (S_{BET}) was 0.3496 and 0.2549 m^2/g for MIP and IIP, respectively. The total pore volumes (V_T) were 0.000384 and 0.000008 cm^3/g, and the corresponding average pore diameters (D_p) were 4.3986 and 0.1180 nm. The relative decrease in S_{BET} in the case of IIP could be precisely attributed to the satisfactory removal of the copper ion template after the MIP leaching process. In addition, the V_T of IIP presented a decrease in comparison to the V_T of MIP because, along the active sites of these adsorbents, the formulated molecular complexes were different, thus showing a specific development of porosity [4,32]. In agreement with the IUPAC definition, the MIP could be considered as a mesoporous material because its pore diameter is around 2–50 nm, and the IIP could be a microporous material because its pore diameter is less than 2 nm [4]. It is very desirable to obtain microporous materials as in this case because the affinity for ion remotion is improved. Copper ion has an ionic radius of 0.087 nm, closer to the IIP pore diameter because the pore of IIP is precisely where templated by the lixiviation of the Cu^{2+}. Then, a high selectivity and chemical functionality of the imprinted polymer could be expected. The specific surface area of NIIP was determined by the relatively easy and inexpensive method of methylene blue adsorption. Then, the specific surface area was defined as the accessible area of the solid surface per unit mass of material [19,20,33]. For comparison, the IIP surface was also determined by this method. The results showed a surface area of IIP higher than NIIP, being 12.60 and 11.30 m^2/g, respectively. By leaching the Cu(II) ions from the polymer matrix, we obtain the formation of specific cavities in the polymer network, resulting in a higher surface area in the IIP than in the NIIP [34].

To improve the adsorption of the superficial area of the imprinted polymers, a porous structure is desirable in which the binding sites at the surface are exposed. In Figure 4, the scanning electron micrographs of MIP and IIP show that their surface is morphological with the presence of meso, as well as micropores across the surface due to the influence of the cross-linking, and the imprinting reaction [4]. This morphological feature of imprinted polymers was also in agreement with the BET results mentioned above. Figure 4a shows a material with irregular particles tending to a smoothly prismatic and spherical surface with an average diameter of 90.73 nm, corresponding to the molecular imprinted polymer. Figure 4b exhibits monoliths of a material with regular particles packed together and a smooth spherical surface with an average diameter of 41.33 nm, corresponding to the ion-imprinted polymer. As shown in Figure 4 and Table S1 (Supplementary Materials), the order of increase of the particle diameter of the prepared materials was IIP < MIP < functional monomer. The presence of many micropores on the spherical surface of the IIP is more beneficial to the fast and homogeneous binding of template ions.

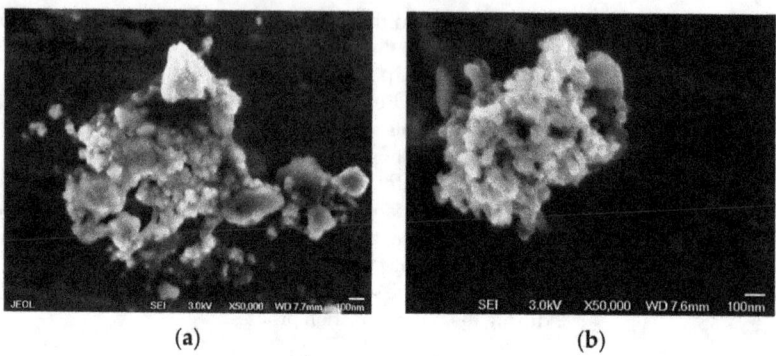

Figure 4. SEM micrographs of the imprinted polymers, (**a**) image at 50,000× of MIP and (**b**) image at 50,000× of IIP.

3.4. Static Adsorption Capacity and Adsorption Isotherm

The maximal concentration of metal ions adsorbed by the ion-imprinted polymer at equilibrium is stated by the adsorption capacity [21,22]. The amount of copper ions adsorbed per unit mass of IIP, almost to saturation, against the initial concentration of Cu(II) is shown in Figure 5. Even though the range of concentrations used in this study was higher than in other studies, a plateau trend was not reached [21,35].

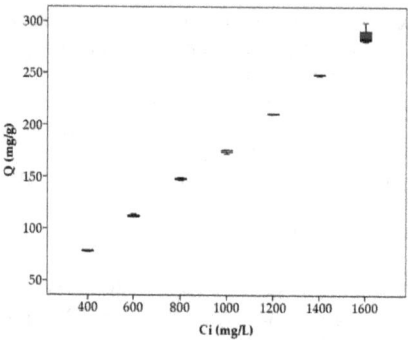

Figure 5. Effect of initial concentration of Cu(II) on adsorption capacity of ion-imprinted polymer.

The maximum static adsorption capacity of the IIP was 287.45 mg/g, being higher with respect to other copper ion-imprinted polymers described in previous publications, with adsorption capacity in the range of 16.55 and 132.77 mg/g [11,12]. A list of maximum adsorption capacities for copper with various adsorbents is shown in Table 3. The adsorbent based on IIP structures has higher sorption capacities for copper than the other materials. The imprinted technology allows us to obtain materials like the MIP and IIP presented in this work, which are highly sensitive materials, with high selectivity and affinity for the analyte, because they have selective molecular recognition properties.

Table 3. Comparison of maximum adsorption capacity of copper(II) ions using different adsorbents.

Adsorbents	Maximum Adsorption Capacity (mg/g)	References
Granular activated carbon	48.22	[8]
Geopolymers	35.88–152.3	[7]
Activated carbon	75.0	[35]
Ion-imprinted polymer	287.45	This work

Heavy metals or adsorbates can interact with the available binding surfaces of the ion-imprinted polymers through an adsorption process. The study of adsorption isotherm is needed to interpret the mechanism of adsorption between IIP and adsorbates and to standardize its application. The adsorption isotherm describes the relationship between the equilibrium of any solute in the solution and the adsorbent. According to the shape of the adsorption curve, the possibilities of the adsorption process occurring can be defined. The Langmuir and Freundlich isotherms are commonly used in water treatment [7,8]. The adsorption isotherm from Figure 6 presents a type of linear adsorption with $y = 0.758x + 0.407$ and $R^2 = 0.981$, where the mass of the solute in the aqueous solution and the mass of solute adsorbed on the solid matrix are kept in equilibrium. There are certain conditions that favor the existence of a linear isotherm; among them, the most important ones are as follows: existence of flexible molecules in the medium due to different degrees of crystallization of the material and a greater affinity of the solute with the substrate than with the solvent. Therefore, the Freundlich model fit better than the Langmuir model when the IIP was employed to remove Cu^{2+}, with a high correlation coefficient compared to the Langmuir model ($R^2 = 0.686$). Due to the Freundlich model's improved fit, it can be deduced that the adsorption of heavy metal ions onto the IIP heterogeneous surface is classified as a multilayer.

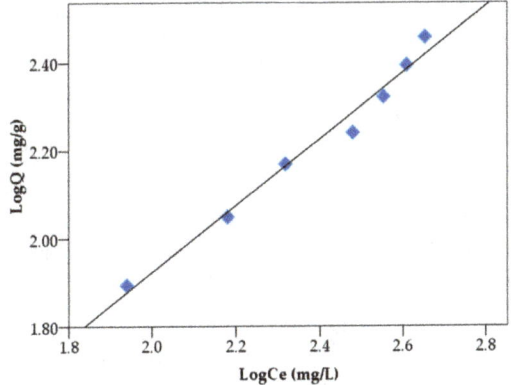

Figure 6. Freundlich plot of copper ions on ion-imprinted polymer. Linear equation $y = 0.758x + 0.407$, $R^2 = 0.981$.

3.5. Selective Recognition

The selective recognition of IIP by Cu(II) against Ni(II) from their solutions was studied. Nickel ion was selected as a competitor sorption of copper, given that the metal ions of the same charge and similar ionic radii as Cu(II) can influence the adsorption process of copper ions. Then, copper and nickel have the same charge of 2+, and the ionic radii are very close, 0.087 for copper and 0.083 nm for nickel. The imprinting effect generated by the template used in the imprinted polymer could be quantified with the selectivity coefficient determination. The results of our experiments indicated a distribution coefficient of 950 and 590 mL/g corresponding to Cu(II) and Ni(II); this is interpreted as the Cu-IIP complex stability being higher than the Ni-IIP complex. Then, the IIP has a good imprinting effect, with a selectivity coefficient of 1.61 [19]. Scheme 2 describes the mechanism of heavy metal adsorption using the ion-imprinted polymer as a multilayer array in agreement with the Freundlich model. In the physical adsorption, the adsorbates are adhered to the available binding surfaces of IIP, filling the micropores of the material, and metal ions can interact with the polymeric matrix through coordination bonds.

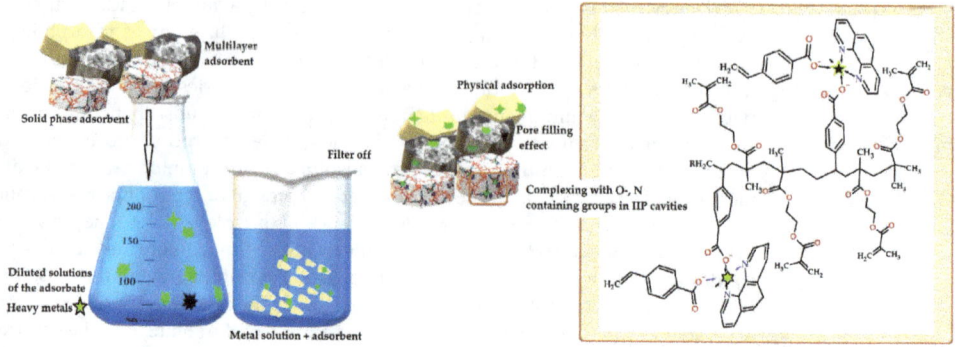

Scheme 2. Adsorption mechanisms of heavy metals by the ion-imprinted polymer.

4. Conclusions

In this work, a Cu(II) ion-imprinted polymer was prepared and characterized. First, the functional monomer of [Cuphen(VBA)$_2$H$_2$O] was synthesized. Then, the molecular imprinted polymer of [Cuphen(VBA)$_2$H$_2$O-co-EGDMA]$_n$ was obtained by free radical polymerization. The copper complex, ethylene glycol dimethacrylate, and 2,2-azobisisobuthyronitrile were used as the functional and cross-linking monomers and the initiator, respectively. The Cu(II) ions were leached in an acidic wash, and the IIP was obtained. The polymers were characterized by some physical and spectroscopical techniques. The morphology of the imprinted materials showed some irregular surface particles for MIP, and the IIP surface area consisted of packed, smoothly spherical particles. Moreover, the MIP is a mesoporous material with a pore diameter of around 4 nm. The IIP with a pore diameter of less than 2 nm was considered as microporous material. The maximum adsorption capacity of IIP was 287.45 mg/g, and this behavior was represented by the Freundlich adsorption isotherm linear model. The Ni(II) was used as the competitive specie against Cu(II); then, the selective recognition of IIP was 1.61 L/g. These results showed that IIP has a high potential for being used for copper or similar chemical species in contaminated water adsorption. In conclusion, we focused on the synthesis of IIP whose preorganized structure came from a metal complex used as a functional monomer. Thus, the relatively facile preparation, cross-linked structure, and low solubility in water rendered the IIP a promising sorbent for selective extraction of metal ion applications in water pollution with specific characteristics like Cu(II) based in an ionic radius, electronegativity, electronic configuration, and geometry. Then, further studies on its selectivity in the presence of other metal ion competitors such as Co(II), Cd(II), Zn(II), Pb(II) would be fundamental. The mechanism of the contaminant removal procedure, regeneration of the material, as well as proceeding with reuse cycles of this IIP are some of the next experiments for future research.

Supplementary Materials: The following supporting information can be downloaded at: https://www.mdpi.com/article/10.3390/polym15051186/s1, Figure S1: FT-IR spectra of the functional monomer; Table S1: SEM micrographs and average particle diameters of the prepared materials, IIP, MIP and functional monomer.

Author Contributions: Conceptualization, E.Y.B.-C.; methodology, E.Y.B.-C.; validation, E.Y.B.-C., A.Z.-N., M.F.-A., J.L.G.-M. and B.M.-O.; formal analysis, E.Y.B.-C and A.Z.-N.; investigation, E.Y.B.-C., A.Z.-N. and B.M.-O.; resources, E.Y.B.-C., M.F.-A. and J.L.G.-M.; data curation, E.Y.B.-C., A.Z.-N. and M.F.-A.; writing—original draft preparation, E.Y.B.-C. and A.Z.-N.; writing—review and editing, E.Y.B.-C., A.Z.-N., J.L.G.-M. and M.F.-A.; visualization, E.Y.B.-C., A.Z.-N. and J.L.G.-M.; supervision, E.Y.B.-C.; project administration, E.Y.B.-C.; funding acquisition, E.Y.B.-C., M.F.-A. and J.L.G.-M. All authors have read and agreed to the published version of the manuscript.

Funding: This research received no external funding.

Article

Application of Molecular Imprinting for Creation of Highly Selective Sorbents for Extraction and Separation of Rare-Earth Elements

Ruslan Kondaurov [1,2], Yevgeniy Melnikov [1,*] and Laura Agibayeva [2]

1. Department of Biochemical Engineering, International Engineering and Technological University, Al-Farabi ave. 93a, Almaty 050060, Kazakhstan
2. Faculty of Chemistry and Chemical Technology, Al-Farabi Kazakh National University, Al-Farabi ave. 71, Almaty 050040, Kazakhstan
* Correspondence: sebas273@mail.ru

Abstract: The aim of the work is to study the effectiveness of a molecular imprinting technique application for the creation of highly selective macromolecular sorbents for selective sorption of light and heavy rare-earth metals (for example, samarium and gadolinium, respectively) with subsequent separation from each other. These sorbents seem to be promising due to the fact that only the target rare-earth metal will be sorbed owing to the fact that complementary cavities are formed during the synthesis of molecularly imprinted polymers. In other words, the advantage of the proposed macromolecules is the absence of accompanying sorption of metals with close chemical properties. Two types of molecularly imprinted polymers (MIP) were synthetized based on methacrylic acid (MAA) and 4-vinylpyridine (4VP) functional monomers. The sorption properties (extraction degree, exchange capacity) of the MIPs were studied. The impact of template removal cycle count (from 20 to 35) on the sorption effectivity was studied. Laboratory experiments on selective sorption and separation of samarium and gadolinium from a model solution were carried out.

Keywords: light and heavy rare-earth metals; selective sorption; separation of rare-earth metals; molecular imprinting technique; functional macromolecular structures; molecularly imprinted polymers

1. Introduction

The market for rare-earth metals (REM) is one of the youngest commodity markets in the world and is growing at an impressive pace compared to other base metals (nickel, copper, iron, gold, etc.): over the last 50 years, the volume of world production and consumption of REMs increased by about 40 times—from 5000 to 200,000 tons per year [1,2]. This was the result of both global economic growth and a change in technological structures based on the innovative development of the world economy. The volumes of production and consumption of REMs are one of the main signs of the development of the national industry of a country and a significant indicator of its manufacturing ability and innovative component. It is rare-earth metals that in the last decade have caused the greatest concern among developed and developing countries due to their strong integration into the production chains of high-tech industries and the level of uncertainty involved in providing this type of raw material [3]. At the same time, in addition to the global geopolitical situation, the rare-earth industry is developing naturally due to scientific and technological progress and free competition [4]. There are new technologies and innovative products (and hence new demand), to which companies respond by modifying and reducing the cost of the production processes of their products. In this regard, an important aspect of the development of a national rare-earth industry is such institutional conditions for suppliers of rare-earth raw materials, its consumers and a state that would protect "their" enterprises and be able to address various crises and the destruction of existing global production chains [5–7].

It is necessary to highlight the following features that determine the relevance and demand for REMs at the present time:

1. REMs in the Earth's crust are not relatively rare; they are more common than, for example, gold, uranium, lead, tin, molybdenum, tungsten, etc. However, deposits with industrial concentrations of rare-earth ores are less common than for most other minerals. According to the report "Strengthening the European rare earths supply-chain", the available REM reserves exceed the current world production by three orders of magnitude [8].

2. Ores are complex in composition. In addition to REMs, they contain elements such as niobium, tantalum, phosphorus, iron, aluminum and others. More than 250 minerals are known that contain REMs, but only 60–65 of them are rare-earth. All rare-earth deposits differ greatly in their specific distribution of metals. As a rule, light rare-earth metals (LREM) make up a much larger proportion of the total content of REMs in the ore than heavy rare-earth metals (HREM). Therefore, in our time, one of the most important tasks is still the development of fundamentally new approaches and technologies for deep and complex processing of complex rare-earth ores that cannot be enriched by traditional physical and mechanical methods. It is for this reason that recently, more and more often in the media, one can find reports on research in the field of search and extraction of REMs from such potential sources as various industrial wastes, tailings and slags (for example, ash and slag dumps, phosphogypsum, red mud). A high value is given to deep-sea rocks and silts from the bottom of the southeastern and central parts of the Pacific Ocean, which, according to various estimates, may contain amounts comparable to or even exceeding continental reserves of REMs [9,10].

3. The third feature is that rare-earth ores contain radioactive thorium and uranium, the concentrations of which are very different for each rare-earth deposit. These elements are considered to be byproducts of mining, and the presence of thorium and uranium in the ore is one of the key factors affecting the attractiveness of a deposit to investors, as these two elements can be the biggest barrier to obtaining permission to mine and process the ore. In this regard, special attention is paid to such issues as radioactive dust and radiation in deposits, radioactive waste management and transportation of rare-earth ore, which must comply with strict regional and international legislative standards [11,12].

4. The fourth feature of the REM sources is that rare-earth elements are often byproducts of mining and processing of ore with elements such as iron, cobalt, manganese, titanium, niobium, tantalum, zirconium and others [13,14]. However, the technologies for capturing and separating associated components are complex and unique for each source of mineral resources; therefore, they have no analogues and are expensive. For this reason, small industrial concentrations of REMs relative to other elements in a deposit may turn out to be untenable during a feasibility study, which will not allow the development and operation of a REM source to begin [14].

5. The fifth feature of REMs that needs to be distinguished is the balance problem, or the balancing problem. As in the case of uranium and thorium, the concentrations of which are very different in the ore object, the specific distributions of metals also differ significantly for each REM deposit [15–18]. Moreover, this distribution does not correspond to the demand of the global market for various types of high-tech products, the production of which requires REMs. The essence of the problem lies in the fact that the mined ore at a deposit is completely processed at the first stages of enrichment into a concentrate without residues and non-selectively. Such a "natural binding" of REMs leads to an excess supply of some of the rare-earth elements and, accordingly, a decrease in prices for them. On the other hand, there is an increased demand for scarce REMs from the market of high-tech products, for the production of which these REM are needed, which leads to an increase in their prices; therefore, surplus REMs are implicitly subsidized by demand at the expense of scarce ones [19].

6. The sixth and last feature of REMs is the change in the dominant area of consumption due to scientific and technological progress, which dramatically changes the demand for REMs and unbalances the market [20–26]. A change in the dominant area of consumption brings the market and industry out of balance, and the demand for individual REMs changes dramatically, which leads to significant changes in prices and supply chains and an increase in uncertainties and risks, including for investors. Therefore, it is important to understand the dynamics of world demand and the structure and distribution of technological chains for the production of high-tech products based on REMs, which will undoubtedly undergo changes in the foreseeable future [27,28].

Figure 1 shows the structure of the technological chain for the production of high-tech products based on REMs [29]. In general, there are three main methods of extracting ore from a deposit: open-pit shallow ore body mining, underground mining and in situ leaching. After the excavation of the rock, due to the complexity of the composition of rare-earth ores, at the second stage, individual (sometimes unique) physical and chemical processing schemes are used, as a result of which various concentrates and intermediate products are obtained at the output. In particular, the initial and intermediate rare-earth products are various concentrates: fluorides, chlorides and carbonates. At the third stage of the production chain, REM oxides are obtained after chemical treatment, from which individual metals are extracted after extraction. Due to the chemical similarity of rare-earth metals, separation into individual metals is a laborious task. Currently, ion exchange and solvent extraction are the two advanced methods for separating concentrates.

Because rare-earth deposits are multicomponent, the technological chain of production "from ore mining to obtaining individual metals" is a multi-stage process. At the same time, each deposit is unique in terms of the composition and content of REEs in the ore, which means that the multi-stage production chain has certain individual technological features for each type of ore. At the same time, there are stages at which similar finished products are obtained: concentrates of different levels (for example, fluorides, chlorides and carbonates of rare-earth metals), oxides or individual metals.

The production chain does not end there, as high-tech goods can be obtained based on REMs and their oxides. It is the possibility of producing such high-tech goods based on rare-earth mineral resources that is a litmus test of the level of a technological structure and the development of a country's industry [1,2,19].

As an alternative to the available sorption technologies based on the use of ion-exchange resins [30], it is possible to use molecular imprinting for the selective sorption and subsequent separation of target REMs. As is known, molecular imprinting is a method for obtaining "molecular imprints" based on the polymerization of functional monomers in the presence of specially introduced target template molecules [31–33]. It is known that molecular recognition is based on the spatial correspondence of structures and non-covalent interactions, namely, electrostatic, hydrophobic, van der Waals, π-π- and cation-π-interactions, as well as hydrogen bonds [34]. At the same time, the combination of the selectivity of specific complexation with strength and the ability to rapidly reverse changes is characteristic of almost all macromolecular structures [35]. Molecular recognition is one of the basic concepts of supramolecular chemistry, which differs from the usual binding between molecules by high selectivity. Molecular recognition is based on the presence on one molecule (the receptor, or "host") of a site of selective binding to another molecule (the ligand, or "guest"). To do this, the receptor and the ligand must show complementarity, that is, structurally and energetically correspond to each other. The concept of complementarity includes the correspondence of the imprint to the template both in size and shape, and in the presence of complementary functional groups in the imprint that are capable of interacting with the functional groups of the template molecule [36–38]. Thus, the ability of molecularly imprinted polymers to recognize a target is based on the conformity of the shape of imprints and the specific functional groups within them to template molecules [39].

The list of modern macroporous sorbents based on synthetic polymers is quite extensive, and most of them are polymethacrylate matrices. The obvious advantages of monolithic sorbents based on synthetic polymers are the relative ease of synthesis and the possibility of varying functional monomers depending on the objectives of the study [40,41].

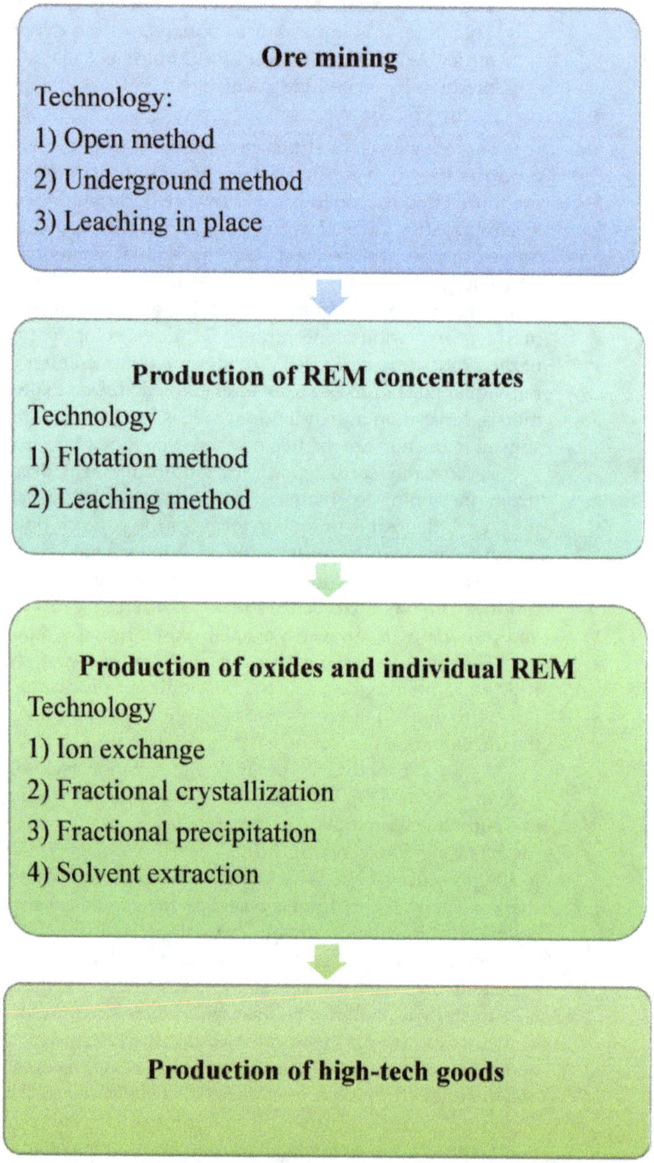

Ore mining
Technology:
1) Open method
2) Underground method
3) Leaching in place

Production of REM concentrates
Technology
1) Flotation method
2) Leaching method

Production of oxides and individual REM
Technology
1) Ion exchange
2) Fractional crystallization
3) Fractional precipitation
4) Solvent extraction

Production of high-tech goods

Figure 1. The structure of the technological chain for the production of high-tech products based on REMs.

The essence of the method for obtaining molecularly imprinted polymers (MIPs) is the formation of a stable prepolymerization complex between a template molecule and a functional monomer (in other words, preorganization occurs) by mixing them in a suitable solvent [42–46]. Next, polymerization is carried out in the presence of a cross-linking agent, during which the pre-polymerization complex is rigidly fixed in the polymer network. At the end of the process, template molecules are removed from the cross-linked polymer matrix. As a result, "imprints" (the so-called "imprint sites") are formed in the polymer, which are cavities that are complementary to the template molecule in size, shape and arrangement of functional groups [47]. Briefly, the process of pore formation in MIPs can be described as follows. After the decomposition of the initiator at the initial stage of polymerization, gel-like oligomeric particles (cores) are formed, which begin to precipitate from the organic phase due to low solubility in porogens. Under such conditions, the monomeric part of the organic phase is the best solvent for nascent polymer chains compared to the porogen phase, which facilitates the penetration of monomers into the precipitated insoluble nuclei and their continued participation in the polymerization process occurring inside the nuclei, which gradually reach the size of microglobules. Growing polymer globules are combined into clusters held by polymer chains penetrating neighboring particles. At the final stage of polymerization, the size of the clusters becomes sufficient for their contact, which leads to the formation of a continuous array inside the polymerizing system. The resulting matrix is gradually strengthened by interglobular cross-links and ongoing polymerization. In this case, the formation of a final porous polymeric material is achieved. At this stage, porogenic solvents are a separate organic phase that fills the voids of the porous polymer mass. The fraction of voids (or macropores) in the final polymer is close to the volume fraction of thoracic solvents in the initial polymerization mixture [48–53].

Thus, an MIP is essentially a solid matrix with artificial receptors of the template molecule capable of repeated highly specific interaction with it or with its analogue [54]. Schematically, the process of obtaining an MIP is shown in Figure 2 [55].

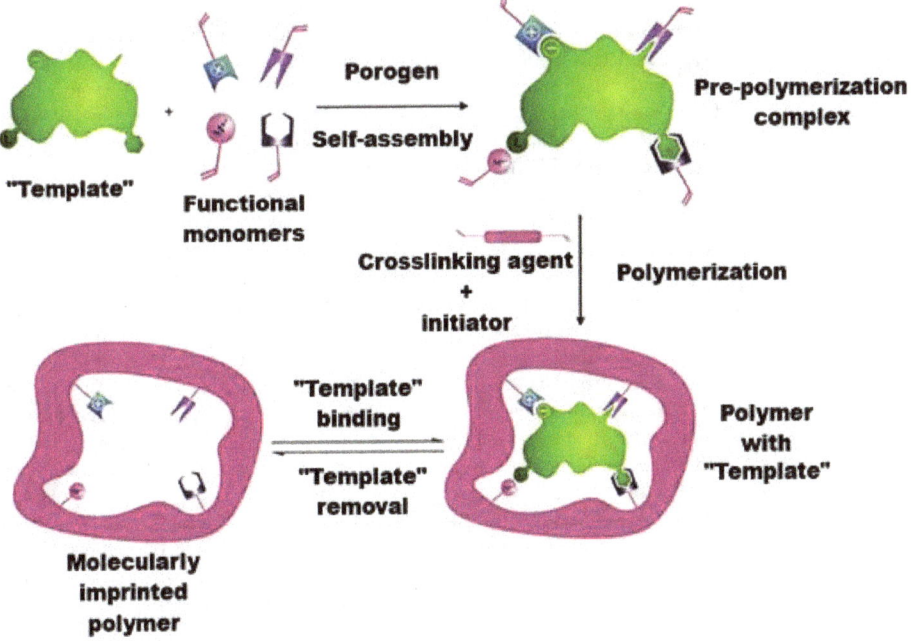

Figure 2. MIP synthesis stages.

In this paper, we propose a variant for the development of polymers with molecular imprints as an alternative to existing sorbents for the purpose of their further application for the selective extraction and separation of rare-earth metals. From the light and heavy REMs, samarium and gadolinium were selected due to their relevance to many areas of modern life. Samarium is widely used in the following spheres: magnetic materials; thermoelectric materials; and production of special luminescent and infrared-absorbing glasses [56–61]. Gadolinium is mainly used in the following areas: creation of storage media with enormous recording density; nuclear energetics; thermoelectric materials; and superconductors [62–65].

2. Materials and Methods

2.1. Materials

The following reactants were used for the synthesis of molecularly imprinted polymers: monomers—methacrylic acid (MAA) and 4-vinylpyridine (4VP); cross-linkers—ethyleneglycol dimethacrylate (EGDMA) and diethyleneglycol dimethacrylate (DEGDMA); initiator—azobisisobutyronitrile (AIBN); porogen—toluene; and stabilizer—hydroxyethylcellulose (HEC). The mentioned reagents were purchased from Sigma-Aldrich (Burlington, MA, USA).

All conducted experiments involved the application of deionized water (χ = 12 µS/cm, pH = 6.95).

Before the experiments the MAA and 4VP monomers were initially purified from inhibitors (monomethyl ether of hydroquinone and hydroquinone) by vacuum distillation.

2.2. Methods

2.2.1. Synthesis of Molecularly Imprinted Polymers

Suspension polymerization was used for the synthesis of molecularly imprinted polymers (MIP). The template molecules were hexahydrate nitrates of gadolinium and samarium. The reactive medium was deionized water. The monomers MAA and 4VP were put into the reactor (containing deionized water) after a template molecule was added. Subsequently, the pre-polymerization complex was added with AIBN, EGDMA (or DEGDMA), tolyene and HEC (polymerizate composition was as follows: template:MAA:4VP:cross-linker = 0.5:1:1:4). Initially the polymerization process occurred at room temperature (for 20 min), but further reaction was carried out at 75 °C in a nitrogen atmosphere with permanent stirring. The obtained imprinted structures are named MIP1 and MIP2, depending on the cross-linker; EGDMA was used in the synthesis of MIP1, and DEGDMA was used in the synthesis of MIP2. The obtained MIPs were crushed into small dispersions and divided by sieving (for further experiments, the particles 200 $\leq$ d $\leq$ 250 µm were taken). After that, the particles were washed firstly with deionized water and after that with acetone to remove impurities and unreacted monomer residues. The subsequent procedure was vacuum drying (for 48 h). Removal of the template from the polymer matrix of the MIP was accomplished by continuous washing with nitric acid (concentration 1 mol/L) for 40 repetitions (each cycle—washing with stirring for 1 h), with further washing and drying for 24 h. For proof that the synthesized MIPs have selectivity for the REMs, non-imprinted structures were synthetized along with the MIPs. The synthesis procedure for the non-MIPs is similar to that mentioned above, except the template molecule is not added to the polymerizate. The scheme of synthesis of the MIPs is presented in Figure 3.

Control of purification of the obtained MIPs and full template removal from the MIP matrix was achieved by using an Expert-002-2-6-p conductometer (Econics, Mocsow, Russia) anda SevenDirect SD50 pH meter (Mettler-Toledo, Columbus, OH, USA) for measurements of specific electric conductivity and pH values. The procedure continued until constant values of specific electric conductivity and pH were reached.

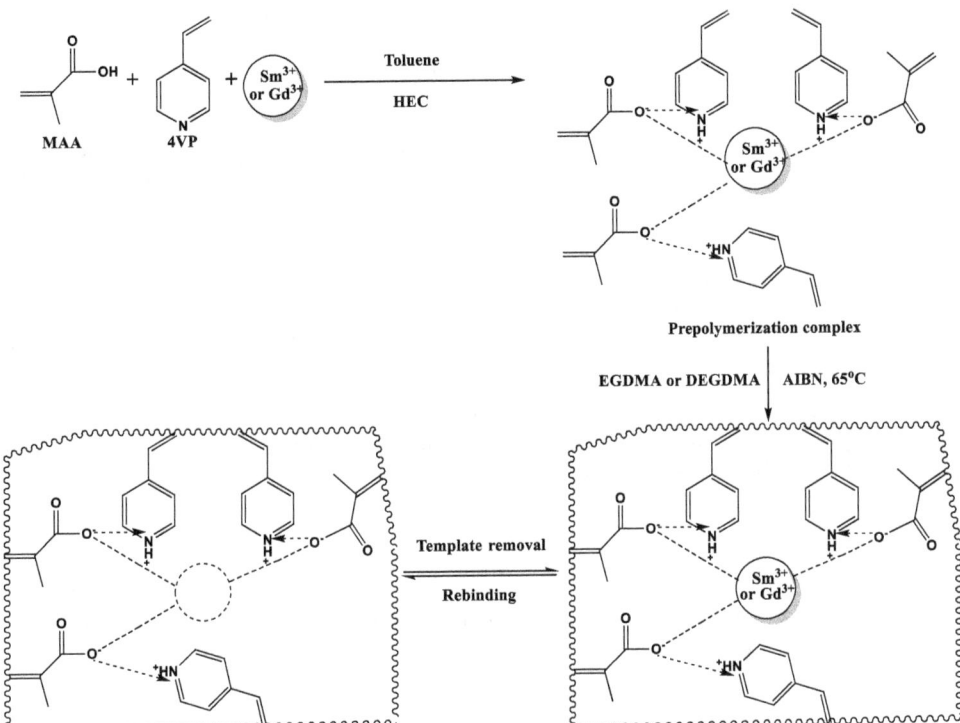

Figure 3. Scheme of the MIP synthesis.

2.2.2. Sorption Experiments

Initially, solutions of REM salts were prepared—hexahydrate nitrates of gadolinium and samarium with concentrations of 100 mg/L. The previously mentioned MIP dispersion (0.12 g) was put into the solutions (200 mL) for 2 days (48 h). The temperature in the laboratory during the sorption experiments was 25 °C. Aliquots of the solutions were sampled at certain intervals, which was necessary for further determination of residual concentrations of REMs.

2.2.3. Study of the Synthetized MIP Selectivity

Laboratory experiments devoted to selective sorption of REMs with their further separation were carried out using a developed installation based on two blocks. Each of the blocks contains cartridges for macromolecular dispersion of MIPs. The first block is filled with a dispersion of MIP-Sm, while the second one is filled with MIP-Gd. The studies were carried out as follows: the model solution (contains Sm and Gd, concentration of each REM is 100 mg/L) is pumped into the 1st block for 48 h (sorption of Sm). Subsequently the solution is pumped into the 2nd block for 48 h (sorption of Gd). During sorption of Sm and Gd, aliquots are sampled at certain intervals. After the sorption/separation process ends, the cartridges can be removed and exchanged with other cartridges containing MIP structures, and the installation is ready to begin a new sorption/separation cycle.

For pumping the model solution, a KNF N 816-3 KT-18 laboratory membrane vacuum pump (KNF, Hamburg, Germany) was used.

2.2.4. Study of Impact of Template Removal Duration on the Sorption Capacity Regeneration Process

The study of the influence of the template removal duration on the efficiency of the sorption capacity regeneration process was carried out as follows:

After sorption of the REMs (Sm and Gd) for 48 h, the structures MIP-Sm and MIP-Gd underwent template removal (desorption) as was described above for the following different times: 20, 25, 30 and 35 h. Subsequently these structures were used for another sorption cycle, and solution aliquots were sampled.

2.2.5. Measurement of Residual Concentrations of Gd and Sm

The residual concentrations of the REMs was determined by the photocolorimetric method and atomic emission spectroscopy; a KFK-3-01 photocolorimeter (ZOMZ, Sergiyev Posad, Russia) and Optima 8300 ICP-OES spectrometer (Perkin-Elmer, Santa Clara, CA, USA) were used.

2.3. Sorption Parameters Calculation

The following sorption parameters are calculated based on the residual concentrations of Sm and Gd in the solution:

(1) Sorption (extraction) degree [66]:

$$\eta = \frac{C_0 - C_e}{C_0} * 100\%$$

where C_0 is the initial concentration of the REM (mg/L); and C_e is the initial (equilibrium) concentration of the REM (mg/L).

(2) Exchange capacity [66]:

$$Q = \frac{m_{sorbed}}{m_{sorbent}}$$

where m_{sorbed} is the mass of the sorbed REM (g); and $m_{sorbent}$ is the mass of the MIP (g).

(3) REM medium sorption efficiency after sorption/desorption cycle:

$$SE_{pi} = \frac{\sum \left(\frac{P_i}{P_0}\right)_m}{n} * 100\%$$

where p is the sorption parameter (sorption degree or exchange capacity); i is the cycles of template removal; m is the time of aliquot taking; n is the number of times aliquots were taken; P_i is the MIP sorption parameter after i cycles of template removal; and P_0 is the initial MIP sorption parameter.

(4) Growth of the sorption parameters depending on amount of template removal cycles:

$$\omega_p = 100\% - SE_{pi}$$

where p is the sorption parameter (sorption degree or exchange capacity); P_i is the MIP sorption parameter (sorption degree or exchange capacity) after i cycles of template removal; and P_0 is the initial MIP sorption parameter.

(5) Sorption parameter medium growth:

$$\omega = \frac{\sum P_i}{i} = \frac{\omega_\eta + \omega_Q}{2}$$

where ω_η is the sorption degree growth (%); and ω_Q is the sorption capacity growth (%).

3. Results and Discussion

The interaction of the synthesized MIPs with samarium and gadolinium salts leads to the sorption of these metals. The sorption character changes with time, and there is the appearance of areas of intense sorption which changes to a slight increase when approaching the equilibrium state between the macromolecular structure and salt solution.

3.1. Sorption of Sm and Gd

Synthetized imprinted structures MIP1-Sm, MIP1-Gd, MIP2-Sm and MIP2-Gd interact with nitrates of samarium and gadolinium, resulting in the sorption of these metals.

Figure 4 presents decreases of the the samarium (a) and gadolinium (b) concentrations during sorption by the imprinted structures. Concentrations of the REMs decrease with time of interaction of the MIPs with the corresponding nitrates. In the case of Sm sorption, the concentration of the metal decreases from 100 mg/L–38.99 mg/L (for MIP1-Sm); for MIP2-Sm, the concentration of Sm decreases from 100 mg/L–46.45 mg/L during first 12 h of interaction. The mean difference of the Sm concentration decrease in this time interval (from 0 h to 12 h) for the structures MIP1-Sm and MIP2-Sm is 9.42 mg/L. The further decrease of the Sm concentration is not so intense for the both the imprinted structures; the mean difference of the Sm concentration decrease in this time interval (from 12 h to 48 h) for the structures MIP1-Sm and MIP2-Sm is 4.06 mg/L. In the case of Gd sorption, a strong decrease of the metal concentration is observed during 12 h after the beginning of the contact. The Gd concentration decreases from 100 mg/L to 42.33 mg/L for MIP1-Gd and from 100 mg/L to 52.57 mg/L for MIP2-Gd. The mean difference of the Gd concentration decrease in this time interval (from 0 h to 12 h) for the structures MIP1-Gd and MIP2-Gd is 10.00 mg/L. The subsequent decrease in the concentration occurs more slightly for the both imprinted structures up to 48 h. The mean difference of the Sm concentration decrease in this time interval (from 12 h to 48 h) for the structures MIP1-Sm and MIP2-Sm is 3.42 mg/L.

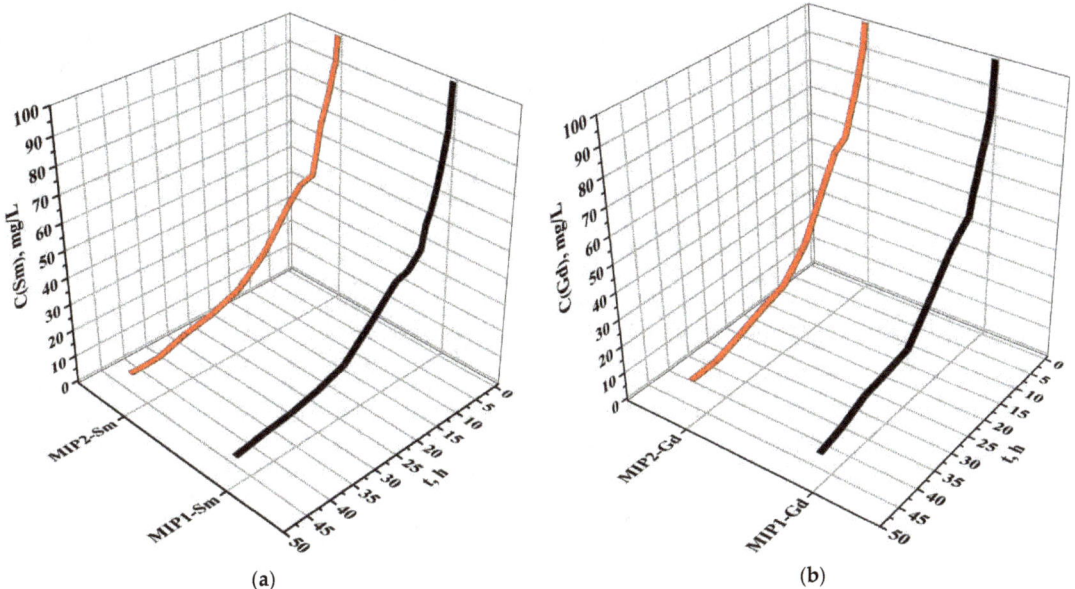

Figure 4. Decrease of samarium (a) and gadolinium (b) concentrations during sorption via molecularly imprinted polymers.

The values of the Sm and Gd concentrations as they decrease during the metals' sorption by the imprinted structures are presented in Table 1.

Table 1. Decreasing values of samarium and gadolinium concentrations via imprinted structures MIP1, MIP2.

t, h	C (Sm) mg/L				C (Gd) mg/L			
	MIP1-Sm	MIP2-Sm	Non-MIP1	Non-MIP2	MIP1-Gd	MIP2-Gd	Non-MIP1	Non-MIP2
0	100.00	100.00	100.00	100.00	100.00	100.00	100.00	100.00
0.5	83.96	90.50	100.00	100.00	84.13	91.99	100.00	100.00
1	79.53	88.35	100.00	100.00	80.09	88.58	100.00	100.00
2	70.75	81.01	100.00	100.00	73.58	82.54	100.00	100.00
3	63.44	75.85	100.00	100.00	69.08	77.02	100.00	100.00
4	57.98	70.70	100.00	100.00	62.72	72.78	100.00	100.00
5	54.33	61.82	100.00	100.00	57.98	67.74	100.00	100.00
6	45.69	53.38	100.00	100.00	50.63	63.53	100.00	100.00
9	40.55	51.91	100.00	100.00	46.95	60.78	100.00	100.00
12	38.99	46.45	100.00	100.00	42.33	52.57	100.00	100.00
18	28.66	32.68	100.00	100.00	30.85	34.41	100.00	100.00
24	19.01	23.87	100.00	100.00	20.05	24.82	100.00	100.00
30	15.37	19.83	100.00	100.00	18.44	21.87	100.00	100.00
36	13.08	17.69	100.00	100.00	16.89	18.46	100.00	100.00
42	11.83	14.28	100.00	100.00	13.01	15.65	100.00	100.00
48	10.01	13.98	100.00	100.00	10.89	15.46	100.00	100.00

The extraction degrees of samarium and gadolinium by the imprinted structures MIP1-Sm, MIP2-Sm, MIP1-Gd and MIP2-Gd are presented in Figures 5a and 5b, respectively. The extraction degree of samarium increases with time, and a strong increase is observed during the first 24 h of interaction for both imprinted structures. At this time of interaction, the sorption degree is 80.99 for MIP1-Sm and 76.13 for MIP2-Sm, with 90.00% and 88.50% of the total samarium amount sorbed, respectively. The further increase (in the interval of time 24–48 h) in the sorption degree is very slight: for MIP1-Sm, it is 84.63%–86.92%–88.17%–89.99%; for MIP2-Sm, it is 80.17%–82.31%–85.72%–86.02%; the time of interaction is 30 h–36 h–42 h–48 h in both cases. In other words, the increase of extraction degree during the subsequent 24 h is over 5–6%. The sorption degree of gadolinium also increases with time, and a strong increase can be observed in the first 24 h of interaction of the imprinted structures with the salt solution. At 24 h the extraction degree is 79.95% for MIP1-Gd and 75.18% for MIP2-Gd, with 88.93% and 89.72% of the total amount of gadolinium sorbed, respectively. There is subsequently a slight growth of the extraction degree up to the 48 h mark for MIP1-Gd and MIP2-Gd. The increase occurs as follows: for MIP1-Gd, it is 81.56%–83.11%–86.99%–89.11%; for MIP2-Gd, it is 78.13%–81.54%–84.35%–84.54%; the interaction time is 30 h–36 h–42 h–48 h in both cases. The parameter increases over 6–7% during the second day. The phenomenon of a slight increase of the sorption degree during the second day points to equilibrium being achieved between MIP structures and the salt solution.

The values of extraction degrees of samarium and gadolinium are presented in Table 2.

Figure 6 shows exchange capacity values (in relation to Sm and Gd) of MIP1-Sm and MIP2-Sm (a) and MIP1-Gd and MIP2-Gd (b). In both cases (sorption of Sm and Gd), a significant increase of the exchange capacity (over 90% of the total values) is observed at 24 h of interaction. The further increase (up to 48 h) of this sorption parameter is insignificant.

The exchange capacity values (in relation to samarium and gadolinium) of the imprinted structures MIP1 and MIP2 are presented in Table 3.

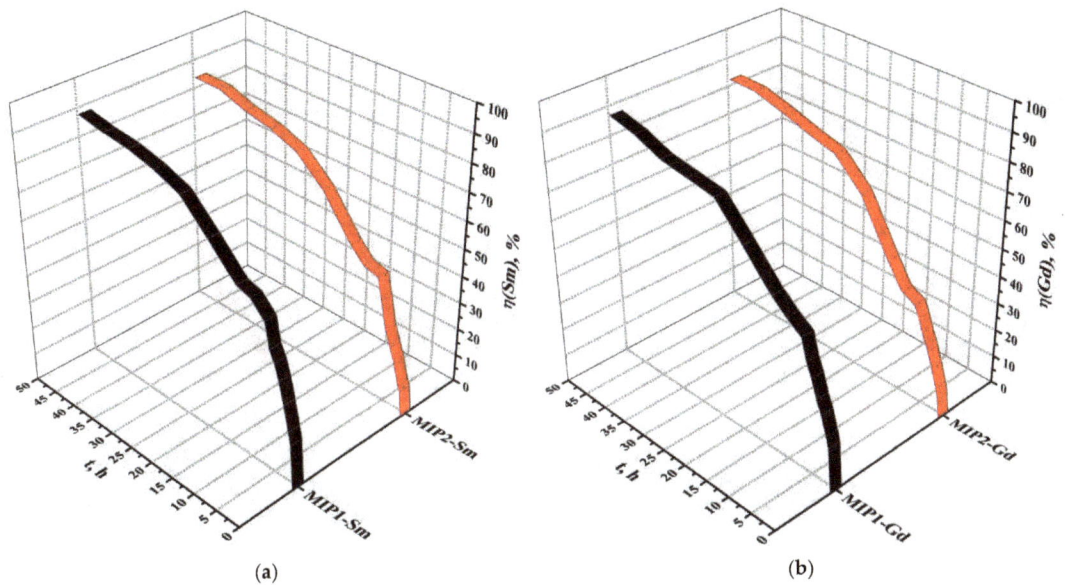

Figure 5. Extraction degrees of samarium (**a**) and gadolinium (**b**) during sorption via molecularly imprinted polymers.

Table 2. Values of extraction degrees of samarium and gadolinium.

t, h	η (Sm), %				η (Gd), %			
	MIP1-Sm	MIP2-Sm	Non-MIP1	Non-MIP2	MIP1-Gd	MIP2-Gd	Non-MIP1	Non-MIP2
0	0	0	0	0	0	0	0	0
0.5	16.04	9.50	0	0	15.87	8.01	0	0
1	20.47	11.65	0	0	19.91	11.42	0	0
2	29.25	18.99	0	0	26.42	17.46	0	0
3	36.56	24.15	0	0	30.92	22.98	0	0
4	42.02	29.30	0	0	37.28	27.22	0	0
5	45.67	38.18	0	0	42.02	32.26	0	0
6	54.31	46.62	0	0	49.37	36.47	0	0
9	59.45	48.09	0	0	53.05	39.22	0	0
12	61.01	53.55	0	0	57.67	47.43	0	0
18	71.34	67.32	0	0	69.15	65.59	0	0
24	80.99	76.13	0	0	79.95	75.18	0	0
30	84.63	80.17	0	0	81.56	78.13	0	0
36	86.92	82.31	0	0	83.11	81.54	0	0
42	88.17	85.72	0	0	86.99	84.35	0	0
48	89.99	86.02	0	0	89.11	84.54	0	0

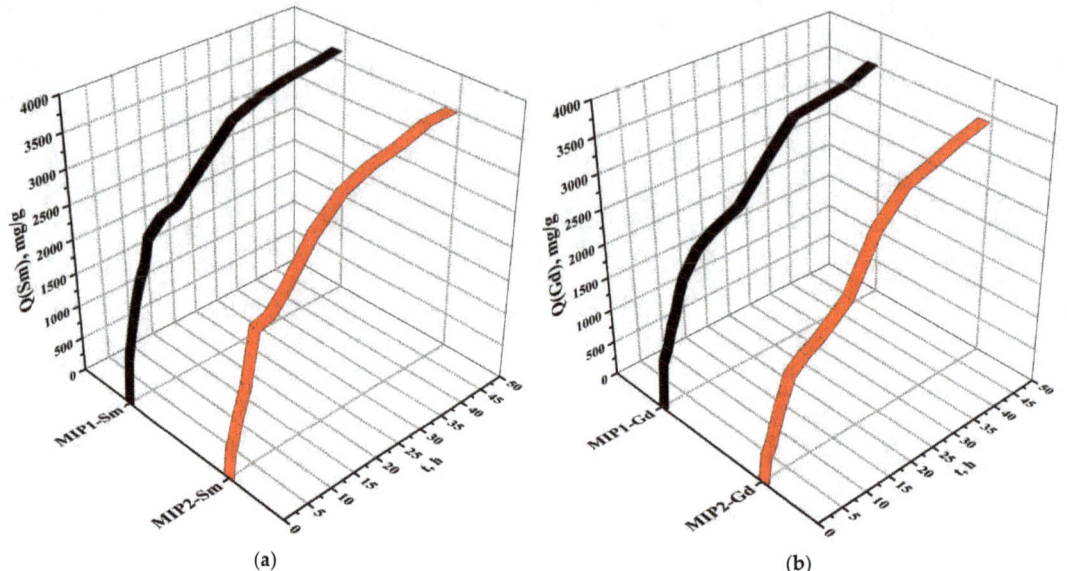

Figure 6. Exchange capacities (in relation to samarium (a) and gadolinium (b)) of imprinted structures MIP1, MIP2.

Table 3. Values of exchange capacity (in relation to samarium and gadolinium) of imprinted structures MIP1 and MIP2.

t, h	Q (Sm), mg/g				Q (Gd), mg/g			
	MIP1-Sm	MIP2-Sm	Non-MIP1	Non-MIP2	MIP1-Gd	MIP2-Gd	Non-MIP1	Non-MIP2
0	0	0	0	0	0	0	0	0
0.5	668.33	395.80	0	0	661.25	333.83	0	0
1	852.92	485.50	0	0	829.58	475.97	0	0
2	1218.75	791.43	0	0	1100.83	727.30	0	0
3	1523.33	1006.37	0	0	1288.33	957.40	0	0
4	1750.83	1220.87	0	0	1553.33	1134.20	0	0
5	1902.92	1590.93	0	0	1750.83	1344.37	0	0
6	2262.92	1942.37	0	0	2057.08	1519.43	0	0
9	2477.08	2003.90	0	0	2210.42	1634.27	0	0
12	2542.08	2231.40	0	0	2402.92	1976.17	0	0
18	2972.50	2805.13	0	0	2881.25	2732.77	0	0
24	3374.58	3172.17	0	0	3331.25	3132.30	0	0
30	3526.25	3340.30	0	0	3398.33	3255.37	0	0
36	3621.67	3429.57	0	0	3462.92	3397.50	0	0
42	3673.75	3571.70	0	0	3624.58	3514.50	0	0
48	3749.58	3584.27	0	0	3712.92	3522.30	0	0

The obtained data show that the non-imprinted structures non-MIP1 and non-MIP2 sorb neither samarium nor gadolinium, which evidences that the synthetized structures MIP1-Sm, MIP2-Sm, MIP1-Gd and MIP2-Gd have selectivity for samarium and gadolinium. The differences in the sorption properties of the structures MIP1 and MIP2 are based on the application of different cross-linking agents during their synthesis. It is supposed that in the case of MIP2, the cross-linking is tighter compared to MIP1, and the sorption process is rather complicated.

3.2. Impact of Amount of Template Removal Cycles on Regeneration of MIP Sorption Efficiency

The sorption efficiency studies are based on the amount of template removal cycles. Herein and after, sorption properties (extraction degree, exchange capacity) at 40 template removal cycles (as described in the synthesis procedure) are taken as 100%.

The values of the sorption properties of the imprinted structures MIP1 and MIP2, dependent on the amount of template removal cycles, are presented in Tables 4–7.

Table 4. Values of extraction degrees of samarium after certain amount of template removal cycles.

t, h/ Number of Cycles	η (Sm), %							
	MIP1-Sm				MIP2-Sm			
	20	25	30	35	20	25	30	35
0	0	0	0	0	0	0	0	0
0.5	10.11	10.91	12.99	14.76	4.84	6.55	7.03	8.36
1	12.90	13.92	16.58	18.83	5.94	8.04	8.62	10.25
2	18.43	19.89	23.69	26.91	9.69	13.11	14.06	16.72
3	23.03	24.86	29.61	33.64	12.32	16.67	17.87	21.25
4	26.47	28.57	34.04	38.66	14.94	20.22	21.68	25.78
5	28.77	31.06	36.99	42.02	19.47	26.35	28.25	33.60
6	34.22	36.93	43.99	49.97	23.77	32.17	34.50	41.02
9	37.45	40.43	48.15	54.69	24.53	33.18	35.59	42.32
12	38.44	41.49	49.42	56.13	27.31	36.95	39.63	47.13
18	44.94	48.51	57.79	65.63	34.33	46.45	49.82	59.24
24	51.02	55.07	65.60	74.51	38.83	52.53	56.34	67.00
30	53.32	57.55	68.55	77.86	40.89	55.32	59.32	70.55
36	54.76	59.11	70.41	79.97	41.98	56.79	60.91	72.43
42	55.55	59.96	71.42	81.12	43.72	59.15	63.43	75.43
48	56.69	61.19	72.89	82.79	43.87	59.36	63.66	75.70

Table 5. Values of exchange capacity (in relation to samarium) after certain amount of template removal cycles.

t, h/ Number of Cycles	Q (Sm), mg/g							
	MIP1-Sm				MIP2-Sm			
	20	25	30	35	20	25	30	35
0	0	0	0	0	0	0	0	0
0.5	421.05	454.47	541.35	614.87	201.86	273.10	292.89	348.30
1	537.34	579.98	690.86	784.68	247.61	335.00	359.27	427.24
2	767.81	828.75	987.19	1121.25	403.63	546.09	585.66	696.46
3	959.70	1035.87	1233.90	1401.47	513.25	694.39	744.71	885.60
4	1103.03	1190.57	1418.18	1610.77	622.64	842.40	903.44	1074.36
5	1198.84	1293.98	1541.36	1750.68	811.38	1097.74	1177.29	1400.02
6	1425.64	1538.78	1832.96	2081.88	990.61	1340.23	1437.35	1709.28
9	1560.56	1684.42	2006.44	2278.92	1021.99	1382.69	1482.89	1763.43
12	1601.51	1728.62	2059.09	2338.72	1138.01	1539.67	1651.24	1963.63
18	1872.68	2021.30	2407.73	2734.70	1430.62	1935.54	2075.80	2468.52
24	2125.99	2294.72	2733.41	3104.62	1617.81	2188.80	2347.40	2791.51
30	2221.54	2397.85	2856.26	3244.15	1703.55	2304.81	2471.82	2939.46
36	2281.65	2462.73	2933.55	3331.93	1749.08	2366.40	2537.88	3018.02
42	2314.46	2498.15	2975.74	3379.85	1821.57	2464.47	2643.06	3143.10
48	2362.24	2549.72	3037.16	3449.62	1827.98	2473.14	2652.36	3154.15

Table 6. Values of extraction degree of gadolinium after certain amount of template removal cycles.

t, h/ Number of Cycles	η (Gd), %							
	MIP1-Gd				MIP2-Gd			
	20	25	30	35	20	25	30	35
0	0	0	0	0	0	0	0	0
0.5	9.36	10.47	13.17	14.28	3.61	5.05	5.69	6.81
1	11.75	13.14	16.53	17.92	5.14	7.20	8.11	9.71
2	15.59	17.44	21.93	23.78	7.85	11.00	12.39	14.84
3	18.24	20.41	25.66	27.83	10.34	14.48	16.31	19.53
4	22.00	24.60	30.94	33.55	12.25	17.15	19.33	23.14
5	24.79	27.73	34.88	37.82	14.52	20.33	22.91	27.43
6	29.13	32.58	40.98	44.43	16.41	22.97	25.89	31.00
9	31.30	35.01	44.03	47.75	17.65	24.71	27.85	33.34
12	34.03	38.06	47.87	51.90	21.34	29.88	33.67	40.31
18	40.80	45.64	57.39	62.24	29.51	41.32	46.57	55.75
24	47.17	52.77	66.36	71.96	33.83	47.36	53.37	63.90
30	48.12	53.83	67.69	73.40	35.16	49.22	55.47	66.41
36	49.03	54.85	68.98	74.80	36.69	51.37	57.89	69.31
42	51.32	57.41	72.20	78.29	37.96	53.14	59.89	71.70
48	52.57	58.81	73.96	80.20	38.04	53.26	60.02	71.85

Table 7. Values of exchange capacity (in relation to gadolinium) after certain amount of template removal cycles.

t, h/ Number of Cycles	Q (Gd), mg/g							
	MIP1-Gd				MIP2-Gd			
	20	25	30	i35	20	25	30	35
0	0	0	0	0	0	0	0	0
0.5	390.14	436.43	548.84	595.13	150.23	210.32	237.02	283.76
1	489.45	547.53	688.55	746.63	214.19	299.86	337.94	404.57
2	649.49	726.55	913.69	990.75	327.29	458.20	516.38	618.21
3	760.12	850.30	1069.32	1159.50	430.83	603.16	679.75	813.79
4	916.47	1025.20	1289.27	1398.00	510.39	714.55	805.28	964.07
5	1032.99	1155.55	1453.19	1575.75	604.97	846.95	954.50	1142.71
6	1213.68	1357.68	1707.38	1851.38	683.75	957.24	1078.80	1291.52
9	1304.15	1458.88	1834.65	1989.38	735.42	1029.59	1160.33	1389.13
12	1417.72	1585.93	1994.42	2162.63	889.28	1244.99	1403.08	1679.74
18	1699.94	1901.63	2391.44	2593.13	1229.75	1721.64	1940.26	2322.85
24	1965.44	2198.63	2764.94	2998.13	1409.54	1973.35	2223.93	2662.46
30	2005.02	2242.90	2820.62	3058.50	1464.92	2050.88	2311.31	2767.06
36	2043.12	2285.53	2874.22	3116.63	1528.88	2140.43	2412.23	2887.88
42	2138.50	2392.23	3008.40	3262.13	1581.53	2214.14	2495.30	2987.33
48	2190.62	2450.53	3081.72	3341.63	1585.04	2219.05	2500.83	2993.96

A comparative analysis of the sorption efficiency of the MIP1 and MIP2 structures, dependent on template removal cycles, is presented in Figure 7. The obtained data show the following sorption efficiency values: for MIP1-Sm, 63% (20 cycles), 68% (25 cycles), 81% (30 cycles) and 92% (35 cycles); for MIP2-Sm, 51% (20 cycles), 69% (25 cycles), 74% (30 cycles) and 88% (35 cycles); for MIP1-Gd, 63% (20 cycles), 59% (25 cycles), 83% (30 cycles) and 90% (35 cycles); and for MIP2-Gd, 45% (20 cycles), 63% (25 cycles), 71% (30 cycles) and 85% (35 cycles). Complete removal of the sorbed REM from the imprinted structure's matrix is a complicated process. Based on the obtained data, it can be concluded that an increase of the template removal cycles by 5 each time provides an average growth of the sorption efficiency of 11.37%.

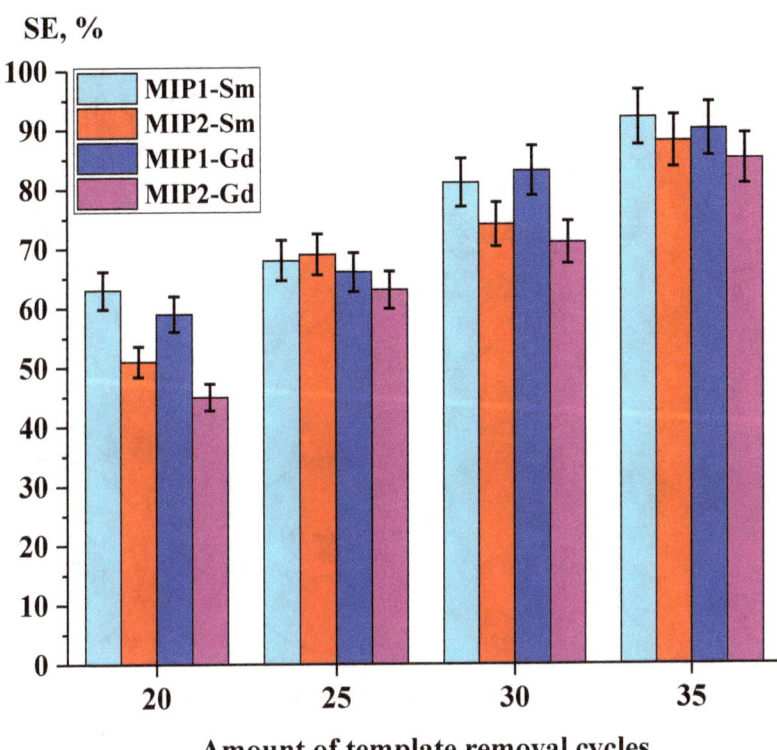

Figure 7. Sorption efficiency of imprinted structures MIP1 and MIP2 depending on amount of template removal cycles.

3.3. Laboratory Experiments on Selective Sorption and Sorption of Sm and Gd

Both imprinted structures (MIP1 and MIP2) are used for the laboratory tests devoted to selective sorption of Sm and Gd. The schematical sorption process during the laboratory tests is shown in Figure 8. As mentioned earlier, laboratory studies on selective sorption of Sm and Gd were carried out with the application of the developed installation (based on two blocks). Each block contains cartridges for the placement of macromolecular dispersions of the imprinted sorbents. The first block is filled with a dispersion of MIP-Sm, and the second one is filled with MIP-Gd. The model solution is pumped into the first block for 48 h (selective sorption of Sm). Subsequently, the solution is pumped into the second block for 48 h (selective sorption of Gd). After the sorption process of both REMs ends, the cartridges with the imprinted structures can be removed and exchanged with other cartridges with MIP structures, and the installation is ready for a new sorption/separation cycle.

The sorption properties of the MIP1 and MIP2 structures during selective sorption of samarium and gadolinium are presented in Tables 8–11.

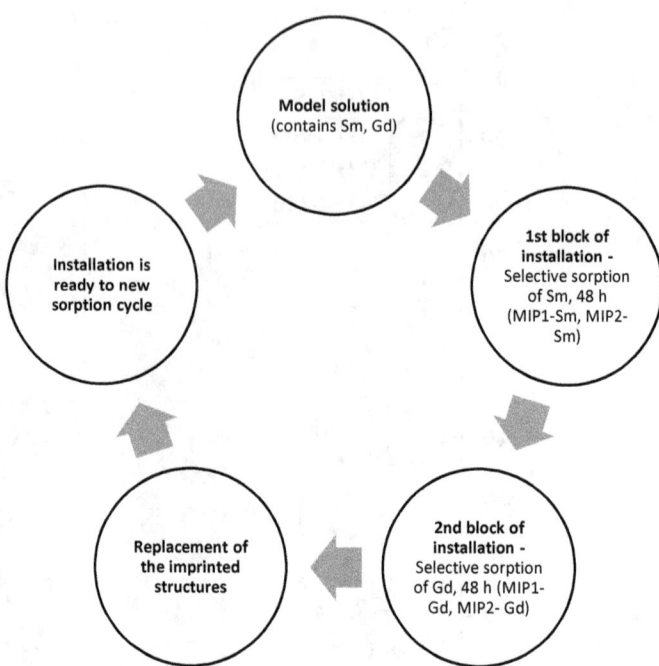

Figure 8. Sorption efficiency of imprinted structures MIP1 and MIP2 in dependence from amount of template removal cycles.

Table 8. Sorption properties of the MIP1-Sm and MIP1-Gd structures during selective sorption of samarium.

t, h	Selective Sorption of Sm			
	MIP1-Sm		MIP1-Gd	
	η (Sm), %	Q (Sm), mg/g	η (Sm), %	Q (Sm), mg/g
0	0	0	0	0
0.5	16.04	668.33	0	0
1	20.47	852.92	0	0
2	29.25	1218.75	0	0
3	36.56	1523.33	0	0
4	42.02	1750.83	0	0
5	45.67	1902.92	0	0
6	54.31	2262.92	0	0
9	59.45	2477.08	0	0
12	61.01	2542.08	0	0
18	71.34	2972.50	0	0
24	80.99	3374.58	0	0
30	84.63	3526.25	0	0
36	86.92	3621.67	0	0
42	88.17	3673.75	0	0
48	89.99	3749.58	0	0

Table 9. Sorption properties of the MIP1-Sm and MIP1-Gd structures during selective sorption of gadolinium.

t, h	Selective Sorption of Gd			
	MIP1-Sm		MIP1-Gd	
	η (Sm), %	Q (Sm), mg/g	η (Sm), %	Q (Sm), mg/g
0	0	0	0	0
0.5	0	0	15.87	661.25
1	0	0	19.91	829.58
2	0	0	26.42	1100.83
3	0	0	30.92	1288.33
4	0	0	37.28	1553.33
5	0	0	42.02	1750.83
6	0	0	49.37	2057.08
9	0	0	53.05	2210.42
12	0	0	57.67	2402.92
18	0	0	69.15	2881.25
24	0	0	79.95	3331.25
30	0	0	81.56	3398.33
36	0	0	83.11	3462.92
42	0	0	86.99	3624.58
48	0	0	89.11	3712.92

Table 10. Sorption properties of the MIP2-Sm and MIP2-Gd structures during selective sorption of samarium.

t, h	Selective Sorption of Sm			
	MIP2-Sm		MIP2-Gd	
	η (Sm), %	Q (Sm), mg/g	η (Sm), %	Q (Sm), mg/g
0	0	0	0	0
0.5	9.50	395.80	0	0
1	11.65	485.50	0	0
2	18.99	791.43	0	0
3	24.15	1006.37	0	0
4	29.30	1220.87	0	0
5	38.18	1590.93	0	0
6	46.62	1942.37	0	0
9	48.09	2003.90	0	0
12	53.55	2231.40	0	0
18	67.32	2805.13	0	0
24	76.13	3172.17	0	0
30	80.17	3340.30	0	0
36	82.31	3429.57	0	0
42	85.72	3571.70	0	0
48	86.02	3584.27	0	0

As seen from the obtained results of the laboratory tests, the molecular imprinting technique seems to be a promising method for the creation of principally new macromolecular sorbents for the selective sorption and separation of rare-earth metals. The developed MIP1 and MIP2 structures showed effectiveness in selective sorption of the target rare-earth metal, wherein the accompanying sorption of another rare-earth metal was absent. Sorption of the target metal occurs due to the presence of complementary cavities in the polymer matrix of the imprinted structures. As was shown earlier, MIP1 structures are more effective for the sorption of REMs due to the fact that their sorption properties are higher (in comparison with MIP2).

Table 11. Sorption properties of the MIP2-Sm and MIP2-Gd structures during selective sorption of gadolinium.

t, h	Selective Sorption of Gd			
	MIP2-Sm		MIP2-Gd	
	η (Sm), %	Q (Sm), mg/g	η (Sm), %	Q (Sm), mg/g
0	0	0	0	0
0.5	0	0	8.01	333.83
1	0	0	11.42	475.97
2	0	0	17.46	727.30
3	0	0	22.98	957.40
4	0	0	27.22	1134.20
5	0	0	32.26	1344.37
6	0	0	36.47	1519.43
9	0	0	39.22	1634.27
12	0	0	47.43	1976.17
18	0	0	65.59	2732.77
24	0	0	75.18	3132.30
30	0	0	78.13	3255.37
36	0	0	81.54	3397.50
42	0	0	84.35	3514.50
48	0	0	84.54	3522.30

4. Conclusions

The developed molecularly imprinted polymers based on MAA and 4VP functional monomers showed good results in the selective sorption/separation of samarium and gadolinium from the common solution. The main advantage of the developed MIPs is the absence of the sorption of accompanying metals (with very close chemical properties) Based on these sorbents, modern effective sorption technologies can be created. As drawbacks, only the complicated procedure of complete desorption of the target metal can be named, along with the relatively high cost of the reactants for MIP synthesis. Despite the mentioned drawbacks, the imprinted sorbents are relevant to be used firstly for modification of existing sorption technologies instead of ion exchangers. Comparison of the synthetized MIPs with the other methods of REM concentration such as extraction of the target REM by phosphoric acid, phosphinic acid and organophosphorus acids [67,68] showed that extraction is high (70–80%), but the drawback that a concentrate of the REM is extracted, and further separation is another complicated step which will make the sorption/separation technology in industry more expensive. The obtained data showed that sorption degree of Sm and Gd by the developed MIPs is almost 90%, and the fact that simultaneous sorption of other metals is absent provides the possibility of using these macromolecular structures as highly effective alternatives to the existing sorption technologies.

Author Contributions: Conceptualization, R.K.; data curation, R.K., Y.M. and L.A.; formal analysis, Y.M. and L.A.; funding acquisition, R.K.; investigation, R.K., Y.M. and L.A.; methodology, R.K. and L.A.; project administration, R.K.; resources, Y.M. and L.A.; supervision, R.K.; validation, Y.M.; visualization, Y.M. and L.A.; writing—original draft, R.K., Y.M. and L.A.; writing—review and editing, R.K. All authors have read and agreed to the published version of the manuscript.

Funding: This research was funded by the Ministry of Science and Higher Education of the Republic of Kazakhstan, grant number AP13067631.

Institutional Review Board Statement: Not applicable.

Informed Consent Statement: Not applicable.

Data Availability Statement: The data presented in this study are available upon request from the corresponding author.

Conflicts of Interest: The authors declare no conflict of interest.

References

1. Hong, F. Rare earth: Production, trade and demand. *J. Iron Steel Res. Int.* **2006**, *13*, 33–38. [CrossRef]
2. Naumov, A.V. Review of the World Market of Rare-Earth Metals. *Russ. J. Non-Ferr. Met.* **2008**, *49*, 14–22. [CrossRef]
3. Zhou, B.L.; Li, Z.X.; Zhao, Y.Q.; Chen, C.C. China's rare earth markets: Value chain and the implications. *Adv. Energy Environ. Res.* **2017**, 85–90. [CrossRef]
4. Pan, A.; Feng, S.S.; Hu, X.Y.; Li, Y.Y. How environmental regulation affects China's rare earth export? *PLoS ONE* **2021**, *16*, e0250407. [CrossRef]
5. Guo, Q.; Mai, Z.S. A Comparative Study on the Export Competitiveness of Rare Earth Products from China, the United States, Russia and India. *Sustainability* **2022**, *14*, 12358. [CrossRef]
6. Klossek, P.; Kullik, J.; van den Boogaart, K.G. A systemic approach to the problems of the rare earth market. *Resour. Policy* **2016**, *50*, 131–140. [CrossRef]
7. Shi, D.F.; Zhang, S.T. Analysis of the Rare Earth Mineral Resources Reserve System and Model Construction Based on Regional Development. *Comput. Intell. Neurosci.* **2022**, *2022*, 9900219. [CrossRef]
8. Goonan, T.G. *Rare Earth Elements—End Use and Recyclability: U.S. Geological Survey Scientific Investigations Report 2011–5094*; U.S. Geological Survey: Reston, VA, USA, 2011. Available online: https://pubs.usgs.gov/sir/2011/5094/pdf/sir2011-5094.pdf (accessed on 5 December 2011).
9. Yu, C.; Higano, Y.; Mizunoya, T.; Ge, J.P.; Lei, Y.L. Analysis of Industrial Correlation Effect of Rare Earth Mining in China: Based on the 2010 Extended Input-output Table of China. *J. Environ. Account. Manag.* **2017**, *5*, 185–199. [CrossRef]
10. Buynovskiy, A.S.; Zhiganov, A.N.; Sofronov, V.L.; Sachkov, V.I.; Daneikina, N.V. Current state of the rare earth industry in Russia and Siberia. *Procedia Chem.* **2014**, *11*, 126–132. [CrossRef]
11. Jurjo, S.; Siinor, L.; Siimenson, C.; Paiste, P.; Lust, E. Two-Step Solvent Extraction of Radioactive Elements and Rare Earths from Estonian Phosphorite Ore Using Nitrated Aliquat 336 and Bis(2-ethylhexyl) Phosphate. *Minerals* **2021**, *11*, 388. [CrossRef]
12. Lima, F.M.; Lovon-Canchumani, G.A.; Sampaio, M.; Tarazona-Alvarado, L.M. Life cycle assessment of the production of rare earth oxides from a Brazilian ore. *Procedia CIRP* **2018**, *69*, 481–486. [CrossRef]
13. Zepf, V. *Rare Earth Elements. A New Approach to the Nexus of Supply, Demand and Use: Exemplified along the Use of Neodymium in Permanent Magnets*; Springer: Berlin/Heidelberg, Germany, 2013; 162p. [CrossRef]
14. Bisaka, K.; Thobadi, I.C.; Pawlik, C. Extraction of rare earths from iron-rich rare earth deposits. *J. South. Afr. Inst. Min. Metall.* **2017**, *117*, 731–739. [CrossRef]
15. Ananyev, Y. Rare-earth element distribution patterns in metasomatites of Eastern Kazakhstan gold-ore deposits. In Proceedings of the Conference on Advanced Engineering Problems in Drilling, Tomsk, Russia, 24–27 November 2014; IOP Publishing: Bristol, UK, 2015; Volume 24, p. 012002. [CrossRef]
16. Manoj, M.C.; Thakur, B.; Prasad, V. Rare earth element distribution in tropical coastal wetland sediments: A case study from Vembanad estuary, southwest India. *Arab. J. Geosci.* **2016**, *9*, 197. [CrossRef]
17. Sivasamandy, R.; Satyanarayanan, M.; Ramesh, R. Distribution of rare earth elements in core sediments of kolakkudi lake, southern india. *Acta Geodyn. Et Geomater.* **2019**, *16*, 473–486. [CrossRef]
18. Chelnokov, G.A.; Kharitonova, N.A.; Bragin, I.V.; Aseeva, A.V.; Bushkareva, K.Y.; Liamina, L.A. The Geochemistry of Rare Earth Elements in Natural Waters and Secondary Mineral Sediments of Thermal Fields of Kamchatka. *Mosc. Univ. Geol. Bull.* **2020**, *75*, 196–204. [CrossRef]
19. Binnemans, K.; Jones, P.T. Rare Earths and the Balance Problem. *J. Sustain. Metall.* **2015**, *1*, 29–38. [CrossRef]
20. Tourre, J.M. Rare earth raw materials and application markets—A balancing act. In Proceedings of the 3rd TMS Symposium on Rare Earths—Science, Technology and Applications, at the TMS 1997 Annual Meeting, Orlando, Fl, USA, 9–13 February 1997; TMS Publishing: Pittsburgh, PA, USA, 1996; pp. 223–234.
21. Li, Y.; Wang, A.J.; Li, J.W.; Chen, Q.S. A Brief Analysis of Global Rare Earth Trade Structure. In Proceedings of the 2nd International Conference on Energy and Environmental Protection (ICEEP 2013), Guilin, China, 19–21 April 2013; Scientific.Net Publishing: Zurich, Switzerland, 2013; Volume 734–737, pp. 3324–3331. [CrossRef]
22. Charalampides, G.; Vatalis, K.; Karayannis, V.; Baklavaridis, A. Environmental defects and economic impact on global market of rare earth metals. In Proceedings of the 20th Innovative Manufacturing Engineering and Energy Conference (IMANEE 2016), Kallithea, Greece, 23–25 September 2016; IOP Publishing: Bristol, England, 2016; Volume 161, p. 012069. [CrossRef]
23. Zhang, L.; Guo, Q.; Zhang, J.B.; Huang, Y.; Xiong, T. Did China's rare earth export policies work?—Empirical evidence from USA and Japan. *Resour. Policy* **2015**, *43*, 82–90. [CrossRef]
24. Gasanov, A.A.; Naumov, A.V.; Yurasova, O.V.; Petrov, I.M.; Litvinova, T.E. Certain Tendencies in the Rare-Earth-Element World Market and Prospects of Russia. *Russ. J. Non-Ferr. Met.* **2018**, *59*, 502–511. [CrossRef]
25. Yan, G.L.; Li, Z.X. International Rare Earth Supply and Demand Forecast Based on Panel Data Analysis. In Proceedings of the 5th International Conference on Energy Equipment Science and Engineering (ICEESE), Harbin, China, 29 November–1 December 2019; IOP Publishing: Bristol, UK, 2020; Volume 461, p. 012011. [CrossRef]
26. Wang, S.J. Rare Earth Metals: Resourcefulness and Recovery. *JOM* **2013**, *65*, 1317–1320. [CrossRef]
27. Fernandez, V. Rare-earth elements market: A historical and financial perspective. *Resour. Policy* **2017**, *53*, 23–45. [CrossRef]
28. Zhdaneev, O.V.; Petrov, Y.I.; Seregina, A.A. Rare and rare-earth metals industry development in Russia and its influence on fourth world energy transition. *Non-Ferr. Met.* **2021**, *2*, 3–8. [CrossRef]

29. Ponomarenko, O.M.; Samchuk, A.I.; Vovk, K.V.; Zaits, O.V.; Kuraeva, I.V. The newest analytical technologies for determining rare-earth elements in granitoids of the ukrainian shield. *Sci. Innov.* **2021**, *17*, 96–102. [CrossRef]
30. Habashi, F. Extractive metallurgy of rare earths. *Can. Metall. Q.* **2013**, *52*, 224–233. [CrossRef]
31. Nicholls, I.A.; Adbo, K.; Andersson, H.S.; Andersson, P.O.; Ankarloo, J.; Hedin-Dahlstrom, J.; Jokela, P.; Karlsson, J.G.; Olofsson, L.; Rosengren, J.; et al. Can we rationally design molecularly imprinted polymers? *Anal. Chim. Acta* **2001**, *435*, 9–18. [CrossRef]
32. Toth, B.; Pap, T.; Horvath, V.; Horvai, G. Which molecularly imprinted polymer is better? *Anal. Chim. Acta* **2001**, *591*, 7–21. [CrossRef]
33. Lai, J.P.; He, X.W.; Guo, H.S.; Liang, H. A review on molecular imprinting technique. *Chin. J. Anal. Chem.* **2001**, *29*, 836–844.
34. Haupt, K.; Linares, A.V.; Bompart, M.; Bernadette, T.S.B. Molecularly Imprinted Polymers. *Mol. Impr.* **2012**, *325*, 1–28. [CrossRef]
35. Yano, K.; Karube, I. Molecularly imprinted polymers for biosensor applications. *Trac-Trends Anal. Chem.* **1999**, *18*, 199–204. [CrossRef]
36. Byrne, M.E.; Park, K.; Peppas, N.A. Molecular imprinting within hydrogels. *Adv. Drug Deliv. Rev.* **2002**, *54*, 149–161. [CrossRef]
37. Cai, W.S.; Gupta, R.B. Molecularly-imprinted polymers selective for tetracycline binding. *Sep. Purif. Technol.* **2004**, *35*, 215–221. [CrossRef]
38. Komiyama, M.; Mori, T.; Ariga, K. Molecular Imprinting: Materials Nanoarchitectonics with Molecular Information. *Bull. Chem. Soc. Jpn.* **2018**, *91*, 1075–1111. [CrossRef]
39. Pohanka, M. Sensors Based on Molecularly Imprinted Polymers. *Int. J. Electrochem. Sci.* **2017**, *12*, 8082–8094. [CrossRef]
40. Chen, W.; Liu, F.; Xu, Y.T.; Li, K.A.; Tong, S.Y. Molecular recognition of procainamide-imprinted polymer. *Anal. Chim. Acta* **2001**, *432*, 277–282. [CrossRef]
41. Cormack, P.A.G.; Elorza, A.Z. Molecularly imprinted polymers: Synthesis and characterization. *J. Chromatogr. B* **2004**, *804*, 173–182. [CrossRef]
42. Allender, C.J.; Richardson, C.; Woodhouse, B.; Heard, C.M.; Brain, K.R. Pharmaceutical applications for molecularly imprinted polymers. *Int. J. Pharm.* **2000**, *195*, 39–43. [CrossRef] [PubMed]
43. Qiao, F.; Sun, H.; Yan, H.; Row, K.H. Molecularly imprinted polymers for solid phase extraction. *Chromatographia* **2006**, *64*, 625–634. [CrossRef]
44. Szatkowska, P.; Koba, M.; Koslinski, P.; Szablewski, M. Molecularly Imprinted Polymers' Applications: A Short Review. *Mini-Rev. Org. Chem.* **2013**, *10*, 400–408. [CrossRef]
45. Saylan, Y.; Yilmaz, F.; Ozgur, E.; Derazshamshir, A.; Yavuz, H.; Denizli, A. Molecular Imprinting of Macromolecules for Sensor Applications. *Sensors* **2017**, *17*, 898. [CrossRef]
46. Lusina, A.; Ceglowski, M. Molecularly Imprinted Polymers as State-of-the-Art Drug Carriers in Hydrogel Transdermal Drug Delivery Applications. *Polymers* **2022**, *14*, 640. [CrossRef]
47. Ceglowski, M.; Kurczewska, J.; Lusina, A.; Nazim, T.; Ruszkowski, P. EGDMA- and TRIM-Based Microparticles Imprinted with 5-Fluorouracil for Prolonged Drug Delivery. *Polymers* **2022**, *14*, 1027. [CrossRef]
48. Ying, T.L.; Gao, M.J.; Zhang, X.L. Highly selective technique-molecular imprinting. *Chin. J. Anal. Chem.* **2001**, *29*, 99–102.
49. Lisichkin, G.V.; Krutyakov, Y.A. Molecularly imprinted materials: Synthesis, properties, applications. *Uspekhi Khimii* **2006**, *75*, 998–1017.
50. Si, B.J.; Chen, C.B.; Zhou, J. New-Generation of Molecular Imprinting Technique. *Prog. Chem.* **2009**, *21*, 1813–1819.
51. Xiao, S.J.; Yu, S.W.; Li, H.X. Synthesis of novel separation materials based on molecular imprinting technology. *Adv. Eng. Mater.* **2012**, *535–537*, 1441. [CrossRef]
52. Erturk, G.; Mattiasson, B. Molecular imprinting techniques used for the preparation of biosensors. *Sensors* **2017**, *17*, 288. [CrossRef]
53. Zhang, N.; Xu, Y.R.; Li, Z.L.; Yan, C.R.; Mei, K.; Ding, M.L.; Ding, S.C.; Guan, P.; Qian, L.W.; Du, C.B. Molecularly imprinted materials for selective biological recognition. *Macromol. Rapid Commun.* **2019**, *40*, 1900096. [CrossRef]
54. Ceglowski, M.; Marien, Y.W.; Smeets, S.; De Smet, L.; D'hooge, D.R.; Schroeder, G.; Hoogenboom, R. Molecularly Imprinted Polymers with Enhanced Selectivity Based on 4-(Aminomethyl)pyridine-Functionalized Poly(2-oxazoline)s for Detecting Hazardous Herbicide Contaminants. *Chem. Mater.* **2022**, *34*, 84–96. [CrossRef]
55. Xie, X.Y.; Bu, Y.S.; Wang, S.C. Molecularly imprinting: A tool of modern chemistry for analysis and monitoring of phenolic environmental estrogens. *Rev. Anal. Chem.* **2016**, *35*, 87–97. [CrossRef]
56. Verma, A.; Verma, P.; Sidhu, R.K. Matrix effect in soft metal-bonded samarium-cobalt (SmCo5) permanent magnets. *Bull. Mater. Sci.* **1996**, *19*, 539–548. [CrossRef]
57. Bailey, G.; Orefice, M.; Sprecher, B.; Onal, M.A.R.; Herraiz, E.; Dewulf, W.; Van Acker, K. Life cycle inventory of samarium-cobalt permanent magnets, compared to neodymium-iron-boron as used in electric vehicles. *J. Clean. Prod.* **2021**, *286*, 125294. [CrossRef]
58. Zhao, X.B.; Zhu, T.J.; Li, W.W.; Wu, Z.T. Electrical properties of iron disilicide thermoelectric materials containing samarium. *Rare Met. Mater. Eng.* **2003**, *32*, 344–347.
59. Li, W.W.; Zhao, X.B.; Zhu, T.J.; Wu, Z.T.; Cao, G.S. Electrical properties of Fe-Sm-Si thermoelectric alloys. *Rare Met. Mater. Eng.* **2004**, *33*, 51–54.
60. Shekhawat, M.S.; Basha, S.K.S.; Rao, M.C. Spectroscopic Studies on Samarium Oxide (Sm_2O_3) Doped Tungsten Tellurite Glasses. In Proceedings of the 2nd International Conference on Condensed Matter and Applied Physics (ICC-2017), Bikaner, India, 24–25 November 2017; AIP Publishing: Woodbury, Long Island, NY, USA, 2018; Volume 1953, p. 090001. [CrossRef]

51. Auwalu, I.A.; Hotoro, M.A.Y.; Jamo, U.H.; Diso, D.G. Effect of Samarium Oxide on Structural and Optical Properties of Zinc Silicate Glass Ceramics from Waste Material. *Nano Hybrids Compos.* **2018**, *22*, 35–46. [CrossRef]
52. Jia, Z.; Misra, R.D.K. Magnetic sensors for data storage: Perspective and future outlook. *Mater. Technol.* **2011**, *26*, 191–199. [CrossRef]
53. Chandrasekaran, S.; Basu, P.; Krishnan, H.; Sivasubramanian, K.; Baskaran, R.; Venkatraman, B. Development of Gadolinium (neutron poison) monitoring system for fuel reprocessing facilities: Computational model and validation with experiments. *Prog. Nucl. Energy* **2018**, *107*, 57–60. [CrossRef]
54. Yang, J.Q.; Xu, B.; Zhang, L.; Liu, Y.D.; Yu, D.L.; Liu, Z.Y.; He, J.L.; Tian, Y.J. Gadolinium filled $CoSb_3$: High pressure synthesis and thermoelectric properties. *Mater. Lett.* **2013**, *98*, 171–173. [CrossRef]
55. Pop, M.; Borodi, G.; Deac, I.G.; Simon, S. Gd substitution effect on the formation of Bi-based superconducting glass ceramics. *Mod. Phys. Lett. B* **2000**, *14*, 59–63. [CrossRef]
56. Sonal, S.; Prakash, P.; Mishra, B.K.; Nayak, G.C. Synthesis, characterization and sorption studies of a zirconium(iv) impregnated highly functionalized mesoporous activated carbons. *RSC Adv.* **2020**, *10*, 13783–13798. [CrossRef]
57. Benedetto, J.S.; Ciminelli, V.S.T.; Neto, J.D. Comparison of extractants in the separation of samarium and gadolinium. *Miner. Eng.* **1993**, *6*, 597–605. [CrossRef]
58. Val'kov, A.V.; Igumnov, S.N.; Ovchinnikov, K.V. Separation of samarium, europium, and gadolinium by extraction with organophosphorus acids. *Theor. Found. Chem. Eng.* **2021**, *55*, 821–828. [CrossRef]

Disclaimer/Publisher's Note: The statements, opinions and data contained in all publications are solely those of the individual author(s) and contributor(s) and not of MDPI and/or the editor(s). MDPI and/or the editor(s) disclaim responsibility for any injury to people or property resulting from any ideas, methods, instructions or products referred to in the content.

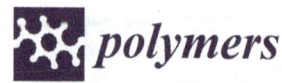

Article

Molecularly Imprinted Ligand-Free Nanogels for Recognizing Bee Venom-Originated Phospholipase A2 Enzyme

Anamaria Zaharia [1,†], Ana-Mihaela Gavrila [1,†], Iuliana Caras [2], Bogdan Trica [1], Anita-Laura Chiriac [1], Catalina Ioana Gifu [1], Iulia Elena Neblea [1], Elena-Bianca Stoica [1], Sorin Viorel Dolana [1] and Tanta-Verona Iordache [1,*]

1 National Institute for Research & Development in Chemistry and Petrochemistry ICECHIM, Advanced Polymer Materials and Polymer Recycling Group, Spl. Independentei 202, 6th District, 060021 Bucharest, Romania
2 National Institute for Medico-Military Research and Development "Cantacuzino", Spl. Independentei 103, 5th District, 011061 Bucharest, Romania
* Correspondence: tanta-verona.iordache@icechim.ro
† These authors contributed equally to this work.

Abstract: In this study, ligand-free nanogels (LFNGs) as potential antivenom mimics were developed with the aim of preventing hypersensitivity and other side effects following massive bee attacks. For this purpose, poly (ethylene glycol) diacrylate was chosen as a main synthetic biocompatible matrix to prepare the experimental LFNGs. The overall concept uses inverse mini-emulsion polymerization as the main route to deliver nanogel caps with complementary cavities for phospholipase A2 (PLA2) from bee venom, created artificially with the use of molecular imprinting (MI) technologies. The morphology and the hydrodynamic features of the nanogels were confirmed by transmission electron microscopy (TEM) and dynamic light scattering (DLS) analysis. The following rebinding experiments evidenced the specificity of molecularly imprinted LFNG for PLA2, with rebinding capacities up to 8-fold higher compared to the reference non-imprinted nanogel, while the in vitro binding assays of PLA2 from commercial bee venom indicated that such synthetic nanogels are able to recognize and retain the targeted PLA2 enzyme. The results were finally collaborated with in vitro cell-viability experiments and resulted in a strong belief that such LFNG may actually be used for future therapies against bee envenomation.

Keywords: molecularly imprinted polymers; ligand-free nanogels; bee venom phospholipase A2; synthetic antivenom; bee envenomation

Citation: Zaharia, A.; Gavrila, A.-M.; Caras, I.; Trica, B.; Chiriac, A.-L.; Gifu, C.I.; Neblea, I.E.; Stoica, E.-B.; Dolana, S.V.; Iordache, T.-V. Molecularly Imprinted Ligand-Free Nanogels for Recognizing Bee Venom-Originated Phospholipase A2 Enzyme. *Polymers* 2022, 14, 4200. https://doi.org/10.3390/polym14194200

Academic Editor: Michał Cegłowski

Received: 19 September 2022
Accepted: 5 October 2022
Published: 7 October 2022

Publisher's Note: MDPI stays neutral with regard to jurisdictional claims in published maps and institutional affiliations.

Copyright: © 2022 by the authors. Licensee MDPI, Basel, Switzerland. This article is an open access article distributed under the terms and conditions of the Creative Commons Attribution (CC BY) license (https://creativecommons.org/licenses/by/4.0/).

1. Introduction

Hymenoptera (bee/wasp/ant) envenomation is not usually lethal for humans and animals if the venom intake is lower than the lethal dose [1]. However, it is well known that the venom from the Hymenoptera insects is a potent neurotoxin and that the main destructive component is the specific secreted phospholipase A2 (PLA2) [2]. Bee venom PLA2 enzyme acts synergistically with the polyvalent cations (toxins) in the venom [3], creating an increased hemolytic effect and quick access of toxins into the blood flow, targeting important organs such as the brain, kidney and liver [4]. This enzyme simply degrades the cellular phospholipidic membranes and in high amounts, as in envenomation, causes decreased blood pressure and thereafter inhibits blood coagulation [5,6]. Therefore, by removing a high amount of PLA2 enzyme from sting/bite zone, the rest of the venom toxins can be locally blocked, and since the phospholipidic membranes are stopped or retarded from degradation, the toxins and other allergens will have limited access into the blood flow. For this reason, the present study targets the development of complementary, or even alternative, antivenom therapies that can reduce the quantity of the toxic PLA2 enzyme intake before the phospholipidic membranes are damaged.

In spite of recent technological developments, no effective and safe therapies are currently available for treating the victims of mass honeybee or wasp attacks [7]. Adrenalin is the first-aid treatment of choice for the systemic allergic response with dyspnea and/or hypertension, while in patients without anaphylaxis, the suggested conservative approach is based on observation and treatment of symptoms [8]. Since 1996, multiple attempts to create antivenom as an emergency treatment for bee envenomation were proposed [9–11]. Antivenom is created by injection of sublethal toxin doses into an animal such as a sheep or horse, followed by harvesting the blood serum of the animal, which contains significant quantities of toxin-recognizing antibodies [12]. However, most of the studies regarding antivenom production suggested that the reason for their ineffectiveness is linked to the low immunogenicity of targeted bee venom toxins [13]. Recently, noteworthy antivenom designs based on monoclonal or oligoclonal antibodies [14,15] have emerged and may contribute to new and effective bee envenomation therapies. Yet, the technology is still young and needs serious efforts to deliver viable antivenom therapies [13]. Meanwhile, a great deal of research was focused on developing nano-sized hydrogels known in the literature under the name of nanogels [16,17]. The most common applications of nanogels, with controlled release properties of various active principles, are found in the tissue engineering field, biomedical implantology and bionanotechnology [16]. Having aside the newest trends in nanogels development [18–20] and access to technologies that allow for creating synthetic antibodies in various polymer networks, i.e., molecular imprinting (MI) [21,22], this work aims to prepare ligand-free nanogels (LFNGs) with complementary specific binding sites for PLA2, named molecularly imprinted polymers (MIPs). Synthetic nanogel antibodies make it possible to directionally modify the molecular size, affinity, specificity, and immunogenicity and effector functions of a natural antibody, as well as to combine antibodies with other functional agents for diagnosing and treating various diseases, particularly using new technologies meant to refine the effector functions of therapeutic antibodies [23]. The advantages of such synthetic antibodies include lower manufacturing costs, a medium level of synthesis complexity and no specific requirements for storage and transportation, as compared to the traditional antivenom. On the other hand, the proposed systems, denoted molecularly imprinted polymer ligand-free nanogels (MIP-LFNGs), are free of template molecules or ligands and can retain various compounds using the matrix itself for targeting [24,25].

Thereby, the novelty introduced by this pioneering work is the exploration of combined methods and concepts of state-of-the-art nanotechnologies and molecular imprinting techniques to deliver novel, efficient and cheaper antivenom variants for bee envenomation. The MIP-LFNG for bee venom-originated PLA2 recognition and retention, as presented in this study, is an original concept and has never been reported as a potential therapy against bee envenomation. To this day, only a few reports dealing with synthetic polymer nanoparticles as plastic antibodies with the capacity to bring and neutralize the hemolytic toxin melittin peptide were reported by Hoshino et al. [26–29]. In this respect, polymer nanoparticles (NPs) were prepared using N-isopropylacrylamide (NIPAm) as a core polymer combined with N-t-butylacrylamide, acrylic acid or N-3-aminopropyl methacrylamide as a functional monomer [26,27,29]. The resulting NPs, with an average size of 50 nm, were able to bind melittin in high amounts, up to 180 µmol/g NP [26]. In some early studies of this group [27,28], NPs based on NIPAm were molecularly imprinted with melittin and labeled with fluorescein o-acrylate to evaluate the binding of melittin and the behavior of the NPs in vivo. Other recent studies on molecularly imprinted nanogels were reported by Takeuchi's group for protein recognition as well [30,31]. In these cases, the authors prepared MIP nanogels of about 45 nm diameter possessing good binding affinity and specificity (F > 20 at 1 mg/mL polymer, but low reaction yields below 1%) capable of protein corona regulation via albumin recognition. Nevertheless, the latter studies have successfully detailed some meaningful insights related to nanoparticle–cell interactions with the emphasis on the cellular uptake mechanism in cancer cells and immune-related cell lines [30], followed by in vivo studies revealing the uptake of albumin in MIP nanogels and their targeting ability for tumor tissue [31].

2. Materials and Methods

2.1. Materials

In order to obtain the MIP-LFNGs, polyethyleneglycol diacrylate (PEGDA, MW = 700 g/mol), sorbitan monooleate (SPAN 80), N,N,N′,N′-tetramethyl ethylenediamine (TMEDA) sodium chloride (>99%), cyclohexane (CHx, 99.5%), phospholipase A2 from bee venom Apis Mellifera (PLA2), 2-amino-2-(hydroxymethyl)-1,3-propanediol (TRIS, 99.8%), hydrochloric acid (HCl, 37%) and acetone (99,92%) were purchased from Sigma-Aldrich (St. Louis, MO, USA). Ammonium persulfate (APS, 98%) was purchased from Peking Chemical Works (Beijing Chemical Works, Beijing, China), polyethylene glycol sorbitan monooleate (TWEEN 80, oleic acid, $\geq$58.0%) was purchased from Sigma-Aldrich (St. Louis, MO, USA) and phosphate-buffered saline (PBS) was purchased from Roti-CEL (Karlsruhe, Germany). Invitrogen EnzChek Phospholipase A2 Assay Kit was purchased from Thermo Scientific LSG (Life Technologies Ltd., Inchinnan Business Park Paisley, UK). Bee venom (BV) was used in the form of lyophilized powder and was purchased from The Research and Development Institute for Beekeeping (Bucharest, Romania). The difunctional macromer polyethyleneglycol diacrylate (PEGDA, MW = 2000 g/mol) was synthesized as previously reported by Radu et al. [32] (results of molecular weight, functionality, structure and thermal stability are provided in the Supplementary Materials, Figures S1–S4).

2.2. Synthesis and Purification of LFNGs for Recognizing and Retaining Bee Venom-Originated PLA2

LFNGs were prepared similarly to the recipe previously described by our group, but with some modifications [33]. The organic phase was prepared by dissolution of the mixed emulsifiers SPAN80 (0.9 mol/L) and TWEEN80 (0.09 mol/L) in a 7:1 ratio (w/w) with cyclohexane (9 mol/L) in a round glass-bottom reactor. This mix was homogenized by magnetic stirring at 600 rpm, degassed and purged with nitrogen for 10 min. Meanwhile, the aqueous phase was prepared by dissolution the difunctional macromers PEGDA 700/PEGDA 2000 (0.4 mol/L), TMEDA (0.04 mol/L) and NaCl (0.2 mol/L) in ultrapure water (2 mL). After 10 min of mechanical stirring (600 rpm), degassing and purging with nitrogen gas, the aqueous phase was added to the organic phase under magnetic stirring (600 rpm). In order to prepare the molecularly imprinted polymer LFNGs (MIP-LFNGs), PLA2 (1:5 molar ratio PLA2:PEGDA) was prepared separately in ultrapure water (1 mL) or TRIS/HCl (pH 8.2; 50 mM), after which it was added to the reaction mixture under quick magnetic stirring (600 rpm). The rationale behind using water and the variant of a buffer solution to solubilize the PLA2 was to maintain as much as possible the conformation of PLA2 that may be found in the venom, in order to increase the specificity of the final nanogels for PLA2. The polymerization reaction was initiated by APS (5% w. relative to PEGDA), after which the stirring rate was lowered to 200 rpm and the temperature was set at 30 °C. The mini-emulsion was maintained in the previously mentioned conditions for 42 h and subsequently centrifuged. The supernatant was removed, and the nanogel phase was washed with cyclohexane, acetone and ultrapure water to remove the continuous media, the emulsifiers, the template PLA2 enzyme and any unreacted macromer. Thus, prepared MIP-LFNGs were lyophilized for 48 h to yield a powder. Furthermore, another nanogel system, called non-imprinted polymer LFNG (NIP-LFNG), was prepared as a reference; in this case, PLA2 was not added during the polymerization. The recipes for preparing the MIP-LFNG W or T (where W stands for water and T for TRIS/HCl solution) and for the reference NIP-LFNG are also summarized in Table 1.

Table 1. The synthesis recipes for LFNGs.

Samples	PEGDA$_{700}$/ PEGDA$_{2000}$ (%)	Span 80/ Tween 60 (%)	Emulsifiers/ Aqueous Phase (%)	Solvent/ Aqueous Phase (%)	PLA2/ PEGDA (Molar Ratio)
NIP-LFNG	75/25	87.5:12.5	3	5.11	0
MIP-LFNG (W)	75/25	87.5:12.5	3	5.11	1/5 (PLA in H$_2$O)
MIP-LFNG (T)	75/25	87.5:12.5	3	5.11	1/5 (PLA in TRIS/HCl)

2.3. Characterization Methods and Instruments

2.3.1. Structural and Morphological Characterization of LFNGs

Fourier transform infrared (FT-IR) spectra, recorded on a Bruker Vertex 70 instrument in the 400–4000 cm^{-1} range with 4 cm^{-1} resolution and 32 scans (on KBr pellets), were useful for highlighting the molecular imprinting effect.

Thermogravimetric analyses (TGA) were carried out by using a Thermal Analysis SDT600 instrument and heating 5–10 mg samples from 30 °C to 1000 °C at a heating rate of 10 °C/min under nitrogen flow.

The particle sizes of LFNGs were determined by dynamic light scattering (DLS) analyses using a Malvern Zetasizer Nano-ZS system equipped with a 4 mW He-Ne laser (633 nm). All measurements were performed in five replicates, and the results are reported as the mean together with the standard deviation.

Transmission electron microscopy (TEM) pictures were taken using a Tecnai G2 F20 TWIN Cryo-TEM. Two protocols were used. The first one consisted of directly sampling the emulsion and placing it on a carbon-film-covered grid. The excess emulsifiers were removed by 5 s immersion of the grid in acetone. The second protocol consisted of the redispersion of the purified nanogels in distilled water and placement of the sample on the same type of grid.

2.3.2. Batch Binding Experiments Assisted by Activity Measurements of PLA2

Binding experiments were performed in order to investigate the specificity and capacity of MIP-LFNG and NIP-LFNG to recognize and rebind PLA2. In this respect, the assays were based on measuring the decrease in activity in PLA2 aqueous solutions or bee venom solutions (U/mL) before and after contact with the nanogels, using the EnzChek Phospholipase A2 Assay Kit, at a plate dilution of 1/2 (U/mL) (or 1/40) initial solution. In brief, the binding experiments involved contacting 10 mg of each LFNG with 1 mL pure PLA2 or bee venom solution of known concentration (0.1 mg/mL PLA2 enzyme and 1 mg/mL bee venom). The supernatants were collected after 15 and 30 min, diluted (1/40) and analyzed by fluorescence for changes in the emission intensity ratio at 490 nm with excitation at 450 nm. To quantify the adsorbed PLA2 (also known as the rebinding capacity, Q (U PLA2/g nanogel)), for MIP-LFNGs and corresponding NIP-LFNG, the study presented the hypothesis that the decrease in activity as measured at 450 nm and 490 nm was due to the nanogels' specific adsorption of PLA2. The method for calculating Q is given in Equation (1), where C_i (U/mL) represents the concentration of PLA2 in the reference aqueous solution or venom solution, C_f (U/mL) represents the concentration of PLA2 after contact with nanogels, m_p (g) is the nanogel weight and V_s (L) is the volume of the feed solution (see also Table S1, Supplementary Materials).

$$Q = (c_i - c_f) \cdot V_S / m_p \tag{1}$$

The imprinting factor, F, expressed by Equation (2), quantified the specificity with which MIP-LFNGs rebind PLA2, compared to the corresponding NIP-LFNG, where QMIP and QNIP are the rebinding capacities of MIP-LFNG (W or T) and NIP-LFNG, respectively. The binding experiments were carried out in duplicate, using fluorescence measurements on a reader for Tecan Infinite M1000 microplates. The results, as mean values of two replicates, were expressed as U/mL after extrapolation on a standard curve made with a standard solution of PLA2.

$$F = Q_{MIP}/Q_{NIP} \qquad (2)$$

2.3.3. Cytotoxicity Study of LFNGs

Extracts from LFNGs were obtained by placing the materials in Dulbecco's Modified Eagle Medium (DMEM), 5 mg/mL, for 24 h at 4 °C and collecting the supernatants by centrifugation (15 min, 10,000× g). Mouse fibroblast cell line L929 (ECACC 85011425) was used to test the cytotoxicity of LFNGs. L929 cells were cultured in DMEM supplemented with 10% fetal bovine serum (FBS), 100 µg/mL penicillin, 100 µg/mL streptomycin and 1 mM L-glutamine in a 5% CO_2 atmosphere incubator at 37 °C. Cells were enzymatically detached and seeded in 96-well culture plates at a density of 1 × 104 cells/well and cultured overnight. Subsequently, the supernatant was discarded and replaced with binary dilutions (of LFNG extracts). After overnight incubation, a 3-(4,5-dimethylthiazol-2-yl)-2,5-diphenyltetrazolium bromide (MTT) assay was performed to evaluate the cell viability. Briefly, the supernatant was discarded and replaced with DMEM containing 0.5 mg/mL MTT. The assay is based on the ability of NAD(P)H-dependent oxidoreductase enzymes in living cells to reduce yellow tetrazolium salts to purple formazan crystals. Cells were incubated for an additional 3 h, and then lysis solution (20% w/v sodium dodecyl sulfate, 50% w/v N,N-dimethylformamide, 0.4% w/v acetic acid, 0.04 M hydrochloric acid) was added to dissolve the insoluble formazan crystals and the resulting-colored solution was quantified by measuring the absorbance at 570 nm using a microplate spectrophotometer (Multiskan FC, Thermo Scientific). The percentage of cell viability for each experimental condition was calculated by setting the control as 100%.

3. Results

3.1. Synthesis of LFNGs

In the current study, we aimed to combine the advantages of more efficient treatments based on nanomaterials and the specificity of MIPs for the development of molecularly imprinted ligand-free nanogels (MIP-LFNGs) for recognizing and retaining bee venom-originated PLA2. The inverse mini-emulsion polymerization system involved the formulation of a stable mixture, composed of droplets of polymer aqueous solution suspended by a mixture of co-surfactants in a continuous organic medium [33,34]. Herein, we investigated (i) the synthesis of LFNG based on poly (ethylene glycol) diacrylate initiated by a redox initiator system at body temperature, in the absence or presence of PLA2 enzyme (according to Scheme 1); (ii) the morphology and structure of LFNGs; (iii) the performance of prepared LFNGs by single-enzyme rebinding experiments and by specific rebinding from bee venom; and the (iv) in vitro cytotoxic effect of LFNGs.

The following work describes the optimized recipes resulting from many variants of nanogel synthesis. In this respect, the same recipes were used to prepare nanogels using either PEDGA 700 or 2000 alone. However, the samples were discarded after performing DLS (Table S1, Supplementary Materials) and TEM analysis (Figure S5, Supplementary Materials) which clearly showed that the systems were not proper for the studied application in terms of average particle size, polydispersity and uniformity.

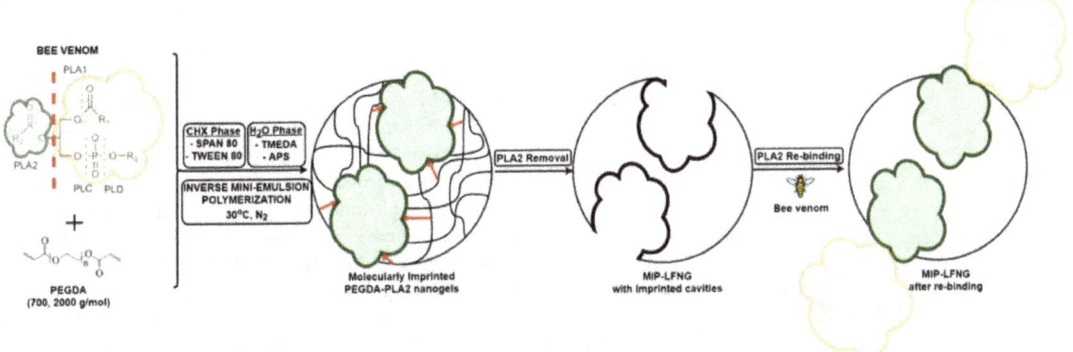

Scheme 1. Preparation of molecularly imprinted ligand-free nanogels (MIP-LFNGs) for recognizing and retaining bee venom-originated PLA2.

3.2. Structural and Morphological Characterization of LFNGs

3.2.1. FT-IR Spectroscopy

The LFNG series were evaluated by FT-IR spectroscopy (Figure 1). In the spectrum of the NIP-LFNG, the bands generated by the C–H stretching vibrations around 2870 cm^{-1}, the carbonyl group (C=O) stretching vibration at 1729 cm^{-1} [35] and C–O stretching vibration at 1097 cm^{-1} can be clearly distinguished. The characteristic band of OH stretching vibrations from poly (ethylene glycol) was registered around 3500–3400 cm^{-1}, while the bands assigned to -C=C- at 1620 cm^{-1} completely disappeared from the LFNG spectrum, indicating the consumption of -C=C- bonds of PEGDA during polymerization [36,37].

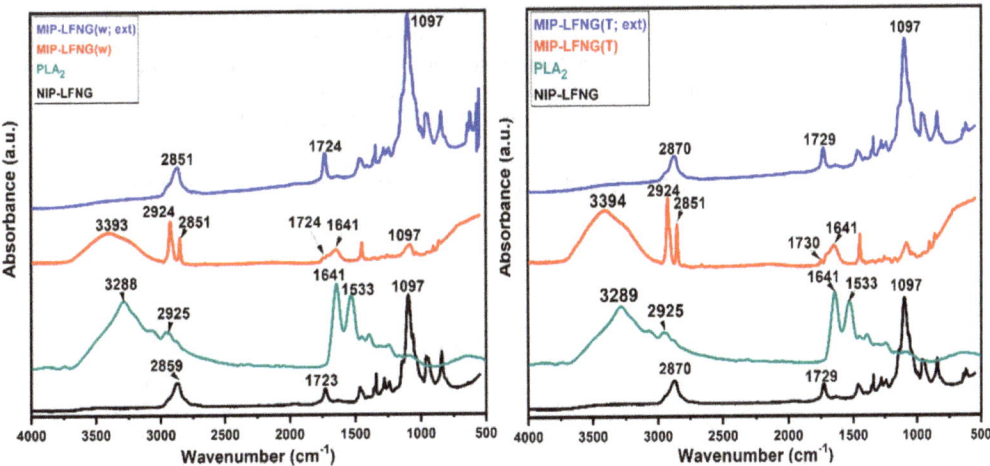

Figure 1. Fourier transform infrared (FT-IR) spectra of NIP-LFNG; PLA2 enzyme (PLA2); and MIP-LFNG (W or T) and MIP-LFNG (W, ext or T, ext) before and after PLA2 extraction, respectively.

The FT-IR spectrum of the PLA2 template presented intense characteristic bands as well. The bands at 3289 and 3072 cm^{-1} correspond to the overlapping O–H stretching vibrations and the N–H stretching vibrations in amide A (more intense) and amide B of proteins. The intense band at 3289 cm^{-1} is the result of resonance between N–H stretching and the overtone of amide II [38,39]. The band at 2924 cm^{-1} is characteristic of symmetric (CH$_2$) groups, while the bands at 1641 and 1533 cm^{-1} (characteristic of the peptide amines and amino acids) correspond to the C=O symmetric stretching vibrations of α-helical structure (amide I) and the N–H in-plane bending and C–N stretching of amino acids (amide II), respectively [40,41].

Interestingly, the spectra of both imprinted nanogels, i.e., MIP-LFNG (T) and MIP-LFNS (W), analyzed before the extraction of the PLA2 template, showed important changes compared to the reference nanogels, NIP-LFNGs, as well as to the same imprinted nanogels analyzed after PLA2 extraction, which indicated an efficient imprinting of PLA2. At a first glance, a broad band between 3000 and 3700 cm^{-1}, similar to that observed in the spectrum of PLA2, was registered for the imprinted nanogels. This band was assigned to the overlapping bands of the O–H stretching vibrations and the N–H stretching vibrations in amide A and amide B of proteins. Furthermore, the characteristic band of -CH$_2$ groups (at 2924 cm^{-1}) and the one associated with the amide I bands (at 1641cm^{-1}) were also spotted in the spectra of both imprinted nanogels before PLA2 extraction.

On the other hand, the FT-IR spectra of the imprinted nanogels analyzed after the extraction of the PLA2 template, named MIP-LFNG (T, ext) and MIP-LFNG (W, ext), are similar to those of the non-imprinted nanogel references, without any characteristic bands of PLA2. This resemblance proves the fact that the chemical structure of the imprinted nanogel matrix was not modified during the imprinting process (this process being non-covalent) and also that a proper extraction of PLA2 from the nanogels was performed with the aim of cleaving the specific binding sites thus created [42].

Thereby, it can be assumed that the imprinting of PLA2 was successful and that specific binding sites were created, considering that the prominent features of PLA2 are present in the spectra of imprinted nanogels before template extraction [43] and disappear after nanogels are thoroughly washed.

3.2.2. TGA Investigation

The thermal stability of LFNGs was highlighted by TGA/DTG analysis provided in Figure 2a,b. The results of TGA and the corresponding derivative curves of NIP-LFNG showed a slight decrease in thermal stability, while the MIP-LFNG (T, ext) revealed a similar thermal stability to that of MIP-LFNG (T), before PLA2 extraction (Figure 2). The NIP-LFNGs presented one small shoulder at 163 °C (Tmax) attributed to polymer lose chain, after which it maintained an expected decomposition trend centered at 392 °C (Tmax) that can be related to the degradation of the polymer backbone chain, with a final weight loss of 95% [44–46].

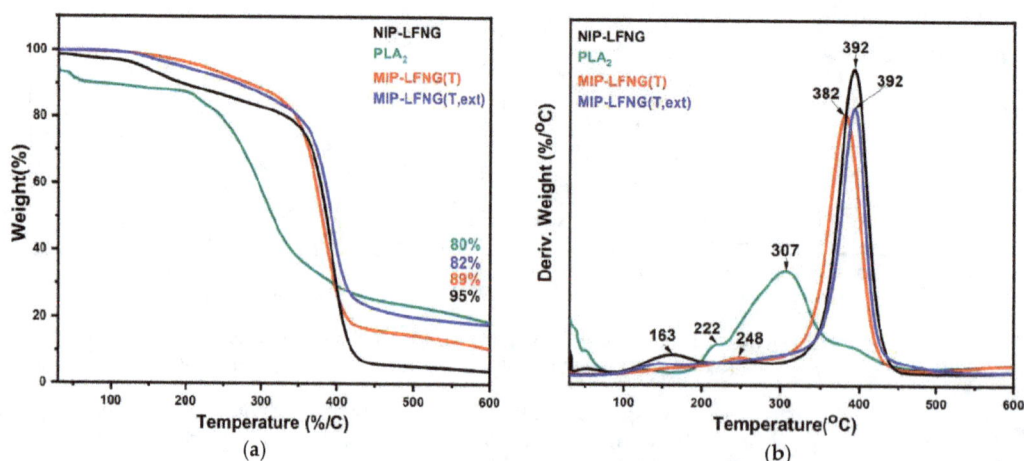

Figure 2. (a)Thermal analysis diagram (TGA) and (b) derivative (DTG) of NIP-LFNG; PLA2 enzyme (PLA2); and MIP-LFNG (T) and MIP-LFNG (T, ext) before and after PLA2 extraction, respectively.

The TGA and derivative curve of the PLA2 enzyme revealed the decomposition of the amino acids in several stages, as follows [47]: the first stage at 222 °C was attributed to the elimination of NH3 and the formation of unsaturated acids; this stage was followed by the release of intramolecular water and the formation of lactams, and ultimately the decarboxylation process occurred, resulting in the formation of amines at 307 °C.

MIP-LFNGs (T, ext) analyzed after the extraction of the PLA2 template exhibited a very small decomposition step in the vicinity of 160 °C followed by the main decomposition step at a maximum temperature of 392 °C (Tmax), with a final weight loss of 82%. The stability of the extracted nanogels, MIP-LFNGs (T, ext), was similar to that of the reference nanogels, NIP-LFNGs, which confirmed the FT-IR observations and conclusions referring to the fact that the chemical structure and composition of the imprinted nanogel matrix was not significantly modified during the imprinting process.

Interestingly, the MIP-LFNG (T) before the extraction of the PLA2 template revealed a small shoulder at 248 °C (Tmax), which was attributed to the very low amounts of PLA2 used for the imprinting process, while the main decomposition process of the nanogel matrix followed a similar trend to that of the NIP-LFNGs, but with a lower maximum temperature for decomposition, at 382 °C (Tmax), and a higher final weight loss of around 89% compared to the extracted MIP-LFNGs (T, ext). The slight decrease in thermal stability of about 10 °C for the nanogels analyzed before PLA2 extraction may be due to the presence of the polymer–template interaction between the amino acid side chains of PLA2 specifically involved in the binding to the functional groups of the nanogel matrix [27,48,49].

3.2.3. DLS Investigation

The particle size distribution and the polydispersity (PDI) of the LFNGs before and after purification were investigated using DLS, given the targeted application. As shown in Table 2, the synthesized LFNGs exhibited low PDI values, and no significant coagulation was observed during polymerization. The PDI of LFNGs should be very low (under 0.5, which means that the nanogels have similar sizes) in order to obtain comparable results in each batch, but also to decrease the potential cytotoxic effects of nanogels given by their uneven size, as demonstrated by other studies as well [33]. The Z-average particle size of the LFNG was registered within the desired range, below 200 nm (and 143 ÷ 198 nm, in this case), and the PDI was below 0.375. The Z-average particle size of the MIP-LFNG (T) and MIP-LFNG (W) before the PLA2 template extraction was approximately 189 nm and 198 nm, respectively, yet slightly bigger than that of NIP-LFNGs. This observation can be linked to the presence of the PLA2 template in the structure of the synthesized LFNGs, which also contributed to the increase in the hydrodynamic volume of the analyzed nanogels. The latter hypothesis was also sustained by the fact that for the MIP-LFNGs analyzed after PLA2 extraction, namely MIP-LFNG (T, ext) and MIP-LFNG (W, ext), the Z-average particle size decreased in the size range of 163–170 nm. It may also be noted that the 20–30 nm difference between the average sizes of extracted MIP-LFNGs and the average size registered for the reference NIP-LFNG may be due to the hydrodynamic volume occupied by the cleaved imprinted cavities specific for PLA2.

Table 2. DLS results of LFNGs before and after PLA2 template extraction.

Sample	Diameter * (nm)	Polydispersity Index (PDI)
NIP-LFNG	143 ± 0.53	0.326
MIP-LFNG (W, ext)	163 ± 2.90	0.251
MIP-LFNG (W)	198 ± 3.91	0.375
MIP-LFNG (T, ext)	170 ± 1.22	0.184
MIP-LFNG (T)	189 ± 3.91	0.322

* Average ± standard deviation of five sequential measurements.

3.2.4. TEM Images

Transmission electron microscopy (TEM) images of nanogels analyzed directly from emulsion and after purification also supported the previously discussed results regarding the average size of nanogels and the imprinting process (Figure 3). The micrographs of the NIP-LFNG taken directly from the final emulsion (Figure 3a,b) revealed the presence of individual spherical nanogels, having dimensions roughly in the range of 60–180 nm. It is important to note the presence of emulsifiers as needle-shaped formations, which form a continuous layer around the synthesized nanogels. After washing, which implied the removal of emulsifiers as well, only the spherical nanogels could be distinguished; they had no significant morphology modifications but were slightly agglomerated and had similar dimensions (Figure 3c).

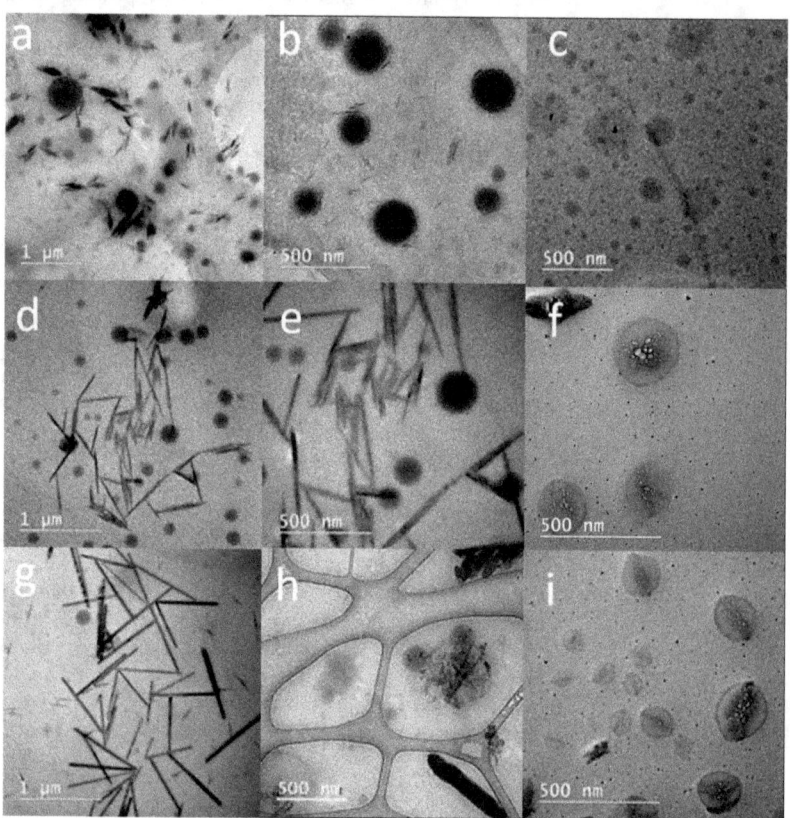

Figure 3. TEM micrograph of the NIP-LFNG ((**a**) 1 μm, (**b**) 500 nm for the final emulsion; (**c**) 500 nm after the washing process), imprinted MIP-LFNG (W) ((**d**) 1 μm and (**e**) 500 nm for the emulsion; (**f**) 500 nm after PLA2 extraction) and MIP-LFNG (T) ((**g**) 1 μm and (**h**) 500 nm for the emulsion; (**i**) 500 nm after PLA2 extraction).

Meanwhile, the micrographs of the MIP-LFNG (W) and MIP-LFNG (T) before and after the PLA2 extraction showed significant morphological changes as compared to the samples analyzed directly from emulsion (Figure 3d–i). Both types of MIP-LFNGs indicate a spherical shape morphology, having dimensions roughly in the range of 90–190 nm. In the case of the LFNGs taken directly from the final emulsion (Figure 3d,e,g,h), the presence of needle-shaped emulsifiers takes an interesting microstructural arrangement in tube form, which can also be due to the presence of the PLA2 enzyme. Thus, on the surface of LFNG spheres, the microcrystals with denser and more homogeneous structures may actually be PLA2 molecules frozen in their crystalline state [50]. These micrographs, probably the first of their kind, show how the PLA2 enzyme binds to the nanogel matrix and, subsequently, leaves marks of its interaction by creating molecularly imprinted cavities (Figure 3f,i). TEM micrographs of the MIP-LFNGs after the PLA2 extraction also showed the presence of multiple spherical zones with different electron densities that sustain the latter affirmation [51]. Therefore, this structural detail is proof of the non-covalent interactions between the template and macromers and can be suggestive and characteristic of the presence of imprinted free nanocavities on the surface of synthesized MIP-LFNGs [52].

3.3. Binding Properties of LFNGs

The binding properties of PLA2 investigated in an aqueous medium in batch mode were determined for the two types of MIP-LFNGs (i.e., MIP-LFNG (T) and MIP-LFNG (W)) and the corresponding blank NIP-LFNG. The specificity for PLA2 uptake was assessed by quantifying the activity of PLA2 in solutions before and after contact with the nanogels (as presented in Table S1, Supplementary Materials). Therefore, the adsorption capacity, Q (U/g$^-$), and imprinting factor, F, of nanogels were also given as units of PLA2 (Figure 4a,b).

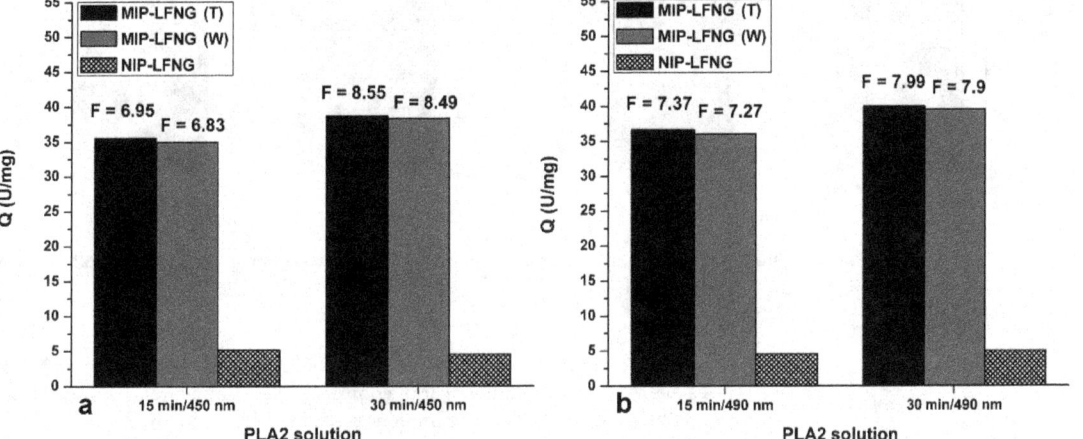

Figure 4. Rebinding capacity, Q, and imprinting factor, F, of LFNGs exposed to PLA2 solution after 15 and 30 min and evaluated at (**a**) 450 nm and (**b**) 490 nm.

Both imprinted nanogels, MIP-LFNG (T) and MIP-LFNG (W), exhibited higher affinity for PLA2 when contacted with an aqueous PLA2 solution than the corresponding NIP-LFNG, leading to impressive rebinding capacities after 30 min of exposure of 39.93 U/mg (490 nm) and 38.66 U/mg (450) for MIP-LFNG (T) and 39.49 U/mg (490 nm) and 38.36 U/mg (450) for MIP-LFNG (W). Due to the low PLA2 amounts adsorbed by the NIP-LFNG after 30 min of exposure, i.e., 5.00 U/mg (490 nm) and 4.52 U/mg (450 nm), the imprinting factor, F, values calculated for MIP-LFNG (T) and MIP-LFNG (W) after 30 min of exposure were close to 8 (Figure 4a,b), meaning that MIP-LFNGs recognize and retain PLA2 about 8-fold more specifically than the reference NIP-LFNG. The resulting values are comparable to the results of other authors related to molecularly imprinted nanogels for peptide recognition [30,31]. It is also important to mention that the two investigated parameters continue to improve with time; an increasing trend was observed from 15 min exposure time to 30 min, indicating that an adsorption equilibrium was surely attained after 30 min of exposure.

The following in vitro experiments of PLA2 binding from bee venom have provided important information regarding the potential of such MIP-LFNGs to retain specifically the enzyme directly from the venom. Although the values of activity measured after contact with the venom (Table S1, Supplementary Materials) were not as spectacular as the ones recorded for the binding assays from PLA2 solutions, a similar trend was observed with regard to the performance of each nanogel system. Figure 5a,b presents the decrease in PLA2 activity (%) for each nanogel system, after exposure to venom at 15 and 30 min, relative to the initial PLA2 activity in the bee venom (100%). However, it was somewhat surprising that the specificity of MIP-LFNGs (W) for PLA2 retention has dropped significantly as compared to the previous assay; their retention capacity for PLA2 this time is close to the retention capacity of NIP-LFNG (see the activity drop of PLA2 after exposure to venom in Figure 5a,b).

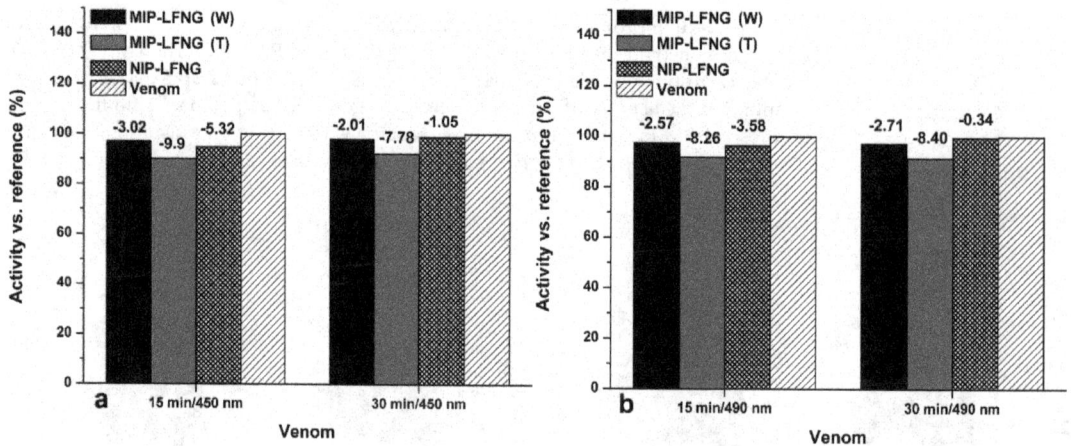

Figure 5. The drop in PLA2 activity after LFNG exposure to venom at 15 and 30 min and evaluated at (**a**) 450 nm and (**b**) 490 nm.

3.4. Cytotoxicity Assay

The potential cytotoxicity of LFNGs was also studied as a result of the targeted application, i.e., as alternatives to traditional antivenom which is administrated intravenously. In this respect, the effect of LFNG concentrations on cell viability was investigated by MTT assay (Figure 6). After 24 h of incubation at different dilutions (1/4, 1/8, 1/16 and 1/32), only a slight reduction in the cell viability was observed, less than 3% (values given in Table S2, Supplementary Materials). The results were, however, significant because the L929 cells remained with high viability ($\geq$97%) as shown in Figure 6, even at high nanogel concentrations of 1/4, in which case a 98.29 $\pm$ 1.33% and 100.8 $\pm$ 1.3% cell viability was registered for MIP-LFNG (T) and MIP-LFNG (W), respectively. Other studies reported similar results when using PEGDA-based nanoparticles [30,53]. A very slight and odd decrease in cell viability was observed at higher dilutions for MIP-LFNG (W), i.e., down to 97.6 $\pm$ 1.7% at 1/16 but, still, very close to the reference. What is also interesting to note is the fact that LFNGs, particularly NIP-LFNG and MIP-LFNG (T), led to an increase in cell viability, especially at higher nanoparticle dilutions, which means that the two systems were also able to induce slight cell proliferation (no more than 9%). Yet, this property may help in the administration of the synthetic antivenom and be of benefit in a secondary activity of cell restoration/proliferation after PLA2 damage to existing viable cells.

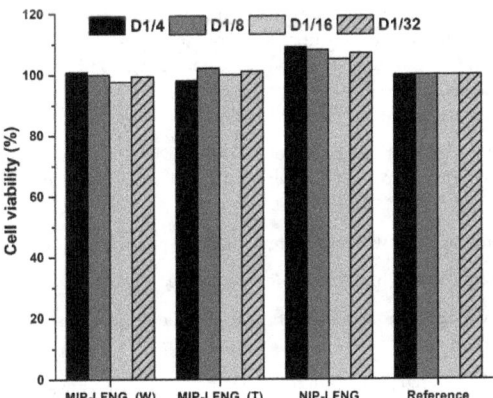

Figure 6. Cell viability of L929 cell line after 24 h exposure to various dilutions of LFNGs compared to a reference (100% viability).

4. Conclusions

In conclusion, this study reports the development of original ligand-free nanogel systems molecularly imprinted with bee venom-originated PLA2 (MIP-LFNGs) as a potential therapy for bee envenomation. In this respect, the nanogels were prepared by a known technique (mini-emulsion polymerization) that can deliver spherical nanosized gels with narrow polydispersity, while the polymer matrix consisted of a mixture between two macromers of PEGDA with two different molecular weights, i.e., 700 and 2000 g/mol. Thus, the nanogels prepared in the presence or absence of the PLA2 template, called MIP-LFNGs and NIP-LFNGs, respectively, were analyzed in terms of structure, composition, morphology and particle size in order to gain a better understanding of their behavior when submitted to rebinding assays of PLA2 from aqueous solution or bee venom and to cytotoxicity investigations. FT-IR, TGA, DLS and TEM analysis have pointed out that specific imprinted cavities for PLA2 retention were created in both nanogel systems, i.e., MIP3-LFNG (T) and MIP-LFNG (W). However, the system denoted MIP-LFNG (T), developed using the PLA2 template solubilized in TRIS/HCl, seems to perform better during the rebinding assays, retaining PLA2 from solution 8.5-fold more specifically than the non-imprinted reference, NIP-LFNG, and attaining a high rebinding capacity of approximately 40 U PLA2/mg of nanogel. The differences between this system and the one denoted MIP-LFNG (W), developed using the PLA2 template solubilized in water, were very small with the exception of the capacity of rebinding the PLA2 from venom, in which case MIP-LFNG (T) reduced the activity of PLA2 in the bee venom by almost 10% (compared to the 3% registered for MIP-LFNG (W)); thus, MIP-LFNG (T) was about 3 times more efficient than MIP-LFNG (W) in recognizing and retaining the PLA2 from the venom. Furthermore, the cytotoxicity of MIP-LFNGs was very low compared to the reference, even at high nanogel concentrations, whereas the lowest values for cell viability were registered for MIP-LFNG (W) at a dilution of 1/16 (97.6 ± 1.7%).

Supplementary Materials: The following supporting information can be downloaded at: https://www.mdpi.com/article/10.3390/polym14194200/s1, Figure S1: Molecular weight distribution by size exclusion chromatography corresponding to the precursor PEG2000 and the synthesized PEGDA2000 macromer (the latter showing a slightly higher molecular weight due to acrylate insertion as end groups to the PEG2000 precursor); Figure S2: H1-NMR spectrum of synthesized PEGDA2000 macromer; Figure S3: FT-IR spectra of synthesized PEGDA2000 monomer; Figure S4: TGA/DTG curves of synthesized PEGDA2000 monomer; Table S1: DLS results corresponding to the optimization study for the synthesis of MIP-LFNGs; Figure S5: TEM images of redispersed NIP7, MIP7 (T) and MIP7 (W); Table S1: PLA2 activity in solution and venom before and after contact with the LFNGs; Table S2: Cell viability of L929 cell line after 24 h exposure to various dilutions of LFNGs.

Author Contributions: Conceptualization, A.Z. and T.-V.I.; formal analysis, A.-L.C., I.E.N., E.-B.S and S.V.D.; funding acquisition, A.Z.; investigation, I.C., B.T., C.I.G. and S.V.D.; methodology, A.Z and A.-M.G.; resources, T.-V.I.; supervision, T.-V.I.; validation, I.C., B.T. and C.I.G.; visualization A.-L.C.; writing—original draft, A.Z. and A.-M.G.; writing—review and editing, T.-V.I. All authors have read and agreed to the published version of the manuscript.

Funding: This work was supported by the Ministry of Research, Innovation and Digitalization CNCS–UEFISCDI, project number PN-III-P1-1.1-TE-2016-1876 and project number PN-III-P1-1.1-TE-2021-1239, within PNCDI III.

Institutional Review Board Statement: Not applicable.

Informed Consent Statement: Not applicable.

Data Availability Statement: The data presented in this study are available on request from the corresponding author.

Conflicts of Interest: The authors declare no conflict of interest.

References

1. Freeman, T.M. Hypersensitivity to Hymenoptera Stings. *N. Engl. J. Med.* **2004**, *351*, 1978–1984. [CrossRef] [PubMed]
2. Schaloske, R.H.; Dennis, E.A. The phospholipase A2 superfamily and its group numbering system. *Biochim. Biophys. Acta* **2006**, *1761*, 1246–1259. [CrossRef] [PubMed]
3. Cajal, Y.; Jain, M.K. Synergism between Mellitin and Phospholipase A2 from Bee Venom: Apparent Activation by Intervesicle Exchange of Phospholipids. *Biochemistry* **1997**, *36*, 3882–3893. [CrossRef] [PubMed]
4. Tuĭchibaev, M.U.; Akhmedova, N.U.; Muksimov, F.A. Hemolytic effect of phospholipase A2 and orientotoxin from venom of the great hornet, *Vespa orientalis*. *Biokhimiia* **1988**, *53*, 434–443. [PubMed]
5. Zambelli, V.O.; Picolo, G.; Fernandes, C.A.H.; Fontes, M.R.M.; Cury, Y. Secreted Phospholipases A2 from Animal Venoms in Pain and Analgesia. *Toxins* **2017**, *9*, 406. [CrossRef]
6. Mingarro, I.; Pérez-Payá, E.; Pinilla, C.; Appel, J.R.; Houghten, R.A.; Blondelle, S.E. Activation of bee venom phospholipase A2 through a peptide-enzyme complex. *FEBS Lett.* **1995**, *372*, 131. [CrossRef]
7. Almeida, R.A.M.B.; Olivo, T.E.T.; Mendes, R.P.; Barraviera, S.R.C.S.; Souza, L.R.; Martins, J.G.; Hashimoto, M.; Fabris, V.E.; Junior, R.S.F.; Barraviera, B. Africanized honeybee stings: How to treat them. *Rev. Soc. Bras. Med. Trop.* **2011**, *44*, 755–761. [CrossRef]
8. Muraro, A.; Roberts, G.; Worm, M.; Bilò, M.B.; Brockow, K.; Rivas, M.F.; Santos, A.F.; Zolkipli, Z.Q.; Bellou, A.; Beyer, K.; et al. Anaphylaxis: Guidelines from the European Academy of Allergy and Clinical Immunology. *Allergy* **2014**, *69*, 1026–1045. [CrossRef]
9. Schumacher, M.J.; Egen, N.B.; Tanner, D. Neutralization of bee venom lethality by immune serum antibodies. *Am. J. Trop. Med. Hyg.* **1996**, *55*, 197–201. [CrossRef]
10. Santos, K.S.; Stephano, M.A.; Marcelino, J.R.; Ferreira, V.M.R.; Rocha, T.; Caricati, C.; Higashi, H.G.; Moro, A.M.; Kalil, J.E.; Malaspina, O.; et al. Production of the first effective hyperimmune equine serum antivenom against Africanized bees. *PLoS ONE* **2013**, *8*, e79971. [CrossRef]
11. Laustsen, A.H.; Gutiérrez, J.M.; Knudsen, C.; Johansen, K.H.; Bermúdez-Méndez, E.; Cerni, F.A.; Jürgensen, J.A.; Ledsgaard, L.; Martos-Esteban, A.; Øhlenschlæger, M.; et al. Pros and cons of different ther-apeutic antibody formats for recombinant antivenom development. *Toxicon* **2018**, *146*, 51–75. [CrossRef] [PubMed]
12. Jones, R.G.A.; Corteling, R.L.; To, H.P.; Bhogal, G.; Landon, J. A novel Fab-based antivenom for the treatment of mass bee attacks. *Am. J. Trop. Med. Hyg.* **1999**, *61*, 361–366. [CrossRef] [PubMed]
13. Pucca, M.B.; Cerni, F.A.; Oliveira, I.S.; Jenkins, T.P.; Argemí, L.; Sørensen, C.V.; Ahmadi, S.; Barbosa, J.E.; Laustsen, A.H. Bee Updated: Current knowledge on bee venom and bee envenoming therapy. *Front. Immunol.* **2019**, *10*, 2090. [CrossRef] [PubMed]
14. Pessenda, G.; Silva, L.C.; Campos, L.B.; Pacello, E.M.; Pucca, M.B.; Martinez, E.Z.; Barbosa, J.E. Human scFv antibodies (Afribumabs) against Africanized bee venom: Advances in melittin recognition. *Toxicon* **2016**, *112*, 59–67. [CrossRef]
15. Jenkins, T.P.; Fryer, T.; Dehli, R.I.; Jürgensen, J.A.; Fuglsang-Madsen, A.; Føns, S.; Laustsen, A.H. Toxin Neutralization Using Alternative Binding Proteins. *Toxins* **2019**, *11*, 53. [CrossRef]
16. Oh, J.K.; Drumright, R.; Siegwart, D.J.; Matyjaszewski, K. The development of microgels/nanogels for drug delivery applications. *Prog. Polym. Sci.* **2008**, *33*, 448–477. [CrossRef]
17. Wani, T.U.; Rashid, M.; Kumar, M.; Chaudhary, S.; Kumar, P.; Mishra, N. Targeting aspects of nano-gels: An overview. *Int. J. Pharm. Sci. Nanotechnol.* **2014**, *7*, 2612–2630. [CrossRef]
18. Patoo, T.S.; Khanday, F.; Qurashi, A. Prospectus of advanced nanomaterials for antiviral properties. *Mater. Adv.* **2022**, *3*, 2960–2970. [CrossRef]
19. Anooj, E.S.; Charumathy, M.; Sharma, V.; Vibala, B.V.; Gopukumar, S.T.; Beema Jainab, S.I.; Vallinayagam, S. Nanogels: An overview of properties, biomedical applications, future research trends and developments. *J. Mol. Struct.* **2021**, *1239*, 130446. [CrossRef]

20. Magadán, S.; Mikelez-Alonso, I.; Borrego, F.; González-Fernández, Á. Nanoparticles and trained immunity: Glimpse into the future. *Adv. Drug Deliv. Rev.* **2021**, *175*, 113821. [CrossRef]
21. Whitcombe, M.J.; Kirsch, N.; Nicholls, I.A. Molecular imprinting science and technology: A survey of the literature for the years 2004–2011. *J. Mol. Recognit.* **2015**, *27*, 297–401. [CrossRef] [PubMed]
22. Ulusoy, M.; Aslıyüce, S.; Keskin, N.; Denizli, A. Beauvericin purification from fungal strain using molecularly imprinted cryogels. *Process Biochem.* **2022**, *113*, 185–193. [CrossRef]
23. Deyev, S.M.; Lebedenko, E.N. Modern Technologies for Creating Synthetic Antibodies for Clinical application. *Acta Nat.* **2009**, *1*, 32–50. [CrossRef]
24. Long, Y.; Li, Z.; Bi, Q.; Deng, C.; Chen, Z.; Bhattachayya, S.; Li, C. Novel polymeric nanoparticles targeting the lipopolysaccharides of *Pseudomonas aeruginosa*. *Int. J. Pharm.* **2016**, *502*, 232–241. [CrossRef]
25. Xu, S.; He, H.; Liu, Z. New Promises of Advanced Molecular Recognition: Bioassays, Single Cell Analysis, Cancer Therapy, and Beyond. *Chin. J. Chem.* **2022**, *40*, 635–650. [CrossRef]
26. Hoshino, Y.; Koide, H.; Furuya, K.; Haberaecker, W.W., III; Lee, S.H.; Kodama, T.; Kanazawa, H.; Oku, N.; Shea, K.J. The rational design of a synthetic polymer nanoparticle that neutralizes a toxic peptide in vivo. *Proc. Natl. Acad. Sci. USA* **2012**, *109*, 33–38. [CrossRef]
27. Hoshino, Y.; Shea, K.J. The evolution of plastic antibodies. *J. Mater. Chem.* **2011**, *21*, 3517–3521. [CrossRef]
28. Hoshino, Y.; Koide, H.; Urakami, T.; Kanazawa, H.; Kodama, T.; Oku, N.; Shea, K.J. Recognition, Neutralization, and Clearance of Target Peptides in the Bloodstream of Living Mice by Molecularly Imprinted Polymer Nanoparticles: A Plastic Antibody. *J. Am. Chem. Soc.* **2010**, *132*, 6644–6645. [CrossRef]
29. Hoshino, Y.; Urakami, T.; Kodama, T.; Koide, H.; Oku, N.; Okahata, Y.; Shea, K.J. Design of Synthetic Polymer Nanoparticles that Capture and Neutralize a Toxic Peptide. *Nano Micro Small* **2009**, *5*, 1562–1568. [CrossRef]
30. Hayakawa, N.; Yamada, T.; Kitayama, Y.; Takeuchi, T. Cellular Interaction Regulation by Protein Corona Control of Molecularly Imprinted Polymer Nanogels Using Intrinsic Proteins. *ACS Appl. Polym. Mater.* **2020**, *2*, 1465–1473. [CrossRef]
31. Takeuchi, T.; Kitayama, Y.; Sasao, R.; Yamada, T.; Toh, K.; Matsumoto, Y.; Kataoka, K. Molecularly Imprinted Nanogels Acquire Stealth In Situ by Cloaking Themselves with Native Dysopsonic Proteins. *Angew. Chem. Int. Ed. Engl.* **2017**, *56*, 1–6. [CrossRef]
32. Radu, A.L.; Gavrila, A.M.; Cursaru, B.; Spatarelu, C.P.; Sandu, T.; Sarbu, A.; Teodorescu, M.; Perrin, F.X.; Iordache, T.V.; Zaharia, A. Poly (ethylene Glycol) Diacrylate-Nanogels Synthesized by Mini-emulsion Polymerization. *Mater. Plast.* **2019**, *56*, 514–519. [CrossRef]
33. Spatarelu, C.P.; Chiriac, A.L.; Cursaru, B.; Iordache, T.V.; Gavrila, A.M.; Cojocaru, C.T.; Botez, R.E.; Trica, B.; Sarbu, A.; Teodorescu, M.; et al. Composite Nanogels Based on Zeolite-Poly (ethylene glycol) Diacrylate for Controlled Drug Delivery. *Nanomaterials* **2020**, *10*, 195. [CrossRef] [PubMed]
34. Raemdonck, K.; Demeester, J.; De Smedt, S. Advanced nanogel engineering for drug delivery. *Soft Matter* **2009**, *5*, 707–715. [CrossRef]
35. Imani, M.; Sharifi, S.; Mirzade, H.; Ziaee, F. Monitoring of Polyethylene Glycol-diacrylate-based Hydrogel Formation by Real Time NMR Spectroscopy. *Iran. Polym. J.* **2007**, *16*, 13–20.
36. Askari, F.; Zandi, M.; Shokrolahi, P.; Tabatabaei, M.H.; Hajirasoliha, E. Reduction in protein absorption on ophthalmic lenses by PEGDA bulk modifcation of silicone acrylate based formulation. *Prog. Biomater.* **2019**, *8*, 169–183. [CrossRef]
37. Bae, M.; Gemeinhart, R.A.; Divan, R.; Suthar, K.J.; Mancini, D.C. Fabrication of Poly (ethylene glycol) Hydrogel Structures for Pharmaceutical Applications using Electron beam and Optical Lithography. *J. Vac. Sci. Technol. B* **2010**, *28*, C6P24–C6P29. [CrossRef]
38. Goormaghtigh, E.; Cabiaux, V.; Ruysschaert, J.-M. Determination of Soluble and Membrane Protein Structure by Fourier Transform Infrared Spectroscopy. In *Physicochemical Methods in the Study of Biomembranes*; Subcellular Biochemistry; Hilderson, H.J., Ralston, G.B., Eds.; Springer: Boston, MA, USA, 1994; Volume 23, pp. 405–450. [CrossRef]
39. Naumann, D. Ft-Infrared and Ft-Raman Spectroscopy in Biomedical Research. *Appl. Spectrosc. Rev.* **2001**, *36*, 239–298. [CrossRef]
40. Kennedy, D.F.; Slotboom, A.J.; de Haas, G.H.; Chapman, D. A Fourier transform infrared spectroscopic (FTIR) study of porcine and bovine pancreatic phospholipase A2 and their interaction with substrate analogues and a transition-state inhibitor. *Biochim. Biophys. Acta (BBA) Protein Struct. Molec. Enzym.* **1990**, *1040*, 317–326. [CrossRef]
41. Barth, A. Infrared spectroscopy of proteins. *Biochim. Biophys. Acta Bioenerg.* **2007**, *1767*, 1073–1101. [CrossRef]
42. Ekomo, V.M.; Branger, C.; Bikanga, R.; Florea, A.-M.; Istamboulie, G.; Calas-Blanchard, C.; Noguer, T.; Sarbu, A.; Brisset, H. Detection of Bisphenol A in aqueous medium by screen printed carbon electrodes incorporating electrochemical molecularly imprinted polymers. *Biosens. Bioelectron.* **2018**, *112*, 156–161. [CrossRef] [PubMed]
43. Barrera, F.N. On 'Fourier transform infrared study of proteins with parallel β-chains' by Heino Susi, D. Michael Byler. *Arch. Biochem. Biophys.* **2022**, *726*, 109114. [CrossRef] [PubMed]
44. Genier, F.S.; Burdin, C.V.; Biria, S.; Hosein, I.D. A novel calcium-ion solid polymer electrolyte based on crosslinked poly (ethylene glycol) diacrylate. *J. Power Sources* **2019**, *414*, 302–307. [CrossRef]
45. Maitlo, I.; Ali, S.; Akram, M.Y.; Shehzad, F.K.; Nie, J. Binary phase solid-state photopolymerization of acrylates: Design, characterization and biomineralization of 3D scaffolds for tissue engineering. *Front. Mater. Sci.* **2017**, *11*, 307–317. [CrossRef]
46. Saimani, S.; Kumar, A. Polyethylene glycol diacrylate and thermoplastic polymers based semi-IPNs for asymmetric membranes. *Compos. Interfaces* **2008**, *15*, 781–797. [CrossRef]

47. Richmond-Aylor, A.; Bell, S.; Callery, P.; Morris, K. Thermal degradation analysis of amino acids in fingerprint residue by pyrolysis GC–MS to develop new latent fingerprint developing reagents. *J. Forensic Sci.* **2007**, *52*, 380–382. [CrossRef]
48. Koch, S.J.; Renner, C.; Xie, X.; Schrader, T. Tuning Linear Copolymers into Protein-Specific Hosts. *Angew. Chem. Int. Ed. Engl.* **2006**, *118*, 6500–6503. [CrossRef]
49. Murakami, M.; Sato, H.; Taketomi, Y. Updating Phospholipase A2 Biology. *Biomolecules* **2020**, *10*, 1457. [CrossRef]
50. Shipolini, R.A.; Callewaert, G.L.; Cottrell, R.C.; Doonan, S.; Vernon, C.A.; Banks, B.E.C. Phospholipase A From Bee Venom. *Eur. J. Biochem.* **1971**, *20*, 459–468. [CrossRef]
51. Hawkins, D.M.; Ellis, E.A.; Stevenson, D.; Holzenburg, A.; Reddy, S.M. Novel critical point drying (CPD) based preparation and transmission electron microscopy (TEM) imaging of protein specific molecularly imprinted polymers (HydroMIPs). *J. Mater. Sci.* **2007**, *42*, 9465–9946. [CrossRef]
52. Shi, H.; Ratner, B.D. Template recognition of protein-imprinted polymer surfaces. *J. Biomed. Mater. Res.* **2000**, *49*, 1–11. [CrossRef]
53. Liu, G.; Li, Y.; Yang, L.; Wei, Y.; Wang, X.; Wang, Z.; Tao, L. Cytotoxicity study of polyethylene glycol derivatives. *RSC Adv.* **2017**, *7*, 18252–18259. [CrossRef]

Article

Molecularly Imprinted Solid Phase Extraction Strategy for Quinic Acid

Sarah H. Megahed [1], Mohammad Abdel-Halim [1], Amr Hefnawy [2], Heba Handoussa [3], Boris Mizaikoff [4,5,*] and Nesrine A. El Gohary [1,*]

1. Pharmaceutical Chemistry Department, Faculty of Pharmacy and Biotechnology, German University in Cairo, Cairo 11835, Egypt
2. Division of Molecular Pharmaceutics and Drug Delivery, College of Pharmacy, University of Texas at Austin, Austin, TX 78712, USA
3. Pharmaceutical Biology Department, Faculty of Pharmacy and Biotechnology, German University in Cairo, Cairo 11835, Egypt
4. Institute of Analytical and Bioanalytical Chemistry, Ulm University, 89081 Ulm, Germany
5. Hahn-Schickard, 89077 Ulm, Germany
* Correspondence: boris.mizaikoff@uni-ulm.de (B.M.); nesrine.elgohary@guc.edu.eg (N.A.E.G.)

Abstract: Quinic acid (QA) and its ester conjugates have been subjected to in-depth scientific investigations for their antioxidant properties. In this study, molecularly imprinted polymers (MIPs) were used for selective extraction of quinic acid (QA) from coffee bean extract. Computational modelling was performed to optimize the process of MIP preparation. Three different functional monomers (allylamine, methacrylic acid (MAA) and 4-vinylpyridine (4-VP)) were tested for imprinting. The ratio of each monomer to template chosen was based on the optimum ratio obtained from computational studies. Equilibrium rebinding studies were conducted and MIP C, which was prepared using 4-VP as functional monomer with template to monomer ratio of 1:5, showed better binding performance than the other prepared MIPs. Accordingly, MIP C was chosen to be applied for selective separation of QA using solid-phase extraction. The selectivity of MIP C towards QA was tested versus its analogues found in coffee (caffeic acid and chlorogenic acid). Molecularly imprinted solid-phase extraction (MISPE) using MIP C as sorbent was then applied for selective extraction of QA from aqueous coffee extract. The applied MISPE was able to retrieve 81.918 ± 3.027% of QA with a significant reduction in the amount of other components in the extract.

Keywords: computational modelling; molecularly imprinted polymers; solid-phase extraction; quinic acid; *Coffea arabica*

1. Introduction

Oxidative stress is the main cause for altering numerous signaling pathways that eventually promote cellular damage. It is considered a key player mediator in the pathophysiology of several health complications [1]. Intracellular antioxidant enzymes and intake of dietary antioxidants may help maintain the utmost antioxidant balance in the body. Epidemiological studies have proven that the consumption of nutraceuticals with potential antioxidant impacts reduces the risk of several diseases, including neurodegenerative diseases, cardiovascular diseases and cancer, through apoptosis-mediated cytotoxicity. They are also known to reduce inflammation by different mechanisms such as the inhibition of pro-inflammatory transcription factor, nuclear factor Kappa B (NF-κB) [2].

Quinic acid (QA) and its ester conjugates (caffeoylquinic acids) are present in various food products [3] and are the major constituents of coffee [4]. Many reports support the efficacy of nutritional QA in the enhancement of several biological processes via its paramount antioxidant effects [5]. QA has been previously reported as a potent antioxidant due to its capability to lower the intracellular ROS levels in H_2O_2 pre-treated cells and inhibition of lipid peroxidation [6]. QA was also found to protect against oxidative stress by increasing the antioxidant capacity as well as decreasing the levels of MDA and nitrite in an in vitro study by Khorasgani et al. [3]. Furthermore, QA upregulated *daf-16 sod-3* expression and downregulated reactive oxygen species (ROS) levels in a *C.elegans* in vivo model [5].

Noteworthily, the ingested QA is used by the human body as a precursor for the synthesis of many important compounds such as nicotinamide [7], which plays a major role in neuronal development and survival [8]. Nicotinamide is also used in the synthesis of two important co-enzymes, nicotinamide adenine dinucleotide (NAD) and nicotinamide adenine dinucleotide phosphate (NADP). Both NAD and NADP are essential in many processes in the human body such as DNA repair, energy production and cell death regulation [9].

Several old techniques have been developed to isolate QA from its original natural source, yet, they are still reports with several limitations [10]. Many extraction protocols and chromatographic methods have been developed to optimize the extraction of QA, such as column chromatography [11], alkaline hydrolysis [12] and liquid–liquid extraction while using an amine as an extractant [13]. However, most of the reported methods are solvent- and time-consuming and none of these methods are considered as having high selectivity towards QA.

Molecular imprinting is a rapidly growing technique used to create synthetic receptors with recognition sites that have the ability to bind specifically to a wide variety of molecules ranging from small drug molecules to large peptides or proteins. The molecularly imprinted polymer (MIP) technique can be described by analogy to the "lock and key model" described by Emil Fischer [14]. The synthesized molecularly imprinted polymers (MIPs) have many cavities complementary to their template molecules in shape, size and chemical functionality, causing them to be particularly selective towards those target molecules [15].

In the past few years, several research articles and reviews have been published, showing the current advances and diversity in the synthesis and application of MIPs [16–18]. These studies show the increasing importance of the use of MIPs in the field of analytical chemistry and their application in sensors, extraction and chromatography [19].

The use of MIPs as a replacement for biological material in optical [20] and electrochemical [21] biosensors has attracted much attention throughout the years. This is attributed to their superiority over the biological components in terms of cost, stability and reusability [17]. They also offer outstanding recognition ability, selectivity, specificity and robustness [17,22]. Accordingly, molecular imprinting has become one of the most important techniques for fabricating synthetic ligands on sensor surfaces [17,19]. MIPs have been successfully coupled to surface plasmon resonance (SPR) sensors [23], quartz crystal microbalance sensors [24], luminescence probes [25] and electrochemical sensors [26,27]. Molecular-imprinted fluorescent sensors (MIFS) have been used for detection of several organic molecules and metal ions including proteins [28], caffeine [29] and cocaine [30] in addition to silver [31] and aluminum ions [32]. Moreover, MIP-based SPR sensors have been applied for detection of biomarkers [33], biomolecules [34], pesticides [35] and banned additives [36].

The use of MIPs as sorbents has become one of the most commonly used methods for SPE [37]. They have attracted much attention owing to their numerous advantages such as high selectivity, ease of preparation, low cost, reusability and their potential application to a wide range of target molecules [38]. MISPE has been applied for the extraction of different analytes from biological fluids such as blood, urine and bile [39] in addition to environmental samples and plant tissues [18].

Different techniques of molecular imprinting have been reported for the extraction of antioxidants from natural resources [40,41] such as the isolation of oleuropein from olive leaf extracts [42] and the concentration of tannins from Brazilian natural sources [43].

MIPs were also applied on different plant extracts, including coffee, for isolation of QA derivatives [12]. In a recent study by Kanao et al. [44], poly(ethylene glycol) hydrogels prepared by molecular imprinting were used for selective extraction of quinic acid gamma-lactone (QAGL) from coffee. The synthesized MIPs successfully removed QAGL from freshly brewed coffee at high speed with high yield, which resulted in better-tasting coffee. Moreover, the prepared MIPS were highly selective towards QAGL, which prevented non-specific adsorption of other components in coffee.

Bulk imprinting is considered the most widely used method for preparation of MIPs [45]. It is the method of choice for the imprinting of small molecules as it allows fast and reversible adsorption and the release of the template molecule [46]. Bulk imprinting has been successfully used for small molecules comparable to QA such as sinapic acid [47], gallic acid [48], caffeic acid and p-hydroxybenzoic acid [49]. However, the preparation of MIP using QA as a template has not been reported before.

In this work, three bulk MIPs were synthesized using three different monomers. Their binding performance and their ability to be used as sorbents for SPE of QA from coffee beans have been examined.

2. Materials and Methods

2.1. Reagents and Materials

Standard QA (98%) was purchased from Alfa Aesar. Caffeic acid (CA) (98%), chlorogenic acid (CLA) (95%), acetonitrile (CAN) (HPLC grade; $\geq$9.99%), methanol (99.8%), absolute ethanol (EtOH) ($\geq$99.5%), formic acid (reagent grade; $\geq$95%), glacial acetic acid ($\geq$99.7%), methacrylic acid (MAA) (stabilized with hydroquinone monomethyl ether; $\geq$90.0%), 4-vinylpyridine (4-VP) (contains 100 ppm hydroquinone as inhibitor; 95%), ethylene glycol and dimethacrylate (EGDMA) (contains 90–100 ppm hydroquinone monomethyl ether as inhibitor; 98%) were obtained from Sigma Aldrich (Darmstadt, Germany). Ethyl acetate (EtOAc) (99.5%) was purchased from Alfa Chemical (India) and a purelab UHQ (ELGA) water purification system (High Wycombe, Buckinghamshire, UK) was used to obtain ultra-pure water. Empty polypropylene SPE 3 mL tubes with PE frits of 20 µm porosity were obtained from Supelco Inc. (Bellefonte, PA, USA).

Green coffee (C. arabica L.) beans were kindly supplied by Misr Coffee (10th of Ramadan Ind. City, Cairo, Egypt Industrial Company). The beans were mechanically ground and milled into size (40 mesh) for extraction and application steps. The obtained granules were completely dried using a hot air oven at a temperature of 38 °C for 2 h.

2.2. Computational Modelling: Monomers Molar Ratio Screening

Gaussian 03 package was used to determine the optimum template to monomer molar ratio for bulk polymers. Gaussview 5.0 software (Gaussian Inc., Pittsburgh, USA) was used first to draw 3D structures of the template, QA, monomers, MAA, allylamine and 4-VP, in addition to template-monomer complexes. All the obtained structures were then optimized to the lowest energy conformation using Hartree-Fock theory with the (6–31 G(D)) basis set. Hartree-Fock is an accurate method for large systems, which makes it easier to screen different monomer ratios for specific templates [50,51]. Different template to monomer molar ratios were screened for each of the used monomers and Equation (1) was used to calculate the binding energies of the complexes.

$$\Delta E = E(\text{template–monomer complex}) - [E(\text{template}) + nE(\text{monomer})] \quad (1)$$

where ΔE refers to the binding energy of the complex and n refers to the monomer number in the template–monomer complexes.

The calculations of the binding energies were conducted in the solvent phase (DMSO) using the polarizable continuum model (PCM) to mimic experimental conditions. In this model, the solvent effect is considered during calculations as it affects the stability and the energy of the template–monomer complexes [52], where the solvent is modelled as a polarizable continuum rather than individual molecules [53].

2.3. Bulk Polymers Preparation

Different bulk MIPs were prepared via the non-covalent approach, introduced by K. Mosbach et al. [54], using thermal free radical polymerization. The reaction was performed in a glass vial, by dissolving 0.5 mmol of QA in 6 mL of the porogen, DMSO. This was followed by the addition of suitable amount of monomer and the pre-polymerization mixture was stirred at room temperature for 30 min. Afterwards, the cross-linker ethylene glycol dimethacrylate (EGDMA) was added and the solution was left to stir for 5 min. Following which, 75 mg of the free radical initiator was added and the solution was purged with argon for 3 min to remove oxygen and create inert conditions. The glass vial was sealed and left in an oil bath at 60 °C for 24 h to allow polymerization. For each MIP, a non-imprinted polymer (NIP) was prepared using the same procedure without adding the template. The glass vials were then smashed to obtain the bulk polymers, which were then subjected to crushing, grinding and sieving. The fraction with a particle size of 40–100 µm was collected. The full composition of the prepared polymers is described in Table 1.

Table 1. Chemical composition of prepared MIPs.

Polymer	Type of Polymerization	Template (T)	Functional Monomer (FM)	Cross-Linker (CL)	T:FM:CL Molar Ratio
A	Bulk	QA	Allylamine	EGDMA	1:6:20
B	Bulk	QA	MAA	EGDMA	1:4:20
C	Bulk	QA	4-VP	EGDMA	1:5:20

2.4. Morphology Characterization

The surface morphology of the MIPs and their corresponding NIPs was examined using FEI Quanta 650 environmental scanning electron microscope (ESEM) under high vacuum at a high voltage of 10 kV with a spot size of 3.5 and working distance set to around 10 mm. N_2 adsorption–desorption isotherms were used to analyze the surface area, pore volume and pore size of all polymers at 77 K via a Quantachrome TouchWin v.1.2 instrument (FL, USA). The polymers were first degassed at 150 °C for 24 h to remove the adsorbed gases and moisture. The specific surface areas were calculated using the Brunauer–Emmett–Teller (BET) method, while the Barrett–Joyner–Halenda (BJH) method was used to calculate the volume and pore size.

2.5. Equilibrium Rebinding Studies

The binding studies were conducted at room temperature by modifying the protocol previously described by Saad et al. [45]. Ten mg of the imprinted and non-imprinted polymers were added to 2 mL of 0.1 mM QA solution prepared in water, methanol or ACN: water (4: 1 v/v). The suspensions were then left to shake at room temperature for 2 h at 200 rpm using a Thermo Scientific™ MaxQ mini 4000 Benchtop Orbital Shaker (Waltham, MA, USA). This was followed by a centrifugation step at 14,000 rpm for 15 min and the supernatants were filtered through 0.22 polytetrafluoroethylene (PTFE) syringe filters. The concentration of the unbound QA was then quantified using UHPLC-MS/MS. The amount of the rebound QA was calculated using Equation (2)

$$B = \frac{(C_i - C_f) \times V \times 1000}{W} \qquad (2)$$

where B is the amount of rebound template in µmol/g polymer, C_i and C_f represent the initial and final concentrations in mM, respectively, V is the volume of the solution in ml and W is the weight of used polymer in mg.

The imprinting factor was then calculated using Equation (3)

$$IF = \frac{B_{MIP}}{B_{NIP}} \qquad (3)$$

where IF is the imprinting factor and B_{MIP} is the amount of template bound in µmol/g of the MIP, while B_{NIP} is the amount of template bound in µmol/g of the NIP.

2.6. Adsorption Kinetics

The uptake profiles of MIP C and its corresponding NIP were studied over 2 h. This was achieved by incubating 10 mg of the polymer with 2 mL of 0.1 mM QA in methanol for 5, 15, 30, 60 and 120 min. This was followed by a centrifugation step and the supernatants were analyzed using UHPLC-MS/MS and the amount of bound template was determined using Equation (2).

The obtained data were further analyzed to determine adsorption kinetics. The pseudo-first order and pseudo-second order kinetics were used to investigate the mechanism of adsorption of MIP C. The pseudo-first order rate can be expressed in Equation (4)

$$ln(q_e - q_t) = lnq_e - K_1 t \tag{4}$$

where q_e and q_t are the binding capacities at equilibrium and at time t (μmol/g), respectively, K_1 is the rate constant of pseudo-first order in min^{-1} and t is time in min [55].

Pseudo-second order is expressed in Equation (5)

$$\frac{1}{q_t} = \frac{1}{K_2 q_e^2} + \frac{t}{q_e} \tag{5}$$

where K_2 is the rate constant of pseudo-second order in g/μmol.min [56].

2.7. Binding Isotherm

Ten mg of MIP C and its corresponding NIP were incubated with 2 mL of QA in methanol over the concentration range of (0.01–0.2 mM) for 2 h. The binding isotherms of both polymers were then obtained by plotting the binding capacity (B) versus the initial QA concentration (C_i). The results were further analyzed using the Freundlich isotherm model [57] expressed by Equation (6).

$$Log\ B = mLog\ C_f + Log\alpha \tag{6}$$

where B represents the binding capacity in μmol/g, m represents the Freundlich constant or heterogenicity factor ranging from 0 to 1, C_f represents the equilibrium concentration in mM and the constant α represents maximum binding capacity in μmol/g [45].

2.8. MISPE Procedure Optimization

MIP C and NIP C were used as sorbent materials for offline-mode solid-phase extraction. Forty mg of each polymer was packed into a 3 mL polypropylene SPE cartridge with a 0.22 PTFE frit placed below the polymer and another similar frit placed above the polymer for secure packing. All trials were performed in triplicate and the analytical measurements were obtained using UHPLC-MS/MS.

A systematic one-factor-at-a-time (OFAT) approach was used to investigate different parameters affecting the extraction procedure including loading amount, loading volume, washing solvent and elution volume. Water: acetic acid (9:1 v/v) was used as the elution solvent in all trials.

2.9. UHPLC-MS/MS Measurements

A new UHPLC-MS/MS method was applied for quantification of QA using ferulic acid as an internal standard. A seven-point calibration curve for QA was prepared in methanol over the concentration range 0.001–0.2 mM.

UHPLC-MS/MS measurements were done using ACQUITY Xevo TQD system (Waters), which is composed of ACQUITY UPLC H-Class system and a XevoTQD triple-quadrupole tandem mass spectrometer with an electrospray ionization (ESI) interface (Waters Corp., Milford, MA, USA). The column used for separation was an Aquity UPLC BEH C_{18} (Waters, Wexford, Ireland), with dimensions of 100 mm × 2.1 mm and stationary phase particle size of 1.7 μm. MassLynx 4.1 software (Waters, Milford, MA, USA) was used for system operation and data acquisition. The TargetLynx quantification program was used to process the acquired data (Waters, Milford, MA, USA). A gradient program was used for chromatographic separation using 0.01% formic acid in water (A) and acetonitrile (B) at a flow rate of 0.3 mL/min, injection volume of 10 µL and column temperature of 40 °C. The gradient was run as follows: 0 min, 90% A, 10% B; 0.75 min, 90% A, 10% B; 2.5 min, 1% A, 99% B; 4 min, 1% A, 99% B; 4.5 min, 90% A, 10% B; 6 min, 90% A, 10% B. The desolvation and cone gas flow rate were 800 and 20 L/h, respectively (nitrogen was used in both cases). The collision gas (argon) was applied at a pressure of 3.67×10^{-3} mbar approx. The MS parameters were as follows: radio frequency (RF) lens voltage 2.5 V, capillary voltage 4 kV, source temperature 150 °C and desolvation gas temperature 300 °C. Cone voltage was 45 V and 28 V for QA and ferulic acid, respectively. The ESI source was operated in negative mode. Quantification was performed using multiple reaction monitoring (MRM) of the transitions of m/z 191 > 85 with collision energy of 18 V for QA and m/z 192.89 > 133.95 with collision energy of 14 V for ferulic acid. Dwell time was set automatically by MassLynx 4.1 software.

2.10. Method Validation

The applied UPLC-MS/MS method was validated according to the ICH guidelines in terms of linearity, limit of detection (LOD), limit of quantification (LOQ), inter- and intra-day precision and accuracy. More details are provided in the supplementary material

2.11. MIP Cartridge Reusability

MIP cartridge reusability was tested over 10 adsorption–desorption cycles, where the SPE cartridge was filled with 40 mg of MIP C. This was followed by a conditioning step using 2 mL of absolute ethanol, then loading with 2 mL of 0.1 mM QA in ethanol, a washing step using 2 mL of acetonitrile and finally an elution step using 2 mL of water: acetic acid (9:1 v/v). After the elution step, the cartridge was subjected to 5 washing steps; 2 steps of washing using 3 mL of water: acetic acid (9:1 v/v) each, then once using 3 mL of water and finally 2 washing steps using 3 mL of absolute ethanol each. The elution fractions were analyzed using the validated UHPLC-MS/MS method and QA recovery percentage was calculated after each elution.

2.12. Selectivity Study

Two mL of equimolar mixture of QA, caffeic acid and chlorogenic acid (Figure 1) (0.05 mM) in ethanol was percolated through SPE cartridges packed with 40 mg of MIP C and NIP C. The cartridges were then washed using 2 mL of acetonitrile. This was followed by the elution step, using 2 mL of 10% acetic acid in UPW.

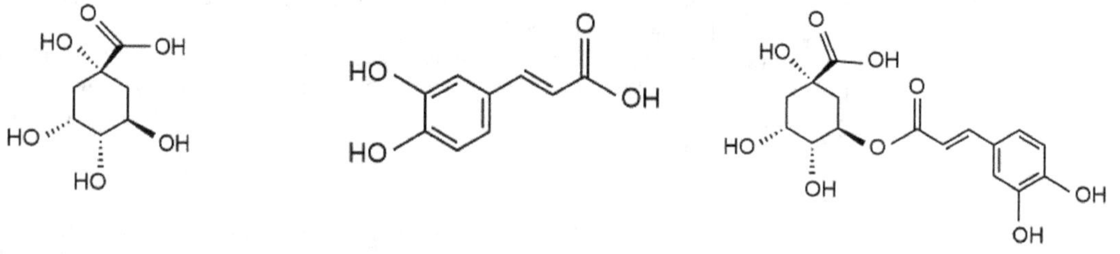

Figure 1. Structures of quinic acid, caffeic acid and chlorogenic acid.

The obtained elution fraction was evaporated and reconstituted in methanol. The amount of QA in the elution solvent was measured using UHPLC-MS/MS, while caffeic acid and chlorogenic acid were quantified using UHPLC-UV at λ_{max} 325 nm.

2.13. MISPE Application on Coffee Extract

2.13.1. UHPLC Method for QA Quantification in Coffee Extract

UHPLC-PDA-ESI- MS and MS/MS analyses were done using the ACQUITY Xevo TQD system (Waters), which is composed of the ACQUITY UPLC H-Class system and a XevoTQD triple-quadrupole tandem mass spectrometer with an electrospray ionization (ESI) interface (Waters Corp., Milford, MA, USA). The column used for separation was an Aquity UPLC BEH C_{18} (Waters, Wexford, Ireland), with dimensions of 100 mm × 2.1 mm and stationary phase particle size of 1.7 µm. MassLynx 4.1 software (Waters, Milford, MA, USA) was used for system operation and data acquisition. The solvent system consisted of 0.01% formic acid in water (A) and acetonitrile (B) by applying the following gradient program: 0 min, 8% B; 30 min, 45% B; 31 min, 8% B; and 33 min, 8% B. The flow rate was 0.2 mL/min and the injection volume was 10 µL. The samples were dissolved in ethanol then filtered through a filter of pore size 0.2 µm. The eluted compounds were detected at mass ranges from 100 to 1000 m/z. The MS scan was carried out at the following conditions: capillary voltage, 3.5 kV; detection at cone voltages, (20 V–95 V); radio frequency (RF) lens voltage, 2.5 V; source temperature, 150 °C and desolvation gas temperature 500 °C. The desolvation and cone gas flow rate were 800 and 20 L/h, respectively (nitrogen was used in both cases). QA was detected through the MRM of the transition m/z 191 > 85 with collision energy of 18 V and cone voltage of 45 V.

2.13.2. Method Validation

The method was validated according to the ICH guidelines in terms of linearity, limit of detection (LOD), limit of quantification (LOQ), inter- and intra-day precision and accuracy. More details are found in the supplementary data.

2.13.3. Preparation of Aqueous Coffee Extract

Fifty grams of roasted coffee beans were subjected to fine grinding and placed in a conical flask, then 1 L of ultrapure water was added. The mixture was heated at 60 °C for 1 h and was left to macerate overnight. This was followed by a centrifugation and a filtration step. Then, the supernatant was concentrated using a rotary vacuum evaporator at 40 °C. The dried residue was stored in an opaque glass bottle for further studies.

2.13.4. Application of MISPE for Extraction of QA from Total Aqueous Coffee Extract

The optimized SPE method was used for the extraction of QA from total aqueous coffee extract. The extract was reconstituted in ethanol: water (97:3 v/v) and 2 concentrations were prepared, 0.25 mg/mL and 0.5 mg/mL. Two ml of each concentration was loaded to SPE cartridge containing 40 mg of MIP C. This was followed by a washing step using 2 mL of acetonitrile and an elution step using 2 mL of 10% acetic acid in water.

3. Results and Discussion

3.1. Computational Modelling: Monomers Molar Ratio Screening

In this study, computational modelling was used to optimize the pre-polymerization complex by determining the most suitable functional monomer molar ratio for each of the chosen monomers, since the self-assembly of the template and functional monomer is the most crucial step in polymer preparation [58]. The study was conducted in the solvent phase and DMSO was the solvent of choice, which was used as the porogen during polymer preparation [15]. The influence of the cross-linker was not considered to simplify the calculations [59]. The three functional monomers used in this study were allylamine, MAA and 4-VP. For each of the chosen monomers, different template: functional monomer ratios were examined. For allylamine, the studied template: monomer ratios were (1:1,

1:2, 1:3, 1:4, 1:5 and 1:6), for MAA, the ratios were (1:1, 1:2, 1:3 and 1:4) and for 4-VP, the studied ratios were (1:1, 1:2, 1:3, 1:4 and 1:5). The optimized structures of QA, functional monomers and pre-polymerization complexes are shown in Figure 2 and Figures S1–S13 in the supplementary data.

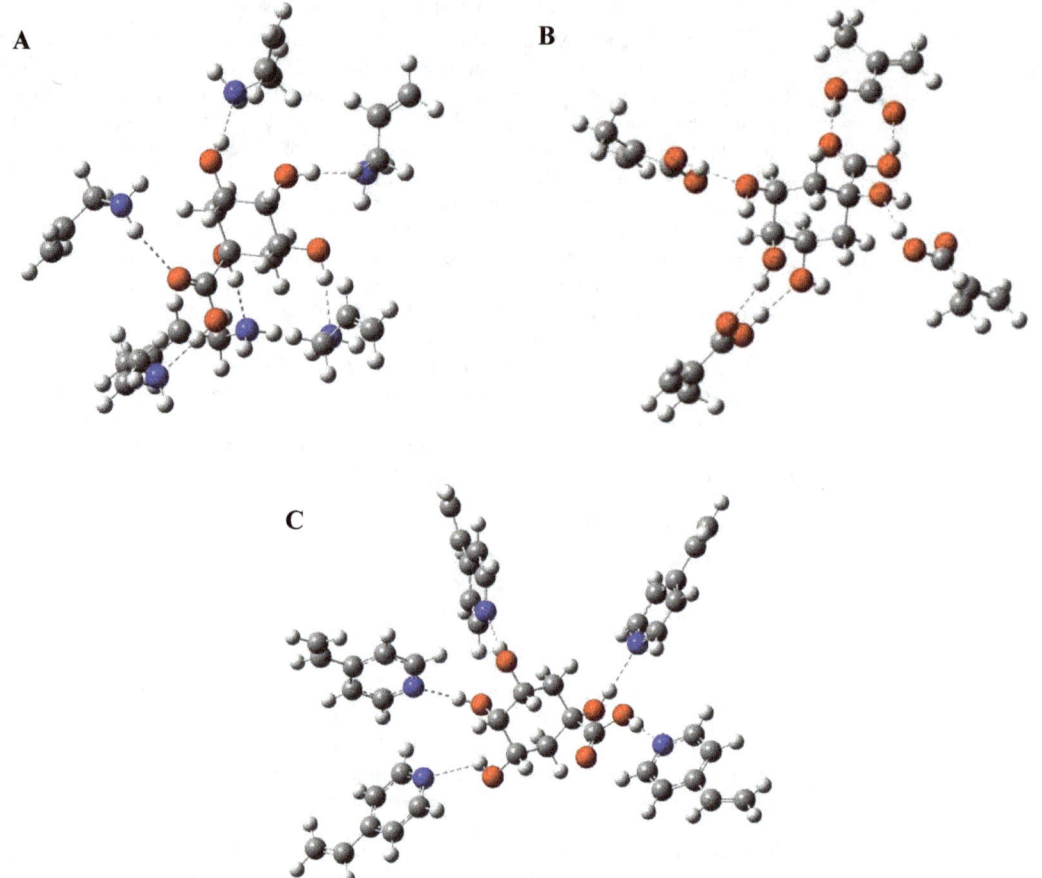

Figure 2. Computer-modelled structures of the best conformations for (**A**) QA–(allylamine)$_6$, (**B**) QA–(MAA)$_4$ and (**C**) QA–(4-VP)$_5$ complexes.

Energies of the most stable conformations were then determined and the binding energies of the formed complexes were calculated according to Equation (1) and the results are shown in supplementary data. Based on the binding energies; the best template: monomer ratio was determined for each functional monomer. Results revealed that by increasing the number of monomers used, the calculated binding energies increased, which indicates the formation of more stable complexes [60]. For allylamine, the best ratio was 1:6 ($E = -175.909$ kJ/mol). For MAA, it was 1:4 ($E = -1633.061$ kJ/mol). Finally, for 4-VP the optimum ratio was 1:5 ($E = -136.5265$ kJ/mol) (Figure 2). Accordingly, these ratios were chosen for the synthesis of MIPs and their corresponding NIPs to be used for further applications.

3.2. Morphology Characterization

The surface morphology of MIPs and their corresponding NIPs of particle size range 40–100 µm were analyzed using scanning electron microscopy (SEM) as shown in Figure 3 and Figures S14 and S15 in the supplementary material. The SEM images showed irregular shapes and sizes, which agrees with the nature of bulk MIPs previously reported in literature [45].

Figure 3. SEM images of (**A**) MIP C and (**B**) NIP C with increasing magnification from left to right.

Nitrogen adsorption–desorption isotherms were performed and BET analysis was used to determine surface areas, while BJH analysis was used to determine the average pore size diameter and pore volume, as these parameters may have a strong impact on the efficiency of adsorption (Figure 4 and Figure S16 in the supplementary material).

The BET results, shown in Table 2 revealed that the MIPs have lower surface areas compared to their corresponding NIPs. This most probably could be attributed to the heterogeneity and roughness of the surface of NIPs which were prepared in the absence of the template, unlike the MIP imprinting process that follows a certain degree of order during the polymerization step [45]. MIP C exhibited the highest surface area (31.41 m^2/g) compared to MIP A and MIP B that exhibited surface areas of 21.51 m^2/g and 23.80 m^2/g, respectively.

The data derived from BJH (Table 2) revealed that all the polymers exhibited a well-developed pore structure. They were all mesoporous with a pore radius of 1.64–1.77 nm, which provides good recognition properties for interaction with the template molecule. These results suggest that the synthesized polymers can be used as sorbents for SPE, since the mesoporous structures are more permeable for solvents compared to micropores and do not require the application of high pressure [61]. All the MIPs and the corresponding NIPs have comparable pore radii, while the pore volumes of all the NIPs are generally larger than the MIPs.

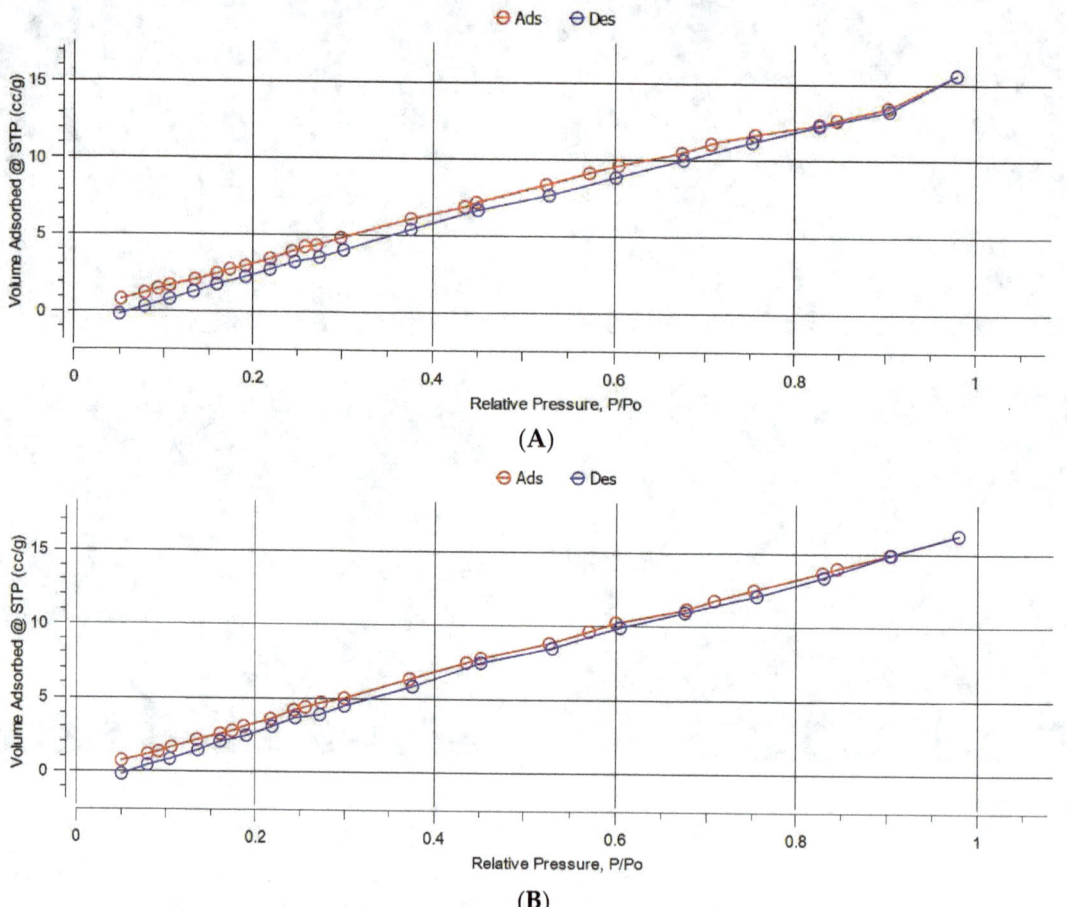

Figure 4. BET isotherms of (**A**) MIP C and (**B**) NIP C.

The overall results reveal that all the NIPs showed higher surface areas and porosities compared to the corresponding MIPs. Thus, it may be concluded that the binding performance of the polymers would be attributed to the imprinting process rather than the surface area and porosity of the particles [45].

Table 2. Surface area, pore volume and pore size of synthesized polymers using BET and BJH methods.

Polymers		A		B		C	
		MIP	NIP	MIP	NIP	MIP	NIP
BET surface area (m²/g)		21.51	37.13	23.80	39.04	31.41	40.90
BJH Pore volume (cc/g)	Adsorption	0.02	0.02	0.02	0.02	0.02	0.02
	Desorption	0.02	0.02	0.02	0.02	0.02	0.02
BJH Pore radius (nm)	Adsorption	1.77	1.64	1.77	1.64	1.77	1.64
	Desorption	1.67	1.67	1.67	1.67	1.67	1.67

3.3. Rebinding Studies

The synthesized polymers were subjected to batch rebinding studies to evaluate their affinity to QA using 0.1 mM QA solution prepared in three different solvents: water, methanol and acetonitrile: water (4:1 v/v) as shown in Table 3.

Table 3. Binding capacities and imprinting factors of bulk polymers in different solvents.

Polymer	FM	Ratio	Water B (µmol/g) ± SD	IF	ACN:Water (4:1) B (µmol/g) ± SD	IF	Methanol B (µmol/g) ± SD	IF
MIP A	Allylamine	1:6:20	2.63 ± 0.48	2.48	2.81 ± 0.57	0.82	7.52 ± 0.64	1.60
NIP A	Allylamine	0:6:20	1.06 ± 0.25		3.45 ± 0.49		4.70 ± 0.29	
MIP B	MAA	1:4:20	2.68 ± 0.39	2.15	0.93 ± 0.27	1.69	3.14 ± 0.36	1.87
NIP B	MAA	0:4:20	1.24 ± 0.22		0.55 ± 0.15		1.68 ± 0.32	
MIP C	4-VP	1:5:20	4.88 ± 0.32	2.14	9.97 ± 0.66	1.60	9.05 ± 0.75	1.62
NIP C	4-VP	0:5:20	2.28 ± 0.38		6.25 ± 0.36		5.58 ± 0.53	

It was observed that when water was used as the rebinding medium, all the polymers showed relatively low binding. This might be attributed to the high solubility of QA in water. This high affinity between QA and water molecules might decrease its interaction with the polymers. Moreover, there was a significant difference between the binding of the MIPs and the corresponding NIPs, pronouncing the specific interaction with the imprinted polymers, where MIP C showed the highest binding capacity of 4.88 ± 0.32 µmol/g, while the binding of its corresponding NIP was 2.28 ± 0.38 µmol/g, with an imprinting factor of 2.14. Although aqueous medium is known to disrupt hydrogen bonding interactions between template and monomer, in the rebinding results of QA a pronounced difference between the binding of QA to the MIPs and the corresponding NIPs was observed. This suggests that hydrogen bonding is not the only factor behind the MIP selectivity towards QA. It can be concluded that, during the imprinting process, different interactions took place based on the size, shape and functionality of the template [62]. During NIP preparation, no proper cavities or recognition sites were formed, therefore the NIP binding to QA was only through non-specific adsorption [63]. As a result, the amount of QA adsorbed by the NIP was lower than by MIP.

It was observed that there was a significant increase in the binding capacity in all polymers when methanol was used as the rebinding medium. This might be because QA has lower solubility in methanol [64], thus a lower affinity to the binding solvent, which increases the chance of interaction between QA and the polymers. It was still observed in this solvent that the binding of QA to MIPs is higher than its binding to the corresponding NIPs, which indicates the success of the imprinting process.

The binding capacities of MIP A, MIP B and MIP C were 7.52 ± 0.64 µmol/g, 3.14 ± 0.36 µmol/g and 9.05 ± 0.80 µmol/g, respectively, while NIP A, NIP B and NIP C showed binding capacities of 4.70 ± 0.28 µmol/g, 1.68 ± 0.32 µmol/g and 5.58 ± 0.53 µmol/g, respectively.

The third binding solvent chosen was ACN, an example of polar aprotic solvent, water was added to acetonitrile with a ratio of ACN:H_2O (4:1) to ensure the solubility of polar QA, which is insoluble in pure acetonitrile. On comparing the results of rebinding to results obtained in UPW, it was found that the use of acetonitrile increased the interaction between QA and the synthesized polymers, causing an increase in the binding capacity in most of the polymers. It could be argued that the addition of a polar aprotic solvent enhances the hydrogen bond formation between QA and the polymers. Additionally, the low solubility of QA in acetonitrile probably enhanced the interaction between QA and the polymers [65]. It was observed that MIP B and its corresponding NIP prepared using MAA as functional monomer showed lower binding in ACN compared to UPW. This most probably indicates that, in the case of this polymer, the hydrophobic interactions are the main interactions that take place between QA- and MAA-based polymers and this type of interaction is more pronounced when using only UPW as a binding solvent.

MIP C synthesized using 4-VP as functional monomer showed the highest binding capacity in all the rebinding solvents. Although the calculated binding energy of the QA–MAA complex was the highest during computational studies, practically, the polymers prepared with 4-VP showed better overall performance. This can be attributed to the extra interaction between the basic monomer and the acidic template, the pyridine ring of the monomer could promote adsorption because it can form both acid–base interactions and strong hydrogen bonds with the template. Thus, a more stable complex between the template and the functional monomer was formed during the imprinting process [55]. This agrees with what was reported in some studies where 4-VP monomer showed superior results during the imprinting of acidic templates when compared to other acidic or neutral monomers. In a study by Zhang et al., salicylic acid was imprinted using 4-VP and acrylamide as functional monomers, where 4-VP showed superior imprinting effect compared to acrylamide [66]. BET results also revealed that MIP C has the highest surface area (31.41 m^2/g) when compared to MIP A and MIP B, which might contribute to its higher binding capacity. The binding of the allylamine MIP was higher than the MAA MIP, which also might be attributed to its basic nature.

3.4. Adsorption Kinetics

In order to study the interactions between the 4-VP polymers and the template during the 2 h equilibrium rebinding study in methanol, an effect of time experiment was conducted on MIP C and NIP C. Ten mg of each polymer was incubated with 0.1 mM QA solution for definite time intervals. The uptake profile shown in Figure 5 revealed that the uptake of both MIP C and NIP C gradually increased during the course of the experiment to reach its maximum after 1 h, after which no significant improvement in the binding performance was observed.

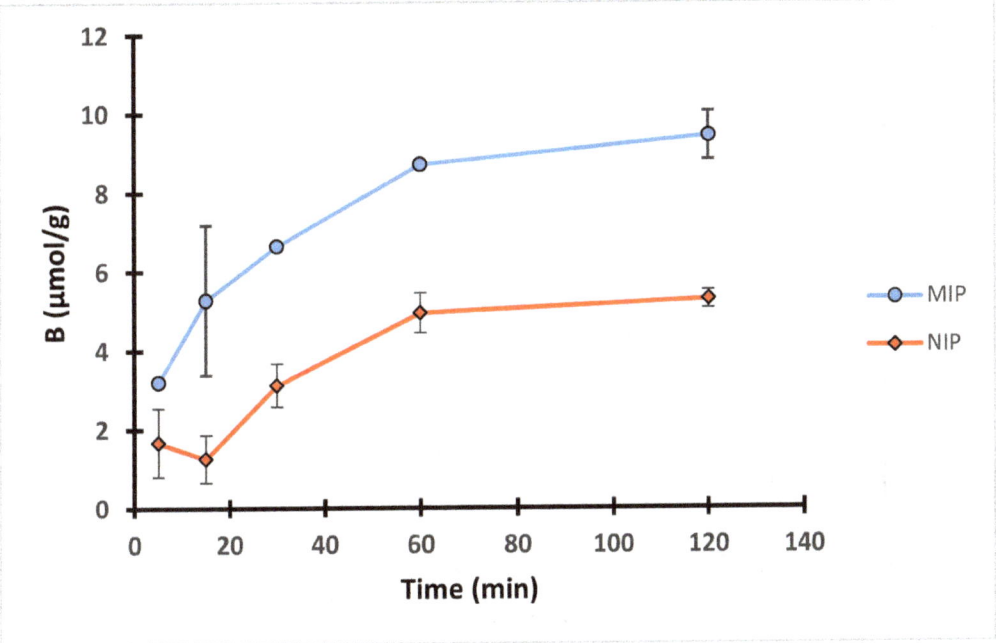

Figure 5. Uptake profiles of MIP C and NIP C over 2 h equilibrium rebinding study in methanol.

The pseudo-first order and pseudo-second order kinetics were used to investigate the mechanism of adsorption of MIP C (Figure S17 in the supplementary material). The results showed better fitting in pseudo-second order equation where R^2 was found to be 0.9967 versus R^2 of 0.8683 obtained from the pseudo-first order equation. This suggests that MIP C binding follows pseudo-second order kinetics with a calculated rate constant of 6.33×10^{-3} in g/µmol min.

3.5. Binding Isotherm

A two-hour equilibrium rebinding study was carried out by incubating 10 mg polymer with 2 mL QA solution in methanol with different concentrations ranging from 0.01 to 0.2 mM. A binding isotherm was conducted by plotting the binding capacity against initial concentration (C_i) as in Figure 6. The amount of QA bound increased with the increase of the initial concentration up to 0.1 mM of QA, while there was only a slight difference in the binding capacity between 0.1 and 0.2 mM QA. The results also revealed that the difference between the binding performance of MIP and NIP was more pronounced in the higher concentration ranges, mostly 0.1 and 0.2 mM.

The obtained results were further analyzed using the Freundlich isotherm for MIP C (Figure S18 in the supplementary material). The model was fitted with a high degree of correlation, R^2 of 0.954. MIP C showed a heterogenicity factor of 0.5389, suggesting heterogenicity of the binding surface, while the α value was 35.3 µmol/g.

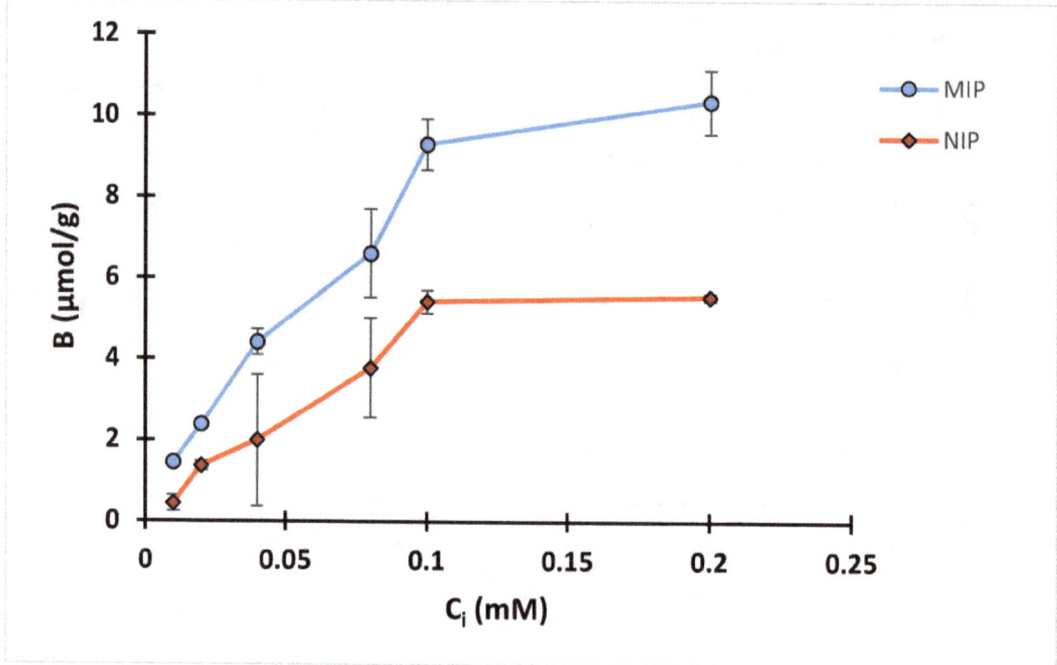

Figure 6. Binding isotherm of MIP C and NIP C.

3.6. MISPE Procedure Optimization

3.6.1. Loading Step Optimization

Different amounts of QA were loaded in SPE cartridges, using methanol as the loading solvent. These amounts were 5, 10, 15 and 20 µmol QA/g polymers. This was followed by a washing step using 2 mL acetonitrile and an elution step using 2 mL of 10% acetic acid in water. The amount of QA in the eluted fraction was calculated and the recovery percent was determined, and the results are shown in Table 4.

The loading amount that showed the highest recovery percent was 5 µmol/g, where the recovery percentage was 72.53 ± 2.68 for the MIP and 56.55 ± 2.77 for the NIP. As the loading amount increased, a decrease in the recovery percentage was attained. This is most probably attributed to occupation of binding sites in the polymer. Therefore, as the loaded amount increased, the fraction of QA that binds to the polymer decreased.

In the following experiment, 5 µmol/g of QA was loaded in different solvents. These solvents were methanol, water, ethanol, ethanol water (1:1 v/v) and acetonitrile: water (4:1 v/v). The recovery percent of QA was then calculated and the results are shown in Table 4. It was observed that when water was used as the loading solvent, there was a significant decrease in the recovery % of QA, which agrees with the previously conducted rebinding studies. This confirms that using water decreases the interaction between QA and the polymers.

3.9. MISPE Selectivity

The selective recognition and retaining properties of the MISPE and NISPE were evaluated. Two compounds were chosen in this study: caffeic acid (CA), having a comparable size to QA, and chlorogenic acid (CLA), an ester of QA and caffeic acid. Both compounds are found in high concentrations in coffee [71]. The results are shown in Figure 8.

Both MISPE and NISPE showed a higher recovery of QA compared to CA and CLA. This difference is probably due to the difference in polarity, where QA is highly hydrophilic with low solubility in acetonitrile, the washing solvent, while CA and CLA exhibit higher solubility in organic solvents. Therefore, some of the CA and CLA may have been removed during the washing step. The recovery percent of QA from MISPE was 82.30 ± 5.58, while it was 53.58 ± 2.77 in case of NISPE. It was observed that for both CA and CLA, the recovery of MISPE was higher than NISPE. For CA, the recovery percent was 23.71 ± 2.85 for MISPE and 14.28 ± 1.84 for NISPE, while for CLA, the recovery percent was 33.41 ± 0.90 for MISPE and 17.46 ± 3.28 for NISPE. This might be due to the structural similarities between QA and the two compounds, where CA is a small molecule with comparable size to QA, it also possesses carboxyl and hydroxyl groups, so it is assumed to bind with some of the functionalities present in the imprinted cavities of the MIP. As for CLA, it has a QA moiety that can fit in the MIP cavities, since CLA is an ester of QA and CA.

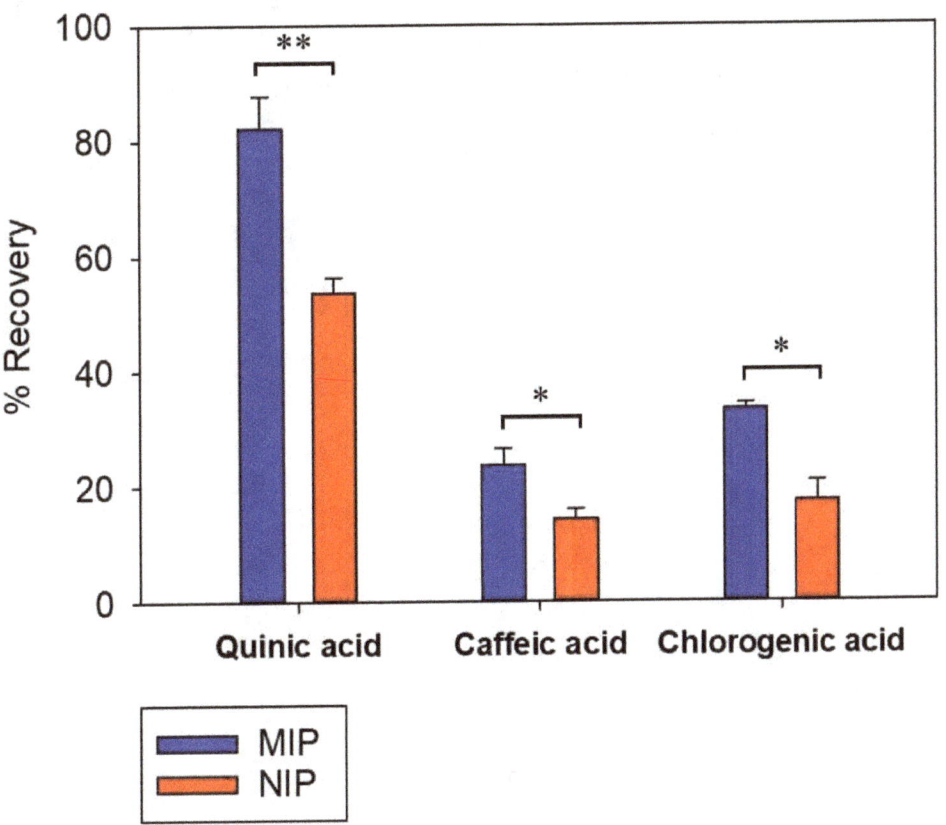

Figure 8. Recovery percentages of QA, CA and CLA upon loading an equimolar mixture of the three compounds to MISPE and NISPE, (n = 3), * indicates p value $\leq$ 0.05 and ** indicates p value $\leq$ 0.01.

3.10. MISPE Application on Coffee Extract
3.10.1. UHPLC-MS/MS Method Validation

The validated method showed good linearity, sensitivity, precision and accuracy. Further details are provided in the supplementary material.

3.10.2. MISPE Application on Coffee Extract

MIP C was tested for its ability to selectively extract QA from coffee extract. The optimized SPE procedure was applied to the aqueous extract of coffee beans. Two mL of coffee extract (0.25 mg/mL and 0.5 mg/mL) was loaded onto the SPE cartridge. This was followed by a washing step using 2 mL of acetonitrile and elution step using 2 mL of 10% acetic acid in water. It was noticed that when 2 mL of 0.5 mg/mL coffee extract was loaded onto MISPE, the recovery percent was only 36.50 ± 1.19 for MISPE and 28.47 ± 1.22 for NISPE. However, decreasing the concentration of the loaded extract to 0.25 mg/mL showed a significant increase in the recovery percent to reach 81.92 ± 3.03 for MISPE while the NISPE showed a much lower recovery percent of 37.26 ± 0.84 using the same concentration of the extract (Figure 9). This concludes that the low recovery percent observed while using a higher concentration of the extract could be attributed to the saturation of the binding cavities within the MIP. These results prove that the optimized MISPE is superior to reported conventional methods for QA isolation, such as liquid–liquid extraction previously reported by Tuyun et al. [13], where the maximum QA recovery was found to be 66.906%. The UV chromatogram for aqueous coffee extract before and after loading to MISPE and NISPE shown in Figure 10 revealed that neither the MISPE nor the NISPE were able to bind significantly to any of the other components of the extract, while there was a significant decrease in the amount of the extract components in the elution fractions of both MISPE and NISPE, compared to the original amounts found in the loaded extract.

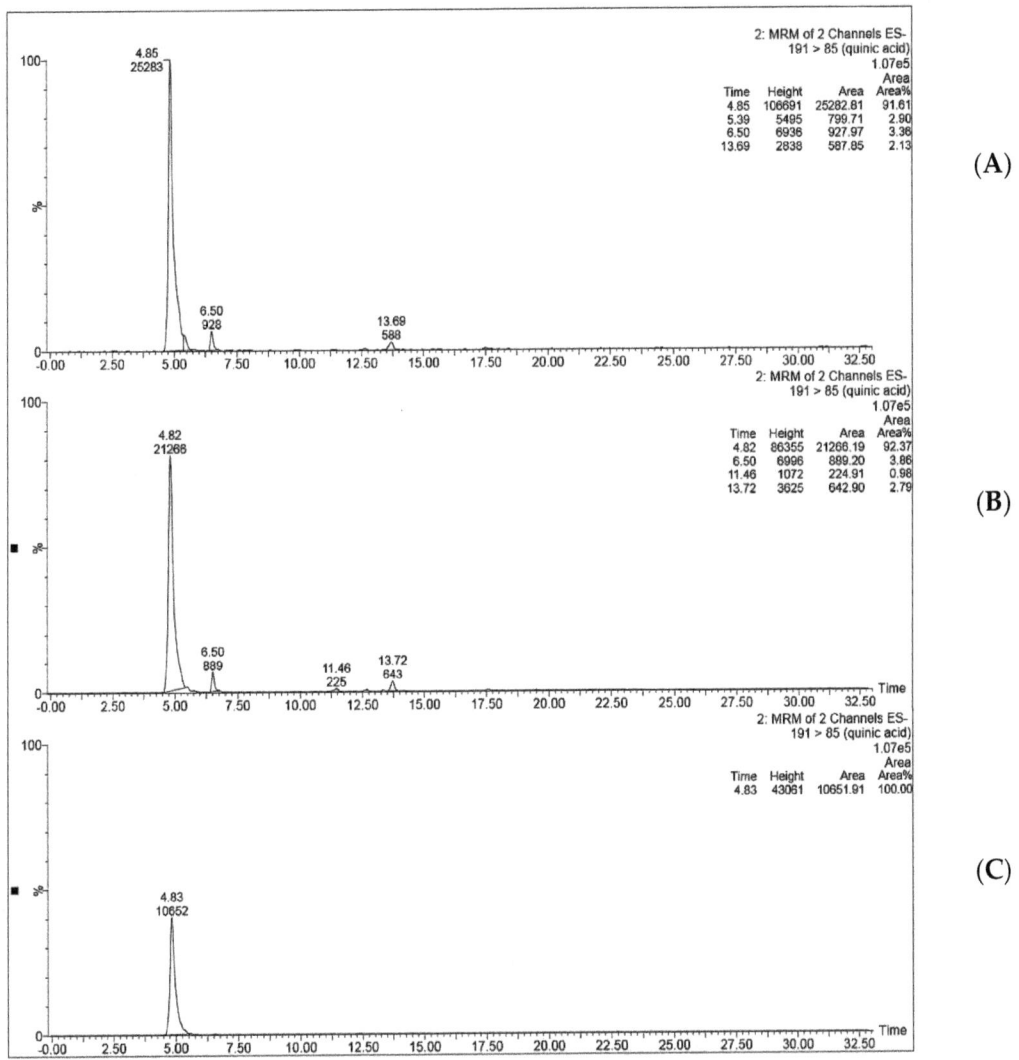

Figure 9. Mass chromatograms of QA (**A**) in 0.25 µg/mL coffee extract dissolved in ethanol: water (97:3 v/v) before loading to MISPE C, (**B**) the elution fraction obtained from MISPE procedure, (**C**) the elution fraction obtained from NISPE procedure.

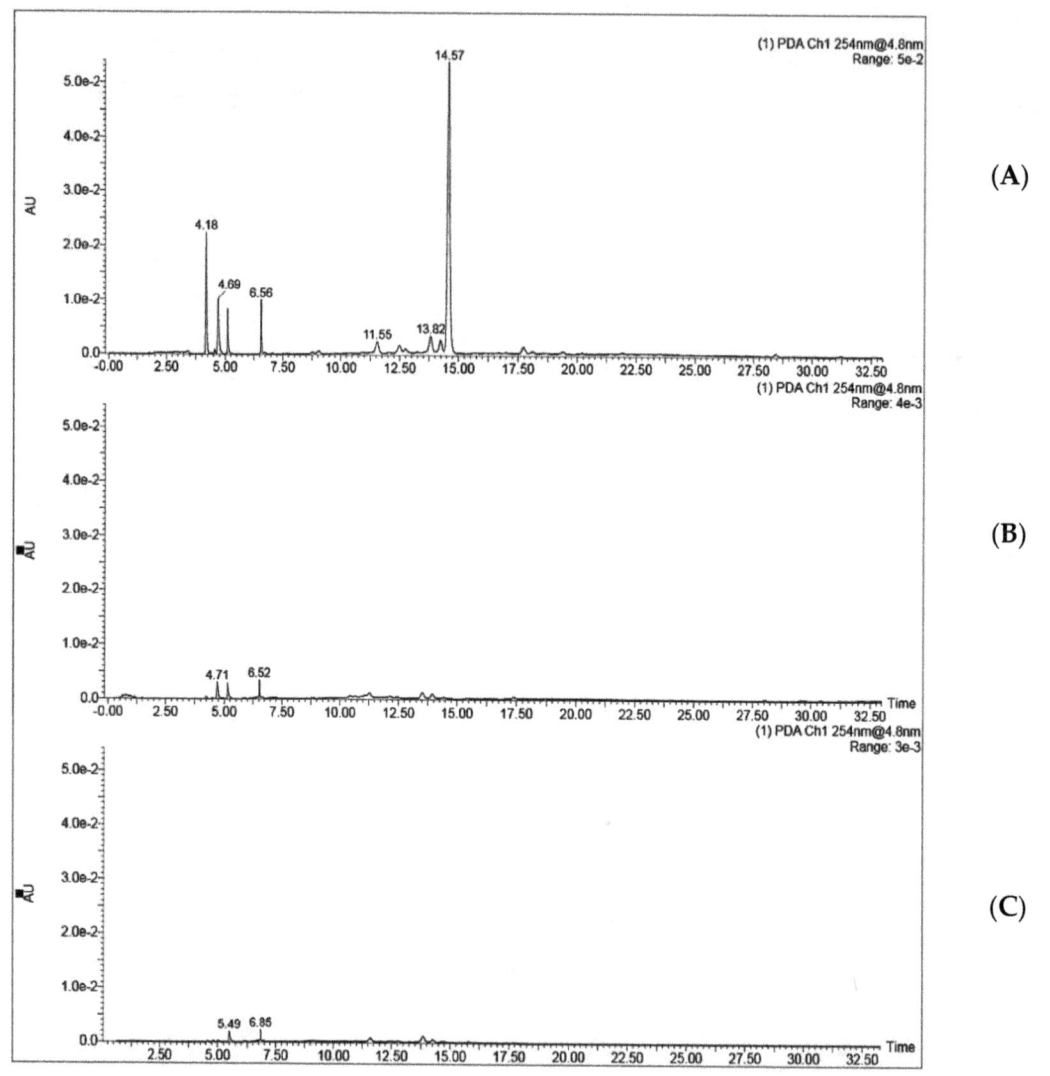

Figure 10. UV chromatograms of QA (**A**) in 0.25 µg/mL coffee extract dissolved in ethanol: water (97:3 v/v) before loading to MISPE C, (**B**) the elution fraction obtained from MISPE procedure, (**C**) the elution fraction obtained from NISPE procedure.

4. Conclusions

The current study represents the use of cheap, selective and simple MISPE procedure for extraction of QA from coffee beans. Three bulk polymers based on three different functional monomers (allylamine, MAA and 4-VP) were synthesized and the molar ratio of each monomer to QA was optimized via computational studies. The 4-VP polymer showed better overall performance in comparison to the other two polymers, thus it was the polymer of choice for SPE application. MIP reusability was tested over ten adsorption–desorption cycles and showed a high recovery of QA (more than 93%) up to the fourth cycle. Selective extraction of QA was observed upon using the optimized MISPE procedure on an equimolar mixture of QA, CA and CGA. The recovery percent of QA was 82.30 ± 5.58, compared to 23.71 ± 2.85 and 33.41 ± 0.90 for CA and CLA, respectively. The application of MISPE for extraction of QA from aqueous coffee extract showed a recovery percent of 81.92 ± 3.03, with a significant reduction in the amounts of other components in the extract. The developed MISPE procedure represents a promising approach for selective extraction of QA from different complex herbal extracts that may be scaled to industrial applications. It can also be applied in the food and beverage industry to decrease the concentration of QA in coffee and enhance its taste. In conclusion, this study succeeded in the isolation of an important nutraceutical in a cost-effective, rapid, robust and reliable method.

Supplementary Materials: The following supporting information can be downloaded at: https://www.mdpi.com/article/10.3390/polym14163339/s1. Figure S1: Computer modeled structures of the best conformations for (a) QA, (b) 4-VP, (c) allylamine, (d)MAA. Figure S2–S13: Computer modeled structures of the best conformations for QA-FM complexes. Figure S14: SEM images of (a) MIP A and (b) NIP A. Figure S15: SEM images of (a) MIP B and (b) NIP B. Figure S16: BET isotherms of (A) MIP A, (B) NIP A, (C) MIP B and (D) NIP B. Figure S17: (A) Pseudo-first order kinetics and (B) pseudo-second order kinetics for MIP C. Figure S18: Freundlich isotherm for MIP C. Figure S19: Calibration curve of QA in methanol over the concentration range of 0.001–0.2 mM. Figure S20: Calibration curve for QA in ethanol over concentration range of 0.2–40 µg/mL. Table S1: The calculated binding energies for complexes prepared in solvent phase. Table S2: Intra-day and Inter-day precision of QA determination in UPLC-MS/MS method. Table S3: Accuracy of QA determination in UHPLC-MS/MS method. Table S4: %RSD of inter-day and intra-day precision assay for UHPLC measurements. Table S5: Recovery % of spiked QA amount 1x, 2x, and 3x the amount of QA present in coffee extract (10, 20, and 30 µg/mL).

Author Contributions: Conceptualization: S.H.M., H.H., M.A.-H., B.M. and N.A.E.G.; Methodology: S.H.M., H.H., M.A.-H., B.M. and N.A.E.G.; Investigation: S.H.M., A.H. and M.A.-H.; Formal Analysis: S.H.M., M.A.-H and N.A.E.G.; Validation: S.H.M., M.A.-H. and N.A.E.G.; Supervision: H.H., M.A.-H., B.M. and N.A.E.G.; Writing—original draft: S.H.M.; Writing—review and editing: S.H.M., H.H., M.A.-H., B.M. and N.A.E.G. All authors have read and agreed to the published version of the manuscript.

Funding: This research did not receive any specific grant from funding agencies in the public, commercial, or not-for-profit sectors.

Conflicts of Interest: The authors declare no conflict of interest.

References

1. Madkour, L. *Oxidative stress and oxidative damage-induced cell death. Reactive Oxygen Species (ROS), Nanoparticles, and Endoplasmic Reticulum (ER) Stress-Induced Cell Death Mechanisms*; Academic Press: Cambridge, MA, USA, 2020; pp. 175–197.
2. Jiang, S.; Liu, H.; Li, C. Dietary Regulation of Oxidative Stress in Chronic Metabolic Diseases. *Foods* **2021**, *10*, 1854. [CrossRef] [PubMed]
3. Toghyani Khorasgani, A.; Amini-Khoei, H.; Shadkhast, M.; Salimian, S.; Majidian, M.; Dehkordi, S.H. Quinic acid through mitigation of oxidative stress in the hippocampus exerts analgesic effect in male mice. *Future Nat. Prod.* **2021**, *7*, 1–11.
4. Magaña, A.; Kamimura, N.; Soumyanath, A.; Stevens, J.; Maier, C. Caffeoylquinic acids: Chemistry, biosynthesis, occurrence, analytical challenges, and bioactivity. *Plant J.* **2021**, *107*, 1299–1319. [CrossRef] [PubMed]
5. Zhang, L.; Zhang, J.; Zhao, B.; Zhao-Wilson, X. Quinic acid could be a potential rejuvenating natural compound by improving survival of Caenorhabditis elegans under deleterious conditions. *Rejuvenation Res.* **2012**, *15*, 573–583. [CrossRef]
6. Kwon, Y.K.; Choi, S.J.; Kim, C.R.; Kim, J.K.; Kim, Y.-J.; Choi, J.H.; Song, S.-W.; Kim, C.-J.; Park, G.G.; Park, C.-S.; et al. Antioxidant and cognitive-enhancing activities of Arctium lappa L. roots in Aβ1-42-induced mouse model. *Appl. Biol. Chem.* **2016**, *59*, 553–565. [CrossRef]
7. Pero, R.W. Health consequences of catabolic synthesis of hippuric acid in humans. *Curr. Clin. Pharmacol.* **2010**, *5*, 67–73. [CrossRef]
8. Fricker, R.A.; Green, E.L.; Jenkins, S.I.; Griffin, S.M. The Influence of Nicotinamide on Health and Disease in the Central Nervous System. *Int. J. Tryptophan Res.* **2018**, *11*, 1178646918776658. [CrossRef]
9. Kansara, K.; Gupta, S.S. Chapter Fourteen-DNA Damage, Repair, and Maintenance of Telomere Length: Role of Nutritional Supplements. In *Mutagenicity: Assays and Applications*; Kumar, A., Ed.; Academic Press: Cambridge, MA, USA, 2018; pp. 287–307.
10. Hulme, A.C. The Isolation of l-Quinic Acid from the Apple Fruit. *J. Exp. Bot.* **1951**, *2*, 298–315. [CrossRef]
11. Yazdi, S.E.; Prinsloo, G.; Heyman, H.M.; Oosthuizen, C.B.; Klimkait, T.; Meyer, J.J.M. Anti-HIV-1 activity of quinic acid isolated from Helichrysum mimetes using NMR-based metabolomics and computational analysis. *S. Afr. J. Bot.* **2019**, *126*, 328–339. [CrossRef]

12. Upadhyay, R.; Rao, L.J.M. An outlook on chlorogenic acids-occurrence, chemistry, technology, and biological activities. *Crit. Rev Food Sci. Nutr.* **2013**, *53*, 968–984. [CrossRef]
13. Tuyun, A.F.; Uslu, H. Extraction of D-(−)-Quinic Acid Using an Amine Extractant in Different Diluents. *J. Chem. Eng. Data* **2012**, *57*, 190–194. [CrossRef]
14. Fischer, E. Einfluss der Configuration auf die Wirkung der Enzyme. *Eur. J. Inorg. Chem.* **1894**, *27*, 2985–2993.
15. Saad, E.M.; Madbouly, A.; Ayoub, N.; el Nashar, R.M. Preparation and application of molecularly imprinted polymer for isolation of chicoric acid from *Chicorium intybus* L. medicinal plant. *Anal. Chim. Acta.* **2015**, *877*, 80–89. [CrossRef] [PubMed]
16. Wackerlig, J.; Schirhagl, R. Applications of Molecularly Imprinted Polymer Nanoparticles and Their Advances toward Industrial Use: A Review. *Anal. Chem.* **2016**, *88*, 250–261. [CrossRef]
17. Uzun, L.; Turner, A.P.F. Molecularly-imprinted polymer sensors: Realising their potential. *Biosens. Bioelectron.* **2016**, *76*, 131–144. [CrossRef] [PubMed]
18. Speltini, A.; Scalabrini, A.; Maraschi, F.; Sturini, M.; Profumo, A. Newest applications of molecularly imprinted polymers for extraction of contaminants from environmental and food matrices: A review. *Anal. Chim. Acta* **2017**, *974*, 1–26. [CrossRef]
19. Malik, M.I.; Shaikh, H.; Mustafa, G.; Bhanger, M.I. Recent Applications of Molecularly Imprinted Polymers in Analytical Chemistry. *Sep. Purif. Rev.* **2019**, *48*, 179–219. [CrossRef]
20. Piletsky, S.; Turner, A. Imprinted polymers and their application in optical sensors. In *Optical Biosensors: Today and Tomorrow*; Elsevier: Amsterdam, The Netherlands, 2008; pp. 543–581.
21. Piletsky, S.A.; Turner, A.P.F. Electrochemical Sensors Based on Molecularly Imprinted Polymers. *Electroanalysis* **2002**, *14*, 317–323. [CrossRef]
22. Elfadil, D.; Lamaoui, A.; della Pelle, F.; Amine, A.; Compagnone, D. Molecularly Imprinted Polymers Combined with Electrochemical Sensors for Food Contaminants Analysis. *Molecules* **2021**, *26*, 4607. [CrossRef]
23. Xu, X.-Y.; Tian, X.-G.; Cai, L.-G.; Xu, Z.-L.; Lei, H.-T.; Wang, H.; Sun, Y.-M. Molecularly imprinted polymer based surface plasmon resonance sensors for detection of Sudan dyes. *Anal. Methods* **2014**, *6*, 3751–3757. [CrossRef]
24. Tsuru, N.; Kikuchi, M.; Kawaguchi, H.; Shiratori, S. A quartz crystal microbalance sensor coated with MIP for "Bisphenol A" and its properties. *Thin Solid Film.* **2006**, *499*, 380–385. [CrossRef]
25. Eskandari, H.; Amirzehni, M.; Hassanzadeh, J.; Vahid, B. Mesoporous MIP-capped luminescent MOF as specific and sensitive analytical probe: Application for chlorpyrifos. *Microchim. Acta* **2020**, *187*, 673. [CrossRef] [PubMed]
26. Antipchik, M.; Reut, J.; Ayankojo, A.G.; Öpik, A.; Syritski, V. MIP-based electrochemical sensor for direct detection of hepatitis C virus via E2 envelope protein. *Talanta* **2022**, *250*, 123737. [CrossRef] [PubMed]
27. Mazouz, Z.; Mokni, M.; Fourati, N.; Zerrouki, C.; Barbault, F.; Seydou, M.; Kalfat, R.; Yaakoubi, N.; Omezzine, A.; Bouslema, A.; et al. Computational approach and electrochemical measurements for protein detection with MIP-based sensor. *Biosens. Bioelectron.* **2020**, *151*, 111978. [CrossRef]
28. Lee, M.-H.; Chen, Y.-C.; Ho, M.-H.; Lin, H.-Y. Optical recognition of salivary proteins by use of molecularly imprinted poly(ethylene-co-vinyl alcohol)/quantum dot composite nanoparticles. *Anal. Bioanal. Chem.* **2010**, *397*, 1457–1466. [CrossRef]
29. Rouhani, S.; Nahavandifard, F. Molecular imprinting-based fluorescent optosensor using a polymerizable 1,8-naphthalimide dye as a florescence functional monomer. *Sens. Actuators B Chem.* **2014**, *197*, 185–192. [CrossRef]
30. Nguyen, T.H.; Hardwick, S.A.; Sun, T.; Grattan, K.T.V. Intrinsic Fluorescence-Based Optical Fiber Sensor for Cocaine Using a Molecularly Imprinted Polymer as the Recognition Element. *IEEE Sens. J.* **2012**, *12*, 255–260. [CrossRef]
31. Sun, H.; Lai, J.-P.; Lin, D.-S.; Huang, X.-X.; Zuo, Y.; Li, Y.-L. A novel fluorescent multi-functional monomer for preparation of silver ion-imprinted fluorescent on–off chemosensor. *Sens. Actuators B Chem.* **2016**, *224*, 485–491. [CrossRef]
32. Ng, S.M.; Narayanaswamy, R. Fluorescence sensor using a molecularly imprinted polymer as a recognition receptor for the detection of aluminium ions in aqueous media. *Anal. Bioanal. Chem.* **2006**, *386*, 1235–1244. [CrossRef]
33. Sener, G.; Ozgur, E.; Rad, A.Y.; Uzun, L.; Say, R.; Denizli, A. Rapid real-time detection of procalcitonin using a microcontact imprinted surface plasmon resonance biosensor. *Analyst* **2013**, *138*, 6422–6428. [CrossRef]
34. Sener, G.; Uzun, L.; Say, R.; Denizli, A. Use of molecular imprinted nanoparticles as biorecognition element on surface plasmon resonance sensor. *Sens. Actuators B Chem.* **2011**, *160*, 791–799. [CrossRef]
35. Zhao, N.; Chen, C.; Zhou, J. Surface plasmon resonance detection of ametryn using a molecularly imprinted sensing film prepared by surface-initiated atom transfer radical polymerization. *Sens. Actuators B Chem.* **2012**, *166–167*, 473–479. [CrossRef]
36. Shaikh, H.; Şener, G.; Memon, N.; Bhanger, M.; Nizamani, S.; Üzek, R.; Denizli, A. Molecularly imprinted surface plasmon resonance (SPR) based sensing of bisphenol A for its selective detection in aqueous systems. *Anal. Methods* **2015**, *7*, 4661–4670. [CrossRef]
37. Hammam, M.A.; Abdel-Halim, M.; Madbouly, A.; Wagdy, H.A.; el Nashar, R.M. Computational design of molecularly imprinted polymer for solid phase extraction of moxifloxacin hydrochloride from Avalox® tablets and spiked human urine samples. *Microchem. J.* **2019**, *148*, 51–56. [CrossRef]
38. Omran, N.H.; Wagdy, H.A.; Abdel-Halim, M.; Nashar, R.M.E. Validation and Application of Molecularly Imprinted Polymers for SPE/UPLC–MS/MS Detection of Gemifloxacin Mesylate. *Chromatographia* **2019**, *82*, 1617–1631. [CrossRef]
39. Stevenson, D.; El-Sharif, H.F.; Reddy, S.M. Selective extraction of proteins and other macromolecules from biological samples using molecular imprinted polymers. *Bioanalysis* **2016**, *8*, 2255–2263. [CrossRef] [PubMed]

40. Pereira, I.; Rodrigues, M.F.; Chaves, A.R.; Vaz, B.G. Molecularly imprinted polymer (MIP) membrane assisted direct spray ionization mass spectrometry for agrochemicals screening in foodstuffs. *Talanta* **2018**, *178*, 507–514. [CrossRef]
41. Voros, V.; Drioli, E.; Fonte, C.; Szekely, G. Process Intensification via Continuous and Simultaneous Isolation of Antioxidants: An Upcycling Approach for Olive Leaf Waste. *ACS Sustain. Chem. Eng.* **2019**, *7*, 18444–18452. [CrossRef]
42. Szekely, G. Valorisation of agricultural waste with adsorption/nanofiltration hybrid process: From materials to sustainable process design. *Green Chem.* **2017**, *19*, 3116–3125.
43. Martins, R.O.; Gomes, I.C.; Telles, A.D.M.; Kato, L.; Souza, P.S.; Chaves, A.R. Molecularly imprinted polymer as solid phase extraction phase for condensed tannin determination from Brazilian natural sources. *J. Chromatogr. A* **2020**, *1620*, 460977. [CrossRef]
44. Kanao, E.; Tsuchiya, Y.; Tanaka, K.; Masuda, H.; Tanigawa, T.; Naito, T.; Sano, T.; Kubo, T.; Otsuka, K. Poly(ethylene glycol) Hydrogels with a Boronic Acid Monomer via Molecular Imprinting for Selective Removal of Quinic Acid Gamma-Lactone in Coffee. *ACS Appl. Polym. Mater.* **2021**, *3*, 226–232. [CrossRef]
45. Saad, E.M.; el Gohary, N.A.; Abdel-Halim, M.; Handoussa, H.; el Nashar, R.M.; Mizaikoff, B. Molecularly imprinted polymers for selective extraction of rosmarinic acid from Rosmarinus officinalis L. *Food Chem.* **2021**, *335*, 127644. [CrossRef] [PubMed]
46. Mujahid, A.; Iqbal, N.; Afzal, A. Bioimprinting strategies: From soft lithography to biomimetic sensors and beyond. *Biotechnol. Adv.* **2013**, *31*, 1435–1447. [CrossRef] [PubMed]
47. Hosny, H.; el Gohary, N.; Saad, E.; Handoussa, H.; el Nashar, R.M. Isolation of sinapic acid from broccoli using molecularly imprinted polymers. *J. Sep. Sci.* **2018**, *41*, 1164–1172. [CrossRef]
48. Nicolescu, T.-V.; Sarbu, A.; Dima, Ş.-O.; Nicolae, C.-A.; Donescu, D. Molecularly imprinted "bulk" copolymers as selective sorbents for gallic acid. *J. Appl. Polym. Sci.* **2013**, *127*, 366–374. [CrossRef]
49. Michailof, C.; Manesiotis, P.; Panayiotou, C. Synthesis of caffeic acid and p-hydroxybenzoic acid molecularly imprinted polymers and their application for the selective extraction of polyphenols from olive mill waste waters. *J. Chromatogr. A* **2008**, *1182*, 25–33. [CrossRef]
50. Li, X.; Shian, Z.; Chen, L.; Whittaker, A. Computer simulation and preparation of molecularly imprinted polymer membranes with chlorogenic acid as template. *Polym. Int.* **2011**, *60*, 592–598. [CrossRef]
51. Liu, Z.; Xu, Z.; Wang, D.; Yang, Y.; Duan, Y.; Ma, L.; Lin, T.; Liu, H. A Review on Molecularly Imprinted Polymers Preparation by Computational Simulation-Aided Methods. *Polymers* **2021**, *13*, 2657. [CrossRef]
52. Roy, E.; Patra, S.; Madhuri, R.; Sharma, P.K. Gold nanoparticle mediated designing of non-hydrolytic sol-gel cross-linked metformin imprinted polymer network: A theoretical and experimental study. *Talanta* **2014**, *120*, 198–207. [CrossRef]
53. Mennucci, B. Polarizable Continuum Model. *Wiley Interdiscip. Rev. Comput. Mol. Sci.* **2012**, *2*, 386–404. [CrossRef]
54. Arshady, R.; Mosbach, K. Synthesis of substrate-selective polymers by host-guest polymerization. *Die Makromol. Chem.* **1981**, *182*, 687–692. [CrossRef]
55. Li, N.; Ng, T.B.; Wong, J.H.; Qiao, J.X.; Zhang, Y.N.; Zhou, R.; Chen, R.R.; Liu, F. Separation and purification of the antioxidant compounds, caffeic acid phenethyl ester and caffeic acid from mushrooms by molecularly imprinted polymer. *Food Chem.* **2013**, *139*, 1161–1167. [CrossRef] [PubMed]
56. Ho, Y.S.; McKay, G. Pseudo-second order model for sorption processes. *Process Biochem.* **1999**, *34*, 451–465. [CrossRef]
57. Langmuir, I. The constitution and fundamental properties of solids and liquids. II. liquids.1. *J. Am. Chem. Soc.* **1917**, *39*, 1848–1906. [CrossRef]
58. Ghani, N.T.A.; el Nashar, R.M.; Abdel-Haleem, F.M.; Madbouly, A. Computational Design, Synthesis and Application of a New Selective Molecularly Imprinted Polymer for Electrochemical Detection. *Electroanalysis* **2016**, *28*, 1530–1538. [CrossRef]
59. El Gohary, N.A.; Madbouly, A.; el Nashar, R.M.; Mizaikoff, B. Synthesis and application of a molecularly imprinted polymer for the voltammetric determination of famciclovir. *Biosens. Bioelectron.* **2015**, *65*, 108–114. [CrossRef]
60. Bakas, I.; Oujji, N.B.; Istambouliè, G.; Piletsky, S.; Piletska, E.; Ait-Addi, E.; Ait-Ichou, I.; Noguer, T.; Rouillon, R. Molecularly imprinted polymer cartridges coupled to high performance liquid chromatography (HPLC-UV) for simple and rapid analysis of fenthion in olive oil. *Talanta* **2014**, *125*, 313–318. [CrossRef]
61. Chrzanowska, A.M.; Poliwoda, A.; Wieczorek, P.P. Characterization of particle morphology of biochanin A molecularly imprinted polymers and their properties as a potential sorbent for solid-phase extraction. *Mater. Sci. Eng. C Mater. Biol. Appl.* **2015**, *49*, 793–798. [CrossRef]
62. Li, Y.; Li, X.; Dong, C.; Li, Y.; Jin, P.; Qi, J. Selective recognition and removal of chlorophenols from aqueous solution using molecularly imprinted polymer prepared by reversible addition-fragmentation chain transfer polymerization. *Biosens. Bioelectron.* **2009**, *25*, 306–312. [CrossRef]
63. Le Noir, M.; Plieva, F.; Hey, T.; Guieysse, B.; Mattiasson, B. Macroporous molecularly imprinted polymer/cryogel composite systems for the removal of endocrine disrupting trace contaminants. *J. Chromatogr. A* **2007**, *1154*, 158–164. [CrossRef]
64. Huang, Q.; Hao, X.; Qiao, L.; Wu, M.; Shen, G.; Ma, S. Measurement and thermodynamic functions of solid–liquid phase equilibrium of d-(−)-quinic acid in H2O, methanol, ethanol and (H2O+methanol), (H2O+ethanol) binary solvent mixtures. *J. Chem. Thermodyn.* **2016**, *100*, 140–147. [CrossRef]
65. Pratiwi, R.; Megantara, S.; Rahayu, D.; Pitaloka, I.; Hasanah, A.N. Comparison of Bulk and Precipitation Polymerization Method of Synthesis Molecular Imprinted Solid Phase Extraction for Atenolol using Methacrylic Acid. *J. Young Pharm.* **2018**, *11*, 12–16. [CrossRef]

66. Zhang, T.; Liu, F.; Wang, J.; Li, N.; Li, K. Molecular recognition properties of salicylic acid-imprinted polymers. *Chromatographia* **2002**, *55*, 447–451. [CrossRef]
67. Song, X.; Turiel, E.; He, L.; Martín-Esteban, A. Synthesis of Molecularly Imprinted Polymers for the Selective Extraction of Polymyxins from Environmental Water Samples. *Polymers* **2020**, *12*, 131. [CrossRef] [PubMed]
68. Jiang, T.; Zhao, L.; Chu, B.; Feng, Q.; Yan, W.; Lin, J.M. Molecularly imprinted solid-phase extraction for the selective determination of 17beta-estradiol in fishery samples with high performance liquid chromatography. *Talanta* **2009**, *78*, 442–447. [CrossRef]
69. Boulanouar, S.; Combes, A.; Mezzache, S.; Pichon, V. Synthesis and application of molecularly imprinted polymers for the selective extraction of organophosphorus pesticides from vegetable oils. *J. Chromatogr. A* **2017**, *1513*, 59–68. [CrossRef]
70. Karasova, G.; Lehotay, J.; Sadecka, J.; Skacani, I.; Lachova, M. Selective extraction of derivates of p-hydroxy-benzoic acid from plant material by using a molecularly imprinted polymer. *J. Sep. Sci.* **2005**, *28*, 2468–2476. [CrossRef]
71. Wang, G.F.; Shi, L.P.; Ren, Y.D.; Liu, Q.F.; Liu, H.F.; Zhang, R.J.; Li, Z.; Zhu, F.H.; He, P.L.; Tang, W.; et al. Anti-hepatitis B virus activity of chlorogenic acid, quinic acid and caffeic acid in vivo and in vitro. *Antivir. Res.* **2009**, *83*, 186–190. [CrossRef]

Article

N-(2-Arylethyl)-2-methylprop-2-enamides as Versatile Reagents for Synthesis of Molecularly Imprinted Polymers

Monika Sobiech ⬤, Dorota Maciejewska and Piotr Luliński *⬤

Department of Organic Chemistry, Faculty of Pharmacy, Medical University of Warsaw, Banacha 1, 02-097 Warsaw, Poland; monika.sobiech@wum.edu.pl (M.S.); dorota.maciejewska@wum.edu.pl (D.M.)
* Correspondence: piotr.lulinski@wum.edu.pl; Tel.: +48-22-5720963

Abstract: The paper describes the formation of six aromatic *N*-(2-arylethyl)-2-methylprop-2-enamides with various substituents in benzene ring, viz., 4-F, 4-Cl, 2,4-Cl$_2$, 4-Br, 4-OMe, and 3,4-(OMe)$_2$ from 2-arylethylamines and methacryloyl chloride in ethylene dichloride with high yields (46–94%). The structure of the compounds was confirmed by ^{1}H NMR, ^{13}C NMR, IR, and HR-MS. Those compounds were obtained to serve as functionalized templates for the fabrication of molecularly imprinted polymers followed by the hydrolysis of an amide linkage. In an exemplary experiment, the imprinted polymer was produced from N-(2-(4-bromophenyl)ethyl)-2-methylprop-2-enamide and divinylbenzene, acting as cross-linker. The hydrolysis of 2-(4-bromophenyl)ethyl residue proceeded and the characterization of material including SEM, EDS, ^{13}C CP MAS NMR, and BET on various steps of preparation was carried out. The adsorption studies proved that there was a high affinity towards the target biomolecules tyramine and L-norepinephrine, with imprinting factors equal to 2.47 and 2.50, respectively, when compared to non-imprinted polymer synthesized from methacrylic acid and divinylbenzene only.

Keywords: *N*-acylation; phenethylamines; molecularly imprinted polymers; semi-covalent imprinting; tyramine

Citation: Sobiech, M.; Maciejewska, D.; Luliński, P. N-(2-Arylethyl)-2-methylprop-2-enamides as Versatile Reagents for Synthesis of Molecularly Imprinted Polymers. *Polymers* **2022**, *14*, 2738. https://doi.org/10.3390/polym14132738

Academic Editor: Michał Cegłowski

Received: 22 May 2022
Accepted: 2 July 2022
Published: 4 July 2022

Publisher's Note: MDPI stays neutral with regard to jurisdictional claims in published maps and institutional affiliations.

Copyright: © 2022 by the authors. Licensee MDPI, Basel, Switzerland. This article is an open access article distributed under the terms and conditions of the Creative Commons Attribution (CC BY) license (https:// creativecommons.org/licenses/by/ 4.0/).

1. Introduction

Molecular-imprinting technology is engaged in searching for advanced selective materials with great potential for environmental, food, or biomedical analyses [1–4]. This technique forms polymers with desired selectivity towards a template, being a result of the interactions between the functional groups of template and monomer(s) prior to the polymerization process. The orientation of molecules is fixed through chemical cross-linking during the polymerization and then the removal of the template is undertaken to obtain a cavity in the molecularly imprinted polymer (MIP). Covalent or non-covalent imprinting strategies employ either chemical bonds or various weak interactions in the formation of template–monomer prepolymerization moieties [5].

The formation of stable prepolymerization structures is a critical step during the imprinting process. The use of a template with covalently bound polymerizable units (a functionalized template) prior to the polymerization resulted in the formation of well-defined binding sites in the polymer matrix. Chemical cleavage is required at the final stage of the process. The fabrication of MIPs, applying the monomer covalently bound to the template, is followed by different rebinding/adsorption approaches, e.g., the rebinding of the target analyte to the polymer matrix by covalent bonds [6] or adsorption of the target analyte via non-covalent intermolecular interactions [7].

The advancement of the use of a functionalized template resulted in a more homogeneous population of binding sites in the resultant MIP and greater binding-site integrity when compared to the MIP synthesized with a non-covalent strategy [8–10]. Hashim and co-workers [8] compared the covalent and non-covalent imprinting strategies for the synthesis of stigmasterol imprinted polymers. In the non-covalent imprinting strategy, stigmasterol was selected as the template and methacrylic acid or 4-vinylpyridine were used as the functional monomers to form different MIPs. In the covalent imprinting strategy stigmasteryl-3-*O*-methacrylate was synthesized prior to its application as the functionalized template. It was found that the non-covalent imprinting method showed insufficient binding affinity and low selectivity towards stigmasterol. In contrast, the application of covalent imprinting in the formation of MIP followed by the chemical cleavage of ester bonds resulted with highly selective imprinted polymer. Similar results were described by Tang and co-workers [9]. Here, the simultaneous reaction of *N*- and *O*-acylation of ractopamine was provided to obtain a novel functionalized template. The results were compared with previously published studies of non-covalent imprinting of ractopamine [10] It was found that MIPs obtained by the covalent imprinting strategy possessed significantly higher binding capacity and selectivity. The homogeneous population of binding sites towards ractopamine in the covalently produced MIP was confirmed by isotherm equilibrium-binding experiments [9], providing a substantial difference between it and the heterogeneous population of binding sites observed in the non-covalently formed MIP [10] More recently, the covalent approach was used to synthesize more advanced and selective materials integrated with metal-organic frameworks [11], magnetic cores [12], surfaces of microtiter plates [13], or dendritic fibrous silica [14].

The current investigations of our group aim to explore the ability of MIPs produced by the covalent imprinting strategy to selectively recognize biogenic amines with 2-phenylethylamine system (phenethylamine system) [15,16]. The group of compounds that contain a backbone of phenethylamine play a very important role in the human nervous system. These molecules act as neurotransmitters and neuromodulators or psychotropic agents, causing neurological disorders related to mood, emotion, attention, and cognition when their levels are irregular or produce hallucinations, illusions, or mental disorders when prolonged or overdosed [17,18]. Moreover, the presence of these compounds, predominantly at low levels in highly complex samples, hampers their analysis. Thus, investigations of selective materials with satisfactory clean-up capabilities for the separation of phenethylamines are completely justified. Here, advanced polymeric materials could improve the detection of the above-mentioned biomolecules as well as contribute to explaining some aspects of neurological diseases or drug addiction.

The aim of the study was to synthesize and characterize various N-(2-arylethyl)-2-methylprop-2-enamides. The potential of those reagents for the fabrication of specific MIPs was proved in an exemplary synthesis of a molecularly imprinted polymer, using one of synthesized compounds as a functionalized template, followed by the analysis and characterization of the resultant material. In this paper, the synthesis of compounds that possess a template fragment of phenethylamine covalently bound with a polymerizable unit is presented, as well as the primary verification of their ability to synthesize an imprinted sorbent for the analysis of phenethylamines.

2. Materials and Methods

2.1. Materials

2.1.1. Reagents

The following are the relevant ethylamines: 2-(4-fluorophenyl)ethylamine (**1a**), 2-(4-chlorophenyl)ethylamine (**1b**), 2-(2,4-dichlorophenyl)ethylamine (**1c**), 2-(4-bromophenyl)ethylamine (**1d**), 2-(4-methoxyphenyl)ethylamine (**1e**), 2-(3,4-dimethoxyphenyl)ethylamine (**1f**). The target analytes are as follows: tyramine; L-norepinephrine or 3,4-dihydroxyphenyl acetic acid; and methacrylic acid, methacryloyl chloride, divinylbenzene were sourced from Sigma-Aldrich (Steinheim, Germany). The polymerization reaction initiator, 1,1′-azobiscyclohexanecarbonitrile, was from Merck (Darmstadt, Germany). The relevant solvents (ethylene dichloride, hexane, petroleum ether, toluene, methanol, hydrochloric acid (36%), triethylamine, and salts used in the synthesis and the post-polymerization treatment) were delivered from POCh (Gliwice, Poland). Ultra-pure water delivered from a Milli-Q purification system (Millipore, France) was used to prepare water solutions.

2.1.2. Synthesis of N-(2-arylethyl)-2-methylprop-2-enamides

The selected amines **1a–1f** (10 mmol) were added to ethylene dichloride (30 mL) and triethylamine (1.67 mL, 12 mmol) under stirring. Subsequently, methacryloyl chloride (1.18 mL, 12 mmol, 20% excess) was slowly added dropwise to the reaction mixture for 15 min. A white precipitate of triethylammonium chloride was formed and the reaction mixture was stirred for 3 h in room temperature. The precipitated salt was filtered off and the remaining organic layer was washed with saturated NaHCO$_3$ and water, prior to drying over anhydrous MgSO$_4$. Following this, the layer was evaporated under vacuum yielding crude solid products which were recrystallized from hexane or petroleum ether to obtain pure products **2a–2f** (Scheme 1).

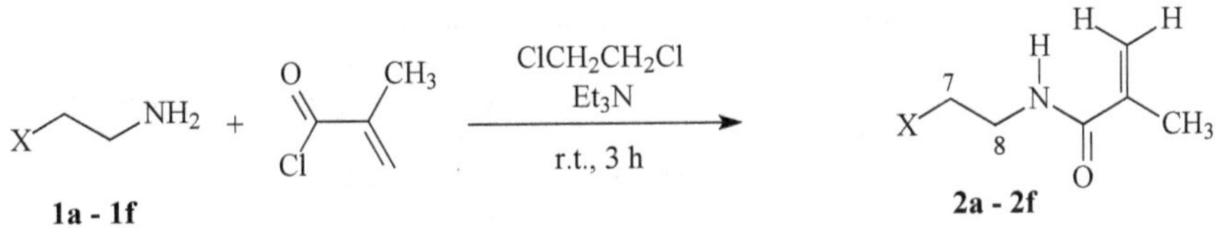

$X = 4\text{-FC}_6\text{H}_4$ (**a**), $4\text{-ClC}_6\text{H}_4$ (**b**), $2,4\text{-Cl}_2\text{C}_6\text{H}_3$ (**c**), $4\text{-BrC}_6\text{H}_4$ (**d**),

$4\text{-MeOC}_6\text{H}_4$ (**e**), $3,4\text{-(MeO)}_2\text{C}_6\text{H}_3$ (**f**)

Scheme 1. Synthesis of N-(2-arylethyl)-2-methylprop-2-enamides.

2.1.3. Preparation of Polymers

The synthesis of the imprinted material, coded MIP$_{ft}$, was carried out. The functionalized template (ft) was used to form an imprinted material. A functionalized template could be defined as the template that possesses one or more polymerizable units that are attached by covalent bonds to form a template–monomer structure by a chemical step independent of polymer formation. In the synthesis of MIP$_{ft}$, the compound **2d** (175.39 mg; 0.8 mmol) was added to toluene (2.056 mL). In the synthesis of NIP, methacrylic acid (69 mg; 0.8 mmol) was added to the same volume of toluene. Following this, divinylbenzene (570 µL; 4 mmol) and 1,1'-azobiscyclohexanecarbonitrile (10 mg) were added to the prepolymerization mixture prior to purging the mixture with nitrogen for 3–5 min. The polymerization process was carried out in 88–92 °C for 24 h. Subsequently, the bulk rigid polymers were ground in a mortar with a pestle and wet-sieved into particles below 45 µm diameter prior to discarding the fine particles by repeated decantation in acetone. Following this, the imprinted particles were treated under reflux with 1 mol L^{-1} hydrochloric acid for 3 h (50 mL) in order to hydrolyze the amide linkage and to remove 4-bromophenylethylamine residue. For comparison, NIP was treated in the same way. Finally, the particles were extensively washed with methanol and were dried prior to the analysis.

2.1.4. Binding Studies

Empty 1 mL solid-phase extraction cartridges were filled with 25 mg of MIP$_{ft}$ or NIP and secured by fiberglass frits. The polymers were then conditioned with methanol–water (85:15 v/v, 1 mL), and loaded with a standard solution of **1d** (conc. 50 µmol L^{-1}, methanol–water, 85:15 v/v, 5 mL) or tyramine (conc. 20 µmol L^{-1}, methanol–water, 85:15 v/v, 5 mL). The unbound amounts of **1d** or tyramine were determined with reference to their respective calibration lines and UV spectroscopy was used for detection. The bound amounts were calculated by subtracting the unbound amount of **1d** or tyramine from initial amount of **1d** or tyramine, respectively. For selectivity tests, L-norepinephrine (conc. 20 µmol L^{-1}) or 3,4-dihydroxyphenylacetic acid (conc. 20 µmol L^{-1}) were used and the analysis was carried out in the same way as described above. The binding capacities and specificity of materials were evaluated [5,19]. The binding capacities (B, µmol g^{-1}) were calculated according to Equation (1):

$$B = \frac{(C_i - C_f)V}{M} \quad (1)$$

where V represents the volume of portion (L) in each loading step, C_i represents the initial analyte concentration (µmol L^{-1}), C_f represents the analyte concentration in solution after adsorption (µmol L^{-1}), and M is the mass of polymer particles (g). The binding capacities of MIP$_{ft}$ and NIP were compared by the determination of the imprinting factor (IF) calculated according to Equation (2):

$$\text{IF} = \frac{B_{MIP}}{B_{NIP}} \quad (2)$$

For isotherm analysis, equilibrium-binding experiments were applied. The polypropylene tubes were filled with 10 mg MIP$_{ft}$ or NIP. Next, a volume of 50 mL of methanol–water (85:15 v/v) standard solution of tyramine was added (concentration range between 10–200 µmol L^{-1}). The tubes were sealed and oscillated by a shaker at room temperature for 24 h prior to centrifugation for 10 min. The aliquots were then used to analyze the unbound amount of tyramine. The amount of tyramine bound to the polymer was calculated according to Equation (1). The detailed analyses of adsorption on MIP$_{ft}$ and NIP were provided using Lineweaver–Burk model [20] represented by Equation (3):

$$\frac{1}{B} = \frac{1}{B_{max}} + \frac{1}{K_L B_{max} F} \quad (3)$$

where B_{max} is the maximum binding capacity and K_L is the equation constant. The Freundlich model [21] represented by Equation (4) was also employed:

$$B = aF^m \quad (4)$$

where B is the adsorbed amount of analyte, F is the unbound amount of analyte, a is the measure of the capacity (B_{max}) and m is a heterogeneity index.

2.1.5. Physicochemical Characterization

The ^{1}H NMR, ^{13}C NMR spectra of **2a–2e** and the ^{13}C CP/MAS NMR spectrum of MIP$_{ft}$ in solid state were recorded with a Bruker Avance DMX 400 spectrometer (Bruker, Germany) at the Faculty of Pharmacy, Medical University of Warsaw, Poland. For the ^{13}C CP/MAS NMR, a powdered sample was contained in 4 mm ZrO$_2$ rotors and was spun at 8 kHz. The 90° pulse length was 2.15 µs. A contact time of 4 ms and a repetition time of 10 s were used for the accumulation of 1900 scans. The chemical shifts, δ ppm, were referenced to tetramethylsilane.

Spectroscopic analyses were carried out using a UV-1605PC spectrophotometer (Shimadzu, Germany). The calibration lines as a function of absorbance (y) versus concentration (x) were constructed at λ$_{max}$ of the analyzed compounds. Each point was measured in triplicate. The linearities of calibration lines were good, with correlation coefficients r^2 > 0.997.

Scanning electron microscopy (SEM) and energy-dispersive X-ray spectrophotometry (EDS) for MIP$_{ft}$ were studied on Merlin FE-SEM (Zeiss, Germany) combined with an EDS X-ray detector (Brucker, Germany). The samples were Au/Pd sputter-coated (SEM) or carbon-coated (EDS) before analysis. The analyses were performed at the Faculty of Chemistry, University of Warsaw, Poland.

The porosity data for MIP$_{ft}$ were determined using the adsorption isotherm of N$_2$ at 77 K (BET) on an ASAP 2420 system (Micromeritics Inc., Norcross, GA, USA) at the Faculty of Chemistry, Maria Curie-Skłodowska University, Lublin, Poland.

3. Results and Discussion

3.1. Synthesis and Identification of N-(2-arylethyl)-2-methylprop-2-enamides

In order to obtain reagents for the fabrication of imprinted polymers, the procedure for the synthesis of six N-(2-arylethyl)-2-methylprop-2-enamides from N-(2-arylethyl)amines with various substituents (methoxy or halogens) in benzene ring was described (Scheme 1). The respective N-(2-arylethyl)-2-methylprop-2-enamides were isolated with high yields (Table 1) and their chemical structures were determined by high-resolution mass spectrometry, IR and NMR analyses.

Table 1. Amines (**1a–1f**) used in synthesis followed by yields and melting points of respective N-(2-arylethyl)-2-methylprop-2-enamides (**2a–2f**).

Amine	Substituents	Product	Yield (%)	M.p. (°C)
1a	X = 4-F-phenyl	2a	90	74–75
1b	X = 4-Cl-phenyl	2b	81	103–104
1c	X = 2,4-diCl-phenyl	2c	92	104–105
1d	X = 4-Br-phenyl	2d	46	114–115
1e	X = 4-H$_3$CO-phenyl	2e	96	74–75
1f	X = 3,4-di(H$_3$CO)-phenyl	2f	63	63–64

The conversion of phenethylamine to phenethylamide is usually a straightforward matter, involving reactions with acyl chloride [22] or acetyl anhydride [23]. Weiner and co-workers [22] used methacryloyl chloride as acylating agent to obtain N-(2-(4-hydroxyphenyl)ethyl)-2-methylprop-2-enamide. The corresponding N-acylations were made in ethyl ether or in benzene, but are inconvenient because of the low boiling point and high toxicity, respectively. Rathelot and co-workers [23] used the N-acylation reaction of phenethylamine to obtain 2-phenylethylacetamide, a substrate in the multistep synthesis of novel antimalarial drugs. In the above-mentioned reaction, acetic anhydride was applied as the N-acylation reagent.

Here, ethylene dichloride was used as a solvent and methacryloyl chloride as acylating agents to obtain derivatives of N-(2-arylethyl)-2-methylprop-2-enamides. The presence of triethylamine provides a satisfactory method of neutralizing the hydrogen halide for amide synthesis. The reaction was finished within 3 h at room temperature. A series of 2-arylethylamines with various substituents in benzene ring, viz., 4-F, 4-Cl, 2,4-Cl$_2$, 4-Br, 4-OMe, 3,4-(OMe)$_2$, **1a–1f** were then used as substrates (Table 1). These amines were selected because of their structural similarity to the biogenic amines or synthetic psychoactive substances used in designer drugs. The conversion of substrates was monitored by TLC. The formation of any other products apart from the main products of the N-acylation reaction was not observed. The structures of compounds **2a–2f** were confirmed by spectroscopy (Table 2).

Table 2. Spectral data of synthesized compounds.

Compound	^{1}H, ^{13}C NMR, IR and MS Data
2a	*N*-(2-(4-fluorophenyl)ethyl)-2-methylprop-2-enamide: ^{1}H NMR (300 MHz, CDCl$_3$) δ (ppm) = 7.19-7.13 (m, 2H, H2, H6), 7.04-6.96 (m, 2H, H3, H5), 5.78 (s, 1H, NH), 5.60 (bt, 1H, H3′a), 5.29 (quint, 1H, H3′b), 3.55 (m, 2H, H8), 2.83 (t, 2H, H7, *J* = 6.9 Hz), 1.92 (dd, 3 H, H4′, *J*$_1$ = 1.2 Hz, *J*$_2$ = 1.5 Hz); ^{13}C NMR (75 MHz, CDCl$_3$) δ (ppm) = 168.86 (C1′), 162.17 (C4), 140.56 (C2′), 135.05 (C1), 130.66 (C2,C6), 119.84 (C3′), 115.94 (C3,C5), 41.32 (C8), 35.31 (C7), 19.09 (C4′); IR (cm^{-1}) 3342 (NH), 1656 (C=O), 1615 (C=C); MS (*m/z*, 70 eV) 207.11 [M$^+$] (calc. 207.20).
2b	*N*-(2-(4-chlorophenyl)ethyl)-2-methylprop-2-enamide: ^{1}H NMR (300 MHz, CDCl$_3$) δ (ppm) = 7.31-7.26 (m, 2H, H3,H5), 7.15-7.11 (m, 2H, H2,H6), 5.81 (bs, 1H, NH), 5.60 (bt, 1H, H3′a), 5.30 (quint, 1H, H3′b, *J* = 1.5 Hz), 3.54 (q, 2H, H8, *J* = 6.9 Hz), 2.83 (t, 2H, H7, *J* = 6.9 Hz), 1.92 (dd, 3H, H4′, *J*$_1$ = 1.2 Hz, *J*$_2$ = 1.5 Hz); ^{13}C NMR (75 MHz, CDCl$_3$) δ (ppm) = 168.87 (C1′), 140.51 (C2′), 137.86 (C1), 132.87 (C4), 130.59 (C3,C5), 129.24 (C2,C6), 119.89 (C3′), 41.15 (C8), 35.48 (C7), 19.09 (C4′); IR (cm^{-1}) 3316 (NH), 1652 (C=O), 1615 (C=C); MS (*m/z*, 70 eV) 223.08 [M$^+$] (calc. 223.66).
2c	*N*-(2-(2,4-dichlorophenyl)ethyl)-2-methylprop-2-enamide: ^{1}H NMR (300 MHz, CDCl$_3$) δ (ppm) = 7.38 (d, 1H, H3, *J* = 1.8 Hz), 7.19 (m, 1H, H5, *J*$_1$ = 8.1 Hz, *J*$_2$ = 1.8 Hz), 7.16 (m, 1H, H6, *J* = 8.1 Hz), 5.96 (bs, 1H, NH), 5.63 (bs, 1H, H3′a), 5.30 (quint, 1H, H3′b, *J* = 1.5 Hz), 3.56 (q, 2H, H8, *J* = 7 Hz), 2.98 (t, 2H, H7, *J* = 7 Hz), 1.93 (dd, 3H, H4′, *J*$_1$ = 1.2 Hz, *J*$_2$ = 1.5 Hz); ^{13}C NMR (75 MHz, CDCl$_3$) δ (ppm) = 168.95 (C1′), 140.39 (C2′), 135.73 (C1), 135.19 (C2), 133.52 (C4), 132.25 (C6), 129.85 (C3), 127.71 (C5), 119.97 (C3′), 39.68 (C8), 33.19 (C7), 19.08 (C4′); IR (cm^{-1}) 3299 (NH), 1654 (C=O), 1616 (C=C); MS (*m/z*, 70 eV) 257.09 [M$^+$] (calc. 258.10).
2d	*N*-(2-(4-bromophenyl)ethyl)-2-methylprop-2-enamide: ^{1}H NMR (300 MHz, CDCl$_3$) δ (ppm) = 7.46-7.41 (m, 2H, H3,H5), 7.10-7.06 (m, 2H, H2,H6), 5.79 (bs, 1H, NH), 5.60 (bt, 1H, H3′a, *J* = 1.5 Hz), 5.30 (bsextet, 1H, H3′b, *J* = 1.5 Hz), 3.54 (bq, 2H, H8, *J* = 7 Hz), 2.82 (t, 2H, H7, *J* = 7 Hz), 1.92 (dd, 3H, H4′, *J*$_1$= 0.9 Hz, *J*$_2$ = 1.5 Hz); ^{13}C NMR (75 MHz, CDCl$_3$) δ (ppm) = 168.88 (C1′), 140.51 (C2′), 138.39 (C1), 132.21 (C3,C5), 131.00 (C2,C6), 120.90 (C4), 119.90 (C3′), 41.10 (C8), 35.55 (C7), 19.10 (C4′); IR (cm^{-1}) 3318 (NH), 1653 (C=O), 1611 (C=C); MS (*m/z*, 70 eV) 267.03 [M$^+$] (calc. 268.12).
2e	*N*-(2-(4-methoxyphenyl)ethyl)-2-methylprop-2-enamide: ^{1}H NMR (300 MHz, CDCl$_3$) δ (ppm) = 7.12 (m, 2H, H2,H6), 6.86 (m, 2H, H3,H5), 5.84 (bs, 1H, NH), 5.60 (bt, 1H, H3′a, *J* = 0.9 Hz), 5.28 (quint, 1H, H3′b, *J* = 1.5 Hz), 3.79 (s, 3H, H9), 3.53 (m, 2H, H8), 2.79 (t, 2H, H7, *J* = 6.9 Hz), 1.92 (dd, 3H, H4′, *J*$_1$ = 0.9 Hz, *J*$_2$ = 1.5 Hz); ^{13}C NMR (75 MHz, CDCl$_3$) δ (ppm) = 168.79 (C1′), 158.76 (C4), 140.59 (C2′), 131.33 (C1), 130.16 (C2,C6), 119.71 (C3′), 114.52 (C3,C5), 55.71 (C9), 41.39 (C8), 35.14 (C7), 19.07 (C4′); IR (cm^{-1}) 3303 (NH), 1652 (C=O), 1615 (C=C); MS (*m/z*, 70 eV) 219.13 [M$^+$] (calc. 219.24).
2f	*N*-(2-(3,4-dimethoxyphenyl)ethyl)-2-methylprop-2-enamide: ^{1}H NMR (300 MHz, CDCl$_3$) δ (ppm) = 6.82 (d, 1H, H5, *J* = 8.4 Hz), 6.74 (dd, 1H, H6, *J*$_1$ = 1.8 Hz, *J*$_2$ = 8.4 Hz), 6.72 (d, 1H, H2, *J* = 1.8 Hz), 5.81 (s, 1H, NH), 5.61 (bt, 1H, H3′a), 5.29 (quint, 1H, H3′b, *J* = 1.5 Hz), 3.87 (s, 6H, H9,H10), 3.55 (q, 2H, H8, *J* = 6.9 Hz), 2.81 (t, 2 H, H7, *J* = 6.9 Hz), 1.92 (bt, 3H, H4′); ^{13}C NMR (75 MHz, CDCl$_3$) δ (ppm) = 168.81 (C1′), 149.56 (C3), 148.21 (C4), 140.61 (C2′), 131.90 (C1), 121.13 (C6), 119.78 (C3′), 112.41 (C2), 111.86 (C5), 56.41 (C10), 56.32 (C9), 41.36 (C8), 35.64 (C7), 19.10 (C4′); IR (cm^{-1}) 3320 (NH), 1650 (C=O), 1618 (C=C); MS (*m/z*, 70 eV) 249.14 [M$^+$] (calc. 249.27).

It has to be underlined that the screening of chemical databases revealed only the presence of *N*-(2-(4-iodophenyl)ethyl)-2-methylprop-2-enamide [24] and *N*-(2-(2-bromophenyl)ethyl)-2-methylprop-2-enamide [25,26].

Hence, it could be assumed that the *N*-acylation reaction of phenethylamines with methacryloyl chloride is effective for phenethylamines with the strong electron-donating group –OCH$_3$ or with the halogens F, Cl, or Br in a benzene ring.

3.2. Preparation of Polymer

In an antecedent paper, Weiner and co-workers [22] described the effect of the attachments of various phenethylamines, viz., phenethylamine, tyramine, ephedrine (2-

(methylamino)-1-phenylpropan-1-ol), and amphetamine into synthetic or natural polymers by an amide or carbamate linkage as a method for increasing their duration of action. However, the imprinting technique was not studied in the above-mentioned paper.

Here, we employed the covalent imprinting approach to synthesize the polymers but the adsorption process on imprinted material was entirely non-covalent in nature [27]. Thus, so-called 'semi-covalent imprinting' was applied. In order to confirm the structure, to analyze the morphology, and to prove the selectivity of the resultant imprinted polymer, the synthesis of MIP_{ft} from N-(2-(4-bromophenyl)ethyl)-2-methylprop-2-enamide, **2d**, was carried out in the presence of divinylbenzene (cross-linker) in toluene (porogen). Radical thermal polymerization was applied to obtain bulk material. The schematic idea of the synthesis, employing a covalent strategy for the imprinting process and an adsorption process on the resultant imprinted polymer that is based on non-covalent interactions of the target analytes, is presented in Figure 1. The NIP was prepared with the employment of methacrylic acid as the functional monomer and divinylbenzene as the cross-linker, omitting the addition of any template, and it was fabricated to compare sorption behavior.

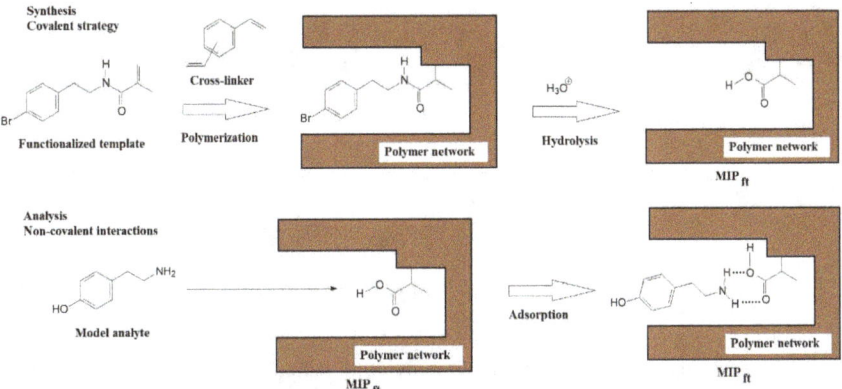

Figure 1. Schematic idea of the synthesis, employing a covalent strategy for the imprinting process and an adsorption process on the resultant imprinted polymer that is based on non-covalent interactions of the target analytes.

The amide linkage, present in the MIP_{ft}, was hydrolyzed prior to the removal of 2-(4-bromophenyl)ethylamine residue in order to form specific cavities in the imprinted polymer. The hydrolysis was carried out in 1 mol L^{-1} hydrochloric acid under reflux. The hydrolysis lasted up to 3 h.

3.3. Characterization of Material

3.3.1. Adsorption Behavior

The binding capacities were determined for MIP_{ft} as well as for NIP in one μmol of **1d** for one gram of polymer particles. The selectivity of the imprinted polymer (imprinting factor, IF) was expressed as the ratio of the amount of **1d** bound to MIP_{ft} in comparison to NIP.

The binding capacities of **1d** and IF were as follows: MIP_{ft}: 8.09 ± 0.08 μmol g^{-1}, NIP: 4.59 ± 0.04 μmol g^{-1}, IF = 1.76. These results proved that the functionalized template **2d** could be used as reagent for covalent imprinting and the resulting MIP_{ft} was possessed of selectivity when compared to NIP.

Sorption behavior is responsible for the effective separation of target analytes on MIPs. In order to prove that the novel imprinted material possessed an affinity towards biogenic amines with a phenethylamine system, the adsorption of tyramine on MIP_{ft} was examined. The binding capacities of tyramine on MIP_{ft} and on NIP were as follows: 2.74 ± 0.03 μmol g^{-1}

and 1.11 ± 0.01 µmol g^{-1}, respectively (IF = 2.47). In order to analyze adsorption data, the Langmuir isotherm was applied [28]. Here, various linearized modifications of the model could be used to determine the most proper fit [29]. It was found that the Lineweaver–Burk modification represented by Equation (3) was characterized by the highest regression coefficients for analyzed data with values of r^2 = 0.974 and r^2 = 0.971 for MIP$_{ft}$ and NIP, respectively. Moreover, to reveal the homogeneity of MIP$_{ft}$, the Freundlich model described by Equation (4) was applied. That model fits well to MIP adsorption data in the low-concentration regions and allows for the surface-homogeneity determination of the material tested. The straight line of log B versus log F is the evidence that adsorption can be described by the Freundlich equation. The correlation coefficients, r^2, for MIP$_{ft}$ and NIP were equal to 0.999 and 0.972, respectively, and the estimated values of m were equal to 0.70 and 0.93 for MIP$_{ft}$ and NIP, respectively (Figure 2).

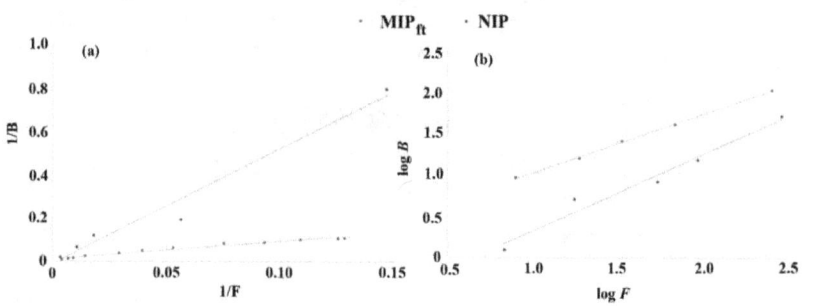

Figure 2. Lineweaver–Burk (**a**) and Freundlich (**b**) models for tyramine adsorption on MIP$_{ft}$ and NIP.

It was found that MIP$_{ft}$ provided affinity towards tyramine. In order to analyze the selectivity, two other biomolecules were used for studies, viz., L-norepinephrine and 3,4-dihydroxyphenylacetic acid. L-norepinephrine possesses a phenethylamine system. In contrast, 3,4-dihydroxyphenylacetic acid (a metabolite in the dopamine system) does not possess a phenethylamine system. The binding capacities for MIP$_{ft}$ and NIP were as follows: for L-norepinephrine, 14.5 ± 1.4 µmol g^{-1} and 5.80 ± 0.59 µmol g^{-1}, respectively (IF = 2.50); and for 3,4-dihydroxyphenylacetic acid, 0.579 ± 0.070 µmol g^{-1} and 0.549 ± 0.060 µmol g^{-1}, respectively (IF = 1.06). The results show the selectivity of MIP$_{ft}$ to L-norepinephrine, a biomolecule with a phenethylamine system, and its lack of selectivity to 3,4-dihydroxyphenylacetic acid, a biomolecule that not possess a phenethylamine system. The higher binding capacity of L-norepinephrine when compared to the binding capacity of tyramine could be explained by the presence of two strong electron-donating hydroxy groups in positions 3 and 4 of the aromatic ring and one hydroxy group in the aliphatic ethylamine chain, enhancing strongly the basicity of L-norepinephrine. Apart from being an exemplary experiment, the results are very promising for the possible application of such reagents in the preparation of sorbents for biomedical purposes.

3.3.2. Morphology Characterization

In order to provide morphological characterization, scanning electron microscopy was employed and the surface of MIP$_{ft}$ was analyzed. Figure 3 presents a micrograph of MIP$_{ft}$ after the hydrolysis process. The particles possessed the morphology of the bulk materials that were composed from spherical entities agglomerated into bigger forms of 10–20 µm. The diameter of single entity varied from 500 nm to 2 µm and was similar for MIP$_{ft}$ and NIP (Figure 3a–d). However, further magnification revealed substantial difference between MIP$_{ft}$ and NIP (Figure 3e–h). The MIP$_{ft}$ was characterized by significant surface extension with numerous macropores clearly detected on the particle's surface. On the contrary, the NIP possessed a smoother surface. The difference could be related to the presence of the functionalized template in the prepolymerization mixture.

3.3.3. Structural Evaluation

In order to confirm the structure of the resultant polymers, EDS was used to prove that the functionalized template was polymerized (Figure 4a). The MIP_{ft} was prepared in order to confirm that the functionalized template was built up into the polymer matrix because the presence of bromine atoms in the structure of **2d** allowed us to detect heteroatom during the analysis of the materials. It has to be underlined that the analysis of the polymer, viz., MIP_{ft}, was carried out omitting the process of the hydrolysis of the amide linkage. The atoms of bromine were detected in the MIP_{ft} structure in the region of 1.50 keV.

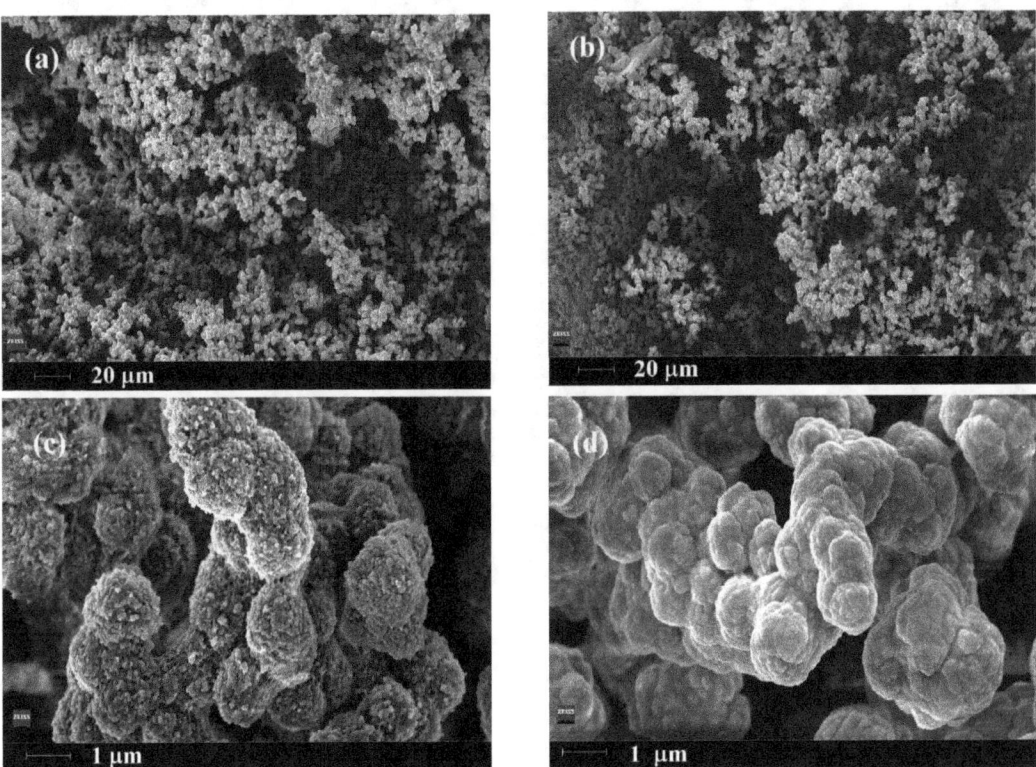

Figure 3. *Cont.*

Figure 3. SEM micrographs of MIP$_{ft}$ (a,c,e,g) and NIP (b,d,f,h).

^{13}C CP/MAS NMR spectroscopy was then applied. This is a versatile tool to confirm the composition of polymer materials. For the purpose of our analysis, the MIP$_{ft}$ was post-treated to remove 2-(4-bromophenyl)ethylamine residue from the polymer matrix (Figure 4b). In the ^{13}C CP/MAS NMR spectrum of MIP$_{ft}$, strong resonances in the aromatic region, representing quaternary benzene C atoms at 137.2 and 144.3 ppm, could be seen. The tertiary –CH atoms at 127.0 ppm originated from the cross-linker. In the aliphatic region, various methyl groups were represented by broad peaks located between 15 and 30 ppm with a narrower sharp peak at 28.7 ppm. Methylene groups in C–CH$_2$–C were found in approximately 44.4 ppm and methylene groups in Φ–CH–C in 39.8 ppm. Carboxyl group, –COOH, atoms were represented by broad resonances in the region of 177.1–182.7 ppm. Low intensity resonance at 111.7 ppm could originate from unreacted double bonds in Φ–(CH=CH$_2$)$_2$.

3.3.4. Porosity Data

Finally, the nitrogen-adsorption isotherm (Brunauer–Emmett–Teller) for MIP$_{ft}$ was analyzed. As it can be seen (Figure 4c) the material revealed physisorption isotherms of type IV with a hysteresis loop. The shape of the hysteresis loop is related to the specific pore structure. Here, type H3 loops characterized MIP$_{ft}$, indicating the slit-shaped structure of its pores. However, the deformation of the desorption-hysteresis line of MIP$_{ft}$ in the region of 0.50 P/P$_o$ could be related to the expulsion of adsorbate from larger-volume mesopores through narrower pore necks. Thus, a more complicated pore system could exist in MIP$_{ft}$. The total specific surface area of MIP$_{ft}$ was equal to 89.88 m^2 g^{-1} and the external surface area was equal to 77.99 m^2 g^{-1}. The plots of pore volume versus diameter for MIP$_{ft}$ showed a peak for a pore diameter of 56 nm (Figure 4d).

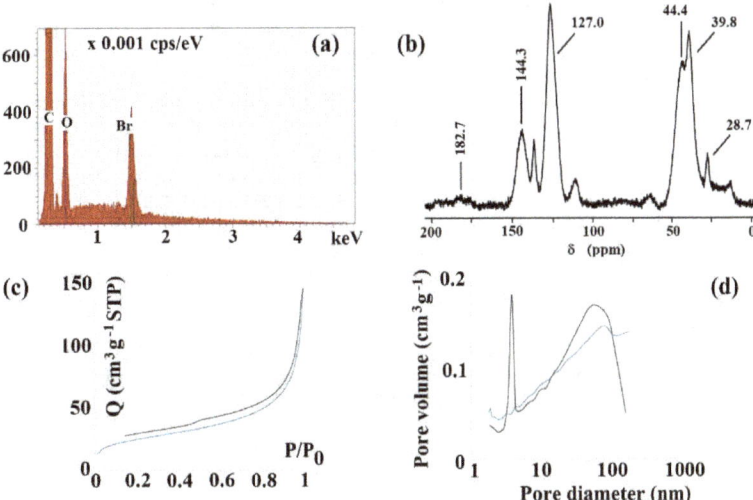

Figure 4. EDS spectrum of MIP$_{ft}$ before hydrolysis of the amide linkage (a), ^{13}C CP MAS NMR spectrum for MIP$_{ft}$ (b), nitrogen-sorption hysteresis (c) and pore-size distributions (d) for MIP$_{ft}$ after hydrolysis of amide linkage.

4. Conclusions

In conclusion, it should be emphasized that a series of compounds, N-(2-arylethyl)-2-methylprop-2-enamides, were obtained with the synthetic procedure presented here with high yields. These compounds possessed fragments of a template covalently bound to polymerizable units and could be used as reagents for the covalent imprinting of polymers. In the control experiment, one of synthesized compounds was used to produce an imprinted polymer (MIP$_{ft}$). Binding-capacity analysis revealed that a molecular imprinting process took place and the polymer, MIP$_{ft}$, possessed selectivity towards biomolecules of tyramine and L-norepinephrine.

Author Contributions: Conceptualization, P.L.; methodology, P.L.; validation, P.L.; formal analysis, M.S. and P.L.; investigation, P.L.; resources, P.L.; data curation, P.L. and D.M.; writing—original draft preparation, P.L., M.S. and D.M.; writing—review and editing, P.L. and M.S.; supervision, P.L. and D.M. All authors have read and agreed to the published version of the manuscript.

Funding: This research received no external funding.

Institutional Review Board Statement: Not applicable.

Informed Consent Statement: Not applicable.

Data Availability Statement: Not applicable.

Conflicts of Interest: The authors declare no conflict of interest.

References

1. Janczura, M.; Luliński, P.; Sobiech, M. Imprinting technology for effective sorbent fabrication: Current state-of-art and future prospects. *Materials* **2021**, *14*, 1850. [CrossRef] [PubMed]
2. Farooq, S.; Wu, H.; Nie, J.; Ahmad, S.; Muhammad, I.; Zeeshan, M.; Khan, R.; Asim, M. Application, advancement and green aspects of magnetic molecularly imprinted polymers in pesticide residue detection. *Sci. Total Environ.* **2022**, *804*, 150293. [CrossRef] [PubMed]

3. Farooq, S.; Nie, J.; Cheng, Y.; Bacha, S.A.S.; Chang, W. Selective extraction of fungicide carbendazim in fruits using β-cyclodextrin based molecularly imprinted polymers. *J. Sep. Sci.* **2020**, *43*, 1145–1153. [CrossRef]
4. Farooq, S.; Nie, J.; Cheng, Y.; Yan, Z.; Li, J.; Bacha, S.A.S.; Mushtaq, A.; Zhang, H. Molecularly imprinted polymers' application in pesticide residue detection. *Analyst* **2018**, *143*, 3971–3989. [CrossRef]
5. BelBruno, J.J. Molecularly imprinted polymers. *Chem. Rev.* **2019**, *119*, 94–119. [CrossRef]
6. Wulff, G.; Sarhan, A.; Zabrocki, K. Enzyme-analogue built polymers and their use for the resolution of racemates. *Tetrahedron Lett.* **1973**, *44*, 4329–4332. [CrossRef]
7. Whitcombe, M.J.; Alexander, C.; Vulfson, E.N. Imprinted polymers: Versatile new tools in synthesis. *Synlett* **2000**, *6*, 911–923.
8. Hashim, S.N.N.S.; Boysen, R.I.; Schwarz, L.J.; Danylec, B.; Hearn, M.T.W. A comparison of covalent and non-covalent imprinting strategies for the synthesis of stigmasterol imprinted polymers. *J. Chromatogr. A* **2014**, *1359*, 35–43. [CrossRef]
9. Tang, Y.-W.; Fang, G.-Z.; Wang, S.; Li, J.-L. Covalent imprinted polymer for selective and rapid enrichment of ractopamine by a non-covalent approach. *Anal. Bioanal. Chem.* **2011**, *401*, 2275–2282. [CrossRef]
10. Hu, Y.; Liu, R.; Li, Y.; Li, G. Investigation of ractopamine-imprinted polymer for dispersive solid-phase extraction of trace β-agonists in pig tissues. *J. Sep. Sci.* **2010**, *33*, 2017–2025. [CrossRef]
11. Hu, X.; Guo, Y.; Wang, T.; Liu, C.; Yang, Y.; Fang, G. A selectivity-enhanced ratiometric fluorescence imprinted sensor based on synergistic effect of covalent and non-covalent recognition units for ultrasensitive detection of ribavirin. *J. Hazard. Mater.* **2022**, *421*, 126748. [CrossRef] [PubMed]
12. Effting, L.; Prete, M.C.; Urbano, A.; Effting, L.M.; Cano Gonzalez, M.E.; Bail, A.; Teixeira Tarley, C.R. Preparation of magnetic nanoparticle-cholesterol imprinted polymer using semi-covalent imprinting approach for ultra-effective and highly selective cholesterol adsorption. *React. Funct. Polym.* **2022**, *172*, 105178. [CrossRef]
13. Tang, Y.; Gao, J.; Liu, X.; Gao, X.; Ma, T.; Lu, X.; Li, J. Ultrasensitive detection of clenbuterol by a covalent imprinted polymer as a biomimetic antibody. *Food Chem.* **2017**, *228*, 62–69. [CrossRef] [PubMed]
14. Zhu, Y.; Pan, Z.; Rong, J.; Mao, K.; Yang, D.; Zhang, T.; Xu, J.; Qiu, F.; Pan, J. Boronate affinity surface imprinted polymers supported on dendritic fibrous silica for enhanced selective separation of shikimic acid via covalent binding. *J. Mol. Liq.* **2021**, *337*, 116408. [CrossRef]
15. Luliński, P.; Maciejewska, D. Effective separation of dopamine from bananas on 2-(3,4-dimethoxyphenyl)ethylamine imprinted polymer. *J. Sep. Sci.* **2012**, *35*, 1050–1057. [CrossRef]
16. Sobiech, M.; Giebułtowicz, J.; Luliński, P. Application of magnetic core–shell imprinted nanoconjugates for the analysis of hordenine in human plasma-preliminary data on pharmacokinetic study after oral administration. *J. Agric. Food Chem.* **2020**, *68*, 14502–14512. [CrossRef]
17. Burchett, S.A.; Hicks, T.P. The mysterious trace amines: Protean neuromodulators of synaptic transmission in mammalian brain. *Prog. Neurobiol.* **2006**, *79*, 223–246. [CrossRef]
18. D'Andrea, G.; Nordera, G.; Pizzolato, G.; Bolner, A.; Colavito, D.; Flaibani, R.; Leon, A. Trace amine metabolism in Parkinson's disease: Low circulating levels of octopamine in early disease stages. *Neurosci. Lett.* **2010**, *469*, 348–351. [CrossRef]
19. Hasanah, A.N.; Safitri, N.; Zulfa, A.; Neli, N.; Rahayu, D. Factors affecting preparation of molecularly imprinted polymer and methods on finding template-monomer interaction as the key of selective properties of the materials. *Molecules* **2021**, *26*, 5612. [CrossRef]
20. Lineweaver, H.; Burk, D. The determination of enzyme dissociation constants. *J. Am. Chem. Soc.* **1934**, *56*, 658–666. [CrossRef]
21. Freundlich, H. Über die Adsorption in Lösungen. *Zeit. Phys. Chem.* **1907**, *57*, 385–470. [CrossRef]
22. Weiner, B.-Z.; Tahan, M.; Zilkha, A. Polymers containing phenethylamines. *J. Med. Chem.* **1972**, *15*, 410–413. [CrossRef] [PubMed]
23. Rathelot, P.; Vanelle, P.; Gasquet, M.; Delmas, F.; Crozet, P.M.; Timon-David, P.; Maldonado, P. Synthesis of novel functionalized 5-nitroisoquinolines and evaluation of in vitro antimalarial activity. *Eur. J. Med. Chem.* **1995**, *30*, 503–508. [CrossRef]
24. Ruowen, W.; Yu, C. Phenoxy-Containing Acryloylphosphoramidite as Well as Preparation Method and Application Thereof. China Patent CN108203446, 26 June 2018.
25. Sharma, U.K.; Sharma, N.; Kumar, Y.; Singh, B.K.; van der Eycken, E.V. Domino carbopalladation/C-H functionalization sequence: An expedient synthesis of bis-heteroaryls through transient alkyl/vinyl–palladium species capture. *Chem. Eur. J.* **2016**, *22*, 481–485. [CrossRef] [PubMed]
26. Vachhani, D.D.; Butani, H.H.; Sharma, N.; Bhoya, U.C.; Shah, A.K.; van der Eycken, E.V. Domino Heck/borylation sequence towards indolinone-3-methyl boronic esters: Trapping of the σ-alkylpalladium intermediate with boron. *Chem. Commun.* **2015**, *51*, 14862–14865. [CrossRef] [PubMed]
27. Mayes, A.G.; Whitcombe, M.J. Synthetic strategies for the generation of molecularly imprinted organic polymers. *Adv. Drug Deliv. Rev.* **2005**, *57*, 1742–1778. [CrossRef]
28. Langmuir, I. The adsorption of gases on plane surfaces of glass, mica and platinum. *J. Am. Chem. Soc.* **1918**, *40*, 1361–1403. [CrossRef]
29. Swenson, H.; Stadie, N.P. Langmuir's Theory of Adsorption: A Centennial Review. *Langmuir* **2019**, *35*, 5409–5426. [CrossRef]

(16 h) incubation at 70 °C, the beads were washed with acetone (4 × 100 mL) and dried at 120 °C for 30 min.

2.3. Immobilisation of Peptide on AHAMTES Glass

AHAMTES-functionalised beads (60 g) were soaked in a large volume of water for 24 h in order to remove multilayers. The beads were then washed with acetone (4 × 100 mL) and dried at 120 °C for 30 min, then incubated with succinimidyl iodoacetate (SIA) (5 mg, 18 µmol) in anhydrous acetonitrile (25 ml) and incubated for 2 h under exclusion of light, before washing with acetonitrile (5 × 50 mL).

Ethylenediaminetetraacetic acid (EDTA) (74 mg, 500 µmol) was dissolved in phosphate buffered saline (PBS, 10 mM, 50 mL) and adjusted to pH 8.2 with sodium hydroxide. SIA-functionalised glass beads (60 g) and EGFR peptide (5 mg) were then added, and the mixture was incubated overnight protected from light. Mercaptoethanol (20 µL, 0.3 mmol) was then added and incubated for 2 h. The beads were then washed with water (3 × 200 mL) and acetone (1 × 100 mL) and allowed to dry.

2.4. Immobilisation of Peptide on IPTMS Glass

EGFR peptide (5 mg) was dissolved in borate buffer (pH 9.2, 30 mM sodium tetraborate, 25 mL), added to IPTMS-functionalised glass beads (60 g) and incubated overnight. Mercaptoethanol (20 µL, 0.3 mmol) was then added and incubated for 2 h. The beads were then washed with water (3 × 200 mL) and acetone (1 × 100 mL) and allowed to dry.

2.5. Peptide Density Measurement

Peptide density in the solid phase was measured using the Pierce Rapid Gold BCA Protein Assay Kit (Thermo Scientific, Waltham, MA, USA). As described in the Thermo Fisher Protein Assay Technical Handbook, Buffer A and Buffer B were mixed in a 50:1 ratio; 400 µL of this solution was mixed with 100 mg of peptide-coated glass beads, and the mixture was incubated at 37 °C with shaking for 30 min. The samples were allowed to cool, and 100 µL of the solution was transferred to each of the three wells in a 96 well microtiter plate. The absorbance of these wells was measured at 562 nm using a Hidex Sense plate reader (LabLogic, Sheffield, UK). A calibration curve was prepared by repeating this measurement using known concentrations of peptides.

2.6. MIP Synthesis

The following monomers were dissolved in water (50 mL): N-isopropyl acrylamide (20 mg, 180 µmol), N-tert-butylacrylamide (16.5 mg, 130 µmol), N,N'-methylene bis(acrylamide) (3 mg, 20 µmol), N-(3-aminopropyl)methacrylamide hydrochloride (3 mg, 17 µmol) and acrylic acid (1.1 µL, 16 µmol). Peptide-functionalised glass beads (60 g) were added to the monomer solution, which was then bubbled with nitrogen for 20 min. Polymerisation was initiated through the addition of ammonium persulphate (30 mg, 0.13 mmol) and N,N,N',N'-tetramethyl ethylenediamine (30 µL, 0.2 mmol) in water (500 µL). The mixture was shaken and incubated for 1 h before being transferred to a solid phase extraction cartridge fitted with a 20 µm polyethylene frit. Unreacted monomers and low affinity polymers were removed from the glass beads by washing with water (5 × 100 mL). High affinity polymers were collected with hot ethanol (65 °C, 2 × 25 mL), reduced to 5 mL under vacuum and dialysed in water for 1 week using 12 kDa cellulose membranes with regular change of water.

2.7. Surface Plasmon Resonance (SPR) Measurement

Binding analysis was performed using a Biacore 3000 instrument (Cytiva, UK) at 25 °C using PBS (0.01 M phosphate buffer, 0.0027 M potassium chloride and 0.137 M sodium chloride, pH 7.4) as the running buffer at a flow of 35 µL min^{-1}. The self-assembled gold sensor chip was plasma-cleaned using a K1050X RF Plasma Etcher/Asher/Cleaner barrel reactor (Quorum Technologies Ltd., Lewes, UK) and placed in a solution of mercaptododecanoic acid in ethanol (1.1 mg mL^{-1}) where they were stored until use. Before assembly, the sensor chip was rinsed with ethanol and water and dried in a stream of air. Each cysteine-containing specific or scrambled peptide was immobilised *in situ* on the chip surface containing carboxyl groups using thiol coupling. First, the surface was activated using an EDC and NHS mixture (0.4 mg and 0.6 mg mL^{-1}, respectively). 2-(2-pyridinyldithio)ethaneamine hydrochloride (PDEA, 80 mM) in 50 mM sodium borate buffer pH 8.5 was injected in order to introduce disulphide bonds; 100 µL of 10 µg mL^{-1} peptide solution in PBS was then injected at a 15 µL mL^{-1} flow rate followed by surface deactivation using cysteine/NaCl solution.

The peptide-specific nanoMIPs were briefly sonicated and diluted with PBS in the concentration range 0.04–1 nM. Sensorgrams were collected sequentially for all analyte concentrations running in KINJECT mode (injection volume—100 µL and dissociation time—120 sec). Dissociation constants (K_d) were calculated from plots of the equilibrium biosensor response using the BiaEvaluation v 4.1 software using a 1:1 Langmuir binding model fitting after subtraction of drift and bulk components.

3. Results and Discussion

This work investigates the replacement of amino silanes for the immobilisation of thiol-presenting templates—a common strategy for the imprinting of peptides and proteins. The immobilisation of peptides is commonly performed via conjugation of a terminal cysteine unit to an amino silane using a linker, such as SIA [7,13,17]. Herein we attempted to replace the amino silane and SIA linker with an iodo silane, (3-iodopropyl)trimethoxy silane (IPTMS) (Scheme 2). This serves three benefits: removal of unwanted side reactions caused by amine groups, reduction in the number of steps necessary for template immobilisation, and the replacement of relatively expensive SIA.

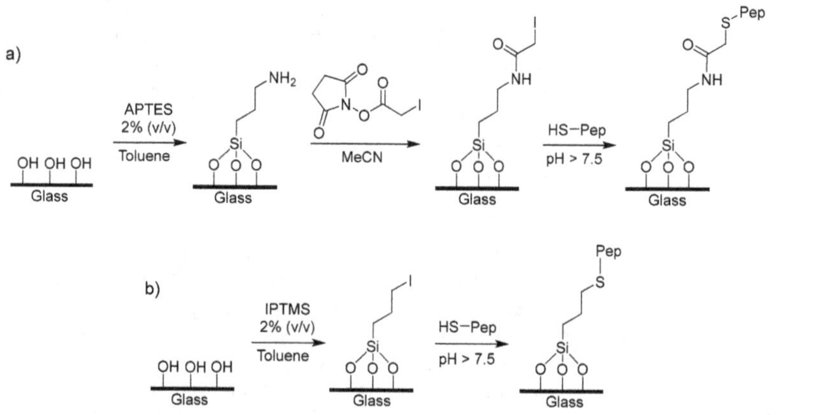

Scheme 2. Immobilisation of peptides on glass beads using; (a) APTES; (b) IPTMS. 'Pep' refers to the EGFR peptide being immobilised.

In order to compare the performance of these two approaches, peptide-functionalised glass beads were prepared using both AHAMTES and IPTMS-based protocols (Table 1). These solid phases were then used for the synthesis of MIPs via a solid phase approach as described by Canfarotta et al. [13]. The peptide selected for imprinting was an epitope of epidermal growth factor receptor (EGFR), a cancer biomarker of clinical interest. This sequence (KLFGTSGQK) was previously identified using a MIP-based epitope mapping technique [8]. The peptide was prepared with a terminal cysteine for immobilisation and an additional glycine to act as a spacer (full sequence CGKLFGTSGQK). A scrambled version of the peptide was also imprinted to act as a control (full sequence CGTKGKQLSGF). The density of peptides on the surface of the glass beads following immobilisation was determined via bicinchoninic acid assay (BCA). The density of immobilised peptide was similar in both cases, at 6.3 nmol peptide/g glass beads using AHAMTES and 6.1 nmol peptide/g glass beads using IPTMS. These values are similar to those previously reported for the immobilisation and imprinting of proteins including trypsin (1.7 nmol/g), pepsin A (2.8 nmol/g) and amylase (2.9 nmol/g) [18]. The difference of a factor of two can be explained by the larger size of these protein templates as compared to the peptides within this study.

Table 1. Comparison of methodologies for AHAMTES and IPTMS-based solid phase preparation.

	AHAMTES	IPTMS
Silanisation	Incubate glass beads (60 g) in 2% (v/v) AHAMTES overnight in dry toluene (60 mL). Wash with acetone (4 × 100 mL). Dry at 120 °C for 30 min.	Incubate glass beads (60 g) in 2% (v/v) IPTMS overnight in dry toluene (60 mL), protected from light. Wash with acetone (4 × 100 mL). Dry at 120 °C for 30 min.
Removal of multilayers	Soak glass beads overnight in a large excess of water. Wash with acetone (4 × 100 mL).	-
Surface activation	Incubate silanised glass beads (60 g) in a solution of succinimidyl iodoacetate (SIA) (5 mg) in anhydrous acetonitrile (25 ml) for 2 h, protected from light. Wash with acetonitrile (5 × 50 mL).	-
Peptide conjugation	Prepare ethylenediaminetetraacetic acid (EDTA) (74 mg, 500 µmol, 5mM) in phosphate buffered saline (PBS, 10 mM, 50 mL), adjust to pH 8.2 with sodium hydroxide. Add SIA-functionalised glass beads (60 g) and EGFR peptide (5 mg), incubate overnight protected from light.	Prepare EGFR peptide (5 mg) in borate buffer (pH 9.2, 30 mM sodium tetraborate, 25 mL). Add IPTMS-functionalised glass beads (60 g) and incubate overnight, protected from light.
Surface quenching	Add mercaptoethanol (20 µL) to a mixture of glass beads and peptide, and incubate for 2 h protected from light. Wash with water (2 × 500 mL) and acetone (100 mL), and allow to dry.	Add mercaptoethanol (20 µL) to a mixture of glass beads and peptide, and incubate for 2 h protected from light. Wash with water (2 × 500 mL) and acetone (100 mL), and allow to dry.

The amount of MIP nanoparticles collected following polymerisation, elution and dialysis was found to be 56 µg MIP/g glass beads using AHAMTES, and 72 µg MIP/g glass beads using IPTMS. The average size of these particles was found to be approximately 60 nm in both cases (Figure 1). The protocol for AHAMTES-functionalisation of glass includes a lengthy (overnight) washing step in order to remove weakly associated silane multilayers. This step is omitted in the IPTMS-based protocol. As shown in the results of elemental analysis, even in the absence of such a washing step the level of silane contamination is lower in the case of IPTMS-MIPs as compared to AHAMTES-MIPs (Tables S2 and S3).

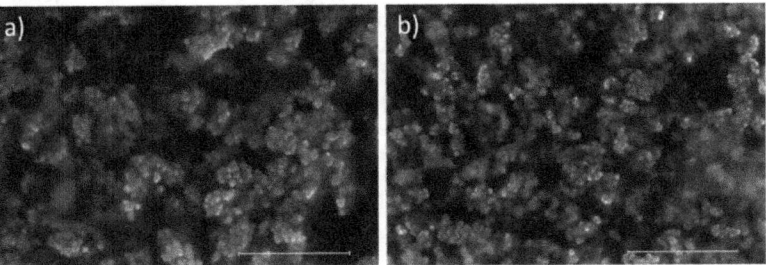

Figure 1. Scanning electron microscopy (SEM) images of MIPs prepared using: (**a**) AHAMTES-based solid phase, (**b**) IPTMS-based solid phase. Scale bar = 2 µm.

The binding performance of MIPs prepared using both silanes was compared via surface plasmon resonance (SPR) measurement (Figure 2, Table 2). The specificity of the resultant MIPs was assessed by comparison of binding affinity with a scrambled version of the same peptide (CGTKGKQLSGF). In both cases, the peptide was immobilised on gold SPR chips, and MIPs prepared using both techniques were injected during SPR measurement.

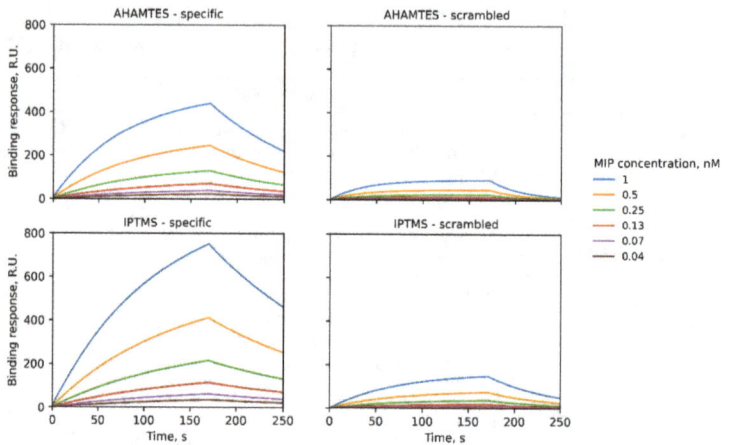

Figure 2. SPR binding curves of MIPs prepared using AHAMTES and IPTMS against specific and scrambled peptides.

Table 2. Dissociation constants (K_d) of MIPs prepared using AHAMTES and IPTMS-functionalised glass against specific and scrambled peptide.

Silane	Peptide	K_d [nM] (χ^2)
AHAMTES	Specific	2.5 (7.14 × 10^{-5})
AHAMTES	Scrambled	676 (1.58 × 10^{-5})
IPTMS	Specific	2.3 (7.14 × 10^{-5})
IPTMS	Scrambled	766 (2.06 × 10^{-5})

Both MIPs prepared using AHAMTES and IPTMS showed excellent binding affinity (K_d of approximately 2 nM) for their specific peptide and significantly lower affinity (K_d of approximately 700 nM) for the scrambled control peptide (Table 2). The dissociation constant for the scrambled control peptide was marginally lower for MIPs prepared using AHAMTES as a solid phase as compared to those made using IPTMS. This may possibly be attributed to the slightly higher level of silane contamination increasing the level of non-specific binding between the MIPs and the gold SPR chip (Tables S2 and S3).

An additional advantage of IPTMS over amino silanes is the lack of residual amines on the surface of the glass beads following MIP formation. Post-synthetic labelling of MIPs is often performed using amine-reactive probes, such as NHS-ester-DyLight and AlexaFluor. In order to preserve binding site efficacy, this tagging can be performed while the MIPs are still bound to the solid phase, preventing the conjugation of probes to functional groups within the binding sites [19]. In this case, the use of amino silanes for template immobilisation would result in probes linking to any remaining amine groups of the solid phase, resulting in wasted (often expensive) probes. The use of iodo silanes circumvents this issue.

Finally, alkyl halides, such as IPTMS are able to react with other nucleophilic groups beyond thiols (notably amines and aromatic alcohols, given basic enough conditions). As a result, the iodo silane-based methodology described herein can be used for the immobilisation of a wide variety of template molecules even in the absence of cysteine groups. However, in the case of templates containing multiple nucleophiles the pH must be carefully controlled or protecting groups employed to ensure immobilisation via the intended functional group to reduce the variability of MIP binding sites.

4. Conclusions

Amino silanes commonly used for solid phase preparation were investigated and found to show undesired side reactions during solid phase synthesis of molecularly imprinted polymers (MIPs), including the formation of silane multilayers and reaction with persulphates. An iodo silane (IPTMS) is presented as an alternative for peptide immobilisation, and was compared to an amino silane (AHAMTES) for the solid phase synthesis of MIP nanoparticles specific for an epitope of epidermal growth factor receptor (EGFR). The iodo silane tested was found to equal or outperform the amino silane by each metric tested. Both protocols produced similar yields of MIPs with excellent affinity ($K_d \approx 2.5$ nM), but with fewer experimental steps and lower cost reagents necessary for the iodo silane-based protocol.

Supplementary Materials: The following supporting information can be downloaded at: https://www.mdpi.com/article/10.3390/polym14081595/s1, S1. Additional Methods; S2. Additional Discussion; Figure S1. Silane density of glass beads coated with APTES and AHAMTES after various levels of washing [20]; Table S1: Contents of vials displayed in Figure S2 and results of gelation experiment [21]; Figure S2: Vials of acrylamide solution in which polymerisation was initiated with APS and various co-initiators; Figure S3: EDS mapping spectra of MIPs prepared using AHAMTES solid phase; Table S2: Elemental analysis of AHAMTES MIPs using EDS; Figure S4: EDS mapping spectra of MIPs prepared using IPTMS solid phase; Table S3: Elemental analysis of IPTMS MIPs using EDS.

Author Contributions: Conceptualisation, S.S.P.; methodology, S.S.P., A.G.C. and E.P.; software, S.S.P., E.P. and A.G.C.; validation, S.S.P., A.G.C. and E.P.; formal analysis, S.S.P., A.G.C. and E.P.; investigation, S.S.P., A.G.C. and E.P.; resources, S.S.P., A.G.C. and E.P.; data curation, S.S.P., A.G.C. and E.P.; writing—original draft preparation, S.S.P.; writing—review and editing, S.S.P.; visualisation, S.S.P.; supervision, S.A.P., E.O.A. and A.C.S.; All authors have read and agreed to the published version of the manuscript.

Funding: This research was funded by the EPSRC, grant number 2133561, the MRC, grant number 2136145 and the Imperial President's PhD Scholarship.

Institutional Review Board Statement: Not applicable.

Informed Consent Statement: Not applicable.

Data Availability Statement: Additional data available upon request from authors.

Conflicts of Interest: The authors declare no conflict of interest.

References

1. Olsen, J.; Martin, P.; Wilson, I.D. Molecular Imprints as Sorbents for Solid Phase Extraction: Potential and Applications. *Anal. Commun.* **1998**, *35*, 13H–14H. [CrossRef]
2. Andersson, L.I. Molecular Imprinting for Drug Bioanalysis: A Review on the Application of Imprinted Polymers to Solid-Phase Extraction and Binding Assay. *J. Chromatogr. B: Biomed. Sci. Appl.* **2000**, *739*, 163–173. [CrossRef]
3. Jodlbauer, J.; Maier, N.M.; Lindner, W. Towards Ochratoxin A Selective Molecularly Imprinted Polymers for Solid-Phase Extraction. *J. Chromatogr. A* **2002**, *945*, 45–63. [CrossRef]
4. Piletsky, S.S.; Cass, A.E.G.; Piletska, E.V.; Czulak, J.; Piletsky, S.A. A Novel Assay Format as an Alternative to ELISA: MINA Test for Biotin. *ChemNanoMat* **2018**, *4*, 1214–1222. [CrossRef]
5. Piletsky, S.S.; Rabinowicz, S.; Yang, Z.; Zagar, C.; Piletska, E.V.; Guerreiro, A.; Piletsky, S.A. Development of Molecularly Imprinted Polymers Specific for Blood Antigens for Application in Antibody-Free Blood Typing. *Chem. Commun.* **2017**, *53*, 1793–1796. [CrossRef]
6. Saylan, Y.; Akgönüllü, S.; Yavuz, H.; Ünal, S.; Denizli, A. Molecularly Imprinted Polymer Based Sensors for Medical Applications. *Sensors* **2019**, *19*, 1279. [CrossRef]

7. Canfarotta, F.; Lezina, L.; Guerreiro, A.; Czulak, J.; Petukhov, A.; Daks, A.; Smolinska-Kempisty, K.; Poma, A.; Piletsky, S.; Barlev, N.A. Specific Drug Delivery to Cancer Cells with Double-Imprinted Nanoparticles against Epidermal Growth Factor Receptor. *Nano Lett.* **2018**, *18*, 4641–4646. [CrossRef]
8. Piletsky, S.S.; Piletska, E.; Poblocka, M.; Macip, S.; Jones, D.J.L.; Braga, M.; Cao, T.H.; Singh, R.; Spivey, A.C.; Aboagye, E.O.; et al. Snapshot Imprinting: Rapid Identification of Cancer Cell Surface Proteins and Epitopes Using Molecularly Imprinted Polymers. *Nano Today* **2021**, *41*, 101304. [CrossRef]
9. Xing, R.; Wen, Y.; He, H.; Guo, Z.; Liu, Z. Recent Progress in the Combination of Molecularly Imprinted Polymer-Based Affinity Extraction and Mass Spectrometry for Targeted Proteomic Analysis. *TrAC Trends Anal. Chem.* **2019**, *110*, 417–428. [CrossRef]
10. Poma, A.; Guerreiro, A.; Whitcombe, M.J.; Piletska, E.; Turner, A.P.F.; Piletsky, S.A. Solid-Phase Synthesis of Molecularly Imprinted Polymer Nanoparticles with a Reusable Template–"Plastic Antibodies". *Adv. Funct. Mater.* **2013**, *23*, 2821–2827. [CrossRef]
11. Medina Rangel, P.X.; Laclef, S.; Xu, J.; Panagiotopoulou, M.; Kovensky, J.; Tse Sum Bui, B.; Haupt, K. Solid-Phase Synthesis of Molecularly Imprinted Polymer Nanolabels: Affinity Tools for Cellular Bioimaging of Glycans. *Sci. Rep.* **2019**, *9*, 3923. [CrossRef] [PubMed]
12. Ambrosini, S.; Beyazit, S.; Haupt, K.; Tse Sum Bui, B. Solid-Phase Synthesis of Molecularly Imprinted Nanoparticles for Protein Recognition. *Chem. Commun.* **2013**, *49*, 6746–6748. [CrossRef] [PubMed]
13. Canfarotta, F.; Poma, A.; Guerreiro, A.; Piletsky, S. Solid-Phase Synthesis of Molecularly Imprinted Nanoparticles. *Nat. Protoc.* **2016**, *11*, 443–455. [CrossRef] [PubMed]
14. Ekpenyong-Akiba, A.E.; Canfarotta, F.; Abd H., B.; Poblocka, M.; Casulleras, M.; Castilla-Vallmanya, L.; Kocsis-Fodor, G.; Kelly, M.E.; Janus, J.; Althubiti, M.; et al. Detecting and Targeting Senescent Cells Using Molecularly Imprinted Nanoparticles. *Nanoscale Horiz.* **2019**, *4*, 757–768. [CrossRef]
15. Zhu, M.; Lerum, M.Z.; Chen, W. How to Prepare Reproducible, Homogeneous, and Hydrolytically Stable Aminosilane-Derived Layers on Silica. *Langmuir* **2012**, *28*, 416–423. [CrossRef] [PubMed]
16. Stayton, P.S.; Olinger, J.M.; Jiang, M.; Bohn, P.W.; Sligar, S.G. Genetic Engineering of Surface Attachment Sites Yields Oriented Protein Monolayers. *J. Am. Chem Soc.* **1992**, *114*, 9298–9299. [CrossRef]
17. Zhang, S.; Shi, Z.; Xu, H.; Ma, X.; Yin, J.; Tian, M. Revisiting the Mechanism of Redox-Polymerization to Build the Hydrogel with Excellent Properties Using a Novel Initiator. *Soft Matter* **2016**, *12*, 2575–2582. [CrossRef]
18. Poma, A.; Guerreiro, A.; Caygill, S.; Moczko, E.; Piletsky, S. Automatic Reactor for Solid-Phase Synthesis of Molecularly Imprinted Polymeric Nanoparticles (MIP NPs) in Water. *RSC Adv.* **2014**, *4*, 4203–4206. [CrossRef]
19. García, Y.; Czulak, J.; Pereira, E.D.; Piletsky, S.A.; Piletska, E. A Magnetic Molecularly Imprinted Nanoparticle Assay (MINA) for Detection of Pepsin. *React. Funct. Polym.* **2022**, *170*, 105133. [CrossRef]
20. Cuoq, F.; Masion, A.; Labille, J.; Rose, J.; Ziarelli, F.; Prelot, B.; Bottero, J.Y. Preparation of Amino-Functionalized Silica in Aqueous Conditions. *Appl. Surf. Sci.* **2013**, *266*, 155–160. [CrossRef]
21. Gómez-Caballero, A.; Elejaga-Jimeno, A.; García del Caño, G.; Unceta, N.; Guerreiro, A.; Saumell-Esnaola, M.; Sallés, J.; Goicolea, M.A.; Barrio, R.J. Solid-Phase Synthesis of Imprinted Nanoparticles as Artificial Antibodies against the C-Terminus of the Cannabinoid CB1 Receptor: Exploring a Viable Alternative for Bioanalysis. *Mikrochim. Acta* **2021**, *188*, 3. [CrossRef] [PubMed]

Article

EGDMA- and TRIM-Based Microparticles Imprinted with 5-Fluorouracil for Prolonged Drug Delivery

Michał Cegłowski [1,*], Joanna Kurczewska [1], Aleksandra Lusina [1], Tomasz Nazim [1] and Piotr Ruszkowski [2]

1 Faculty of Chemistry, Adam Mickiewicz University, 61-614 Poznan, Poland; asiaw@amu.edu.pl (J.K.); aleksandra.lusina@amu.edu.pl (A.L.); tomasz.nazim@amu.edu.pl (T.N.)
2 Department of Pharmacology, Poznan University of Medical Sciences, 61-614 Poznan, Poland; pruszkowski@gmail.com
* Correspondence: michal.ceglowski@amu.edu.pl; Tel.: +48-61-8291-799

Abstract: Imprinted materials possess designed cavities capable of forming selective interactions with molecules used in the imprinting process. In this work, we report the synthesis of 5-fluorouracil (5-FU)-imprinted microparticles and their application in prolonged drug delivery. The materials were synthesized using either ethylene glycol dimethacrylate (EGDMA) or trimethylolpropane trimethacrylate (TRIM) cross-linkers. For both types of polymers, methacrylic acid was used as a functional monomer, whereas 2-hydroxyethyl methacrylate was applied to increase the final materials' hydrophilicity. Adsorption isotherms and adsorption kinetics were investigated to characterize the interactions that occur between the materials and 5-FU. The microparticles synthesized using the TRIM cross-linker showed higher adsorption properties towards 5-FU than those with EGDMA. The release kinetics was highly dependent upon the cross-linker and pH of the release medium. The highest cumulative release was obtained for TRIM-based microparticles at pH 7.4. The IC_{50} values proved that 5-FU-loaded TRIM-based microparticles possess cytotoxic activity against HeLa cell lines similar to pure 5-FU, whereas their toxicity towards normal HDF cell lines was ca. three times lower than for 5-FU.

Keywords: molecularly imprinted polymers; microspheres; drug delivery; 5-fluorouracil

Citation: Cegłowski, M.; Kurczewska, J.; Lusina, A.; Nazim, T.; Ruszkowski, P. EGDMA- and TRIM-Based Microparticles Imprinted with 5-Fluorouracil for Prolonged Drug Delivery. *Polymers* **2022**, *14*, 1027. https://doi.org/10.3390/polym14051027

Academic Editor: Beom Soo Kim

Received: 31 December 2021
Accepted: 2 March 2022
Published: 4 March 2022

Publisher's Note: MDPI stays neutral with regard to jurisdictional claims in published maps and institutional affiliations.

Copyright: © 2022 by the authors. Licensee MDPI, Basel, Switzerland. This article is an open access article distributed under the terms and conditions of the Creative Commons Attribution (CC BY) license (https://creativecommons.org/licenses/by/4.0/).

1. Introduction

Drug delivery systems (DDS) are sophisticated technologies that allow targeted delivery of particular pharmaceutical or controlled therapeutic agents' release. DDS application is essential for therapeutic agents limited due to their low solubility, drug administration issues, or very fast metabolism. Moreover, the application of DDS allows to optimize drug efficiency and improve administration, while lowering the possibility of causing adverse side effects. As a result, the development of new DDS occurs in parallel with creating new therapeutic agents. Whenever a new drug has a limited therapeutic effect in a free form, the combination of drugs and DDS may overcome this problem. The design, synthesis, and production of new substances and materials that can be applied to prepare new DDS have become a major topic of many research groups. Their goal is to develop a DDS which would allow drug administration to a specific site with a known quantity and for a precise amount of time. It is, of course, necessary to produce DDS that, as well as their metabolites, show no toxicity and are easily removed from the human body [1–3].

Various materials have been used to generate DDS. The most frequently used are lipids [4,5], chitosan [6,7], silica [8–10], halloysite [11,12], functional polymers [13–19], and dendrimers [20,21]. Among the described materials, functional polymers are frequently used due to their high synthetic versatility, a broad range of final properties, and various possible applications [22,23]. Sophisticated polymeric DDS should guarantee drug delivery in a predesigned manner. Functional polymers can be designed to fulfill the most sophisticated criteria of DDS. They can be synthesized to become triggered-release DDS or release a drug following a particular kinetic profile. As a result, it is possible to obtain passive or active targeting, which means that the whole system is responsible for the therapeutic benefit. Molecularly imprinted polymers (MIPs) represent a group of functional polymers that can be easily tuned to possess a defined affinity to a drug molecule, and thus they often find application in producing new DDS [24–26].

MIPs are synthesized by copolymerizing a cross-linker mixed with a complex formed by functional monomers and template molecules. The cross-linker initiates the formation of the bulk polymer structure and entrapment of the template molecules inside, whereas functional monomers form stable, non-covalent interactions with template molecules. Finally, it results in the synthesis of MIPs selective towards the template used, which can be reversibly bound by cavities formed during the reaction. As a result, they have found many applications in fields such as solid-phase extraction (SPE) [27–29], development of sensors [30–32], and drug delivery [33,34].

Many researchers find the application of MIPs in drug delivery particularly interesting due to the possibility of producing materials that alter their interaction strength with drug molecules influenced by a specific change in the environment. MIPs possess an enhanced affinity to the drug template, which increases the residence time of the drug. Moreover, MIPs' properties allow for reaching high drug loading, and the fact that the drug is entrapped within the polymeric network results in its higher stability and durability against harsh conditions. MIPs can be synthesized as smart materials that can be effectively used to produce DDS, considering that they can be programmed to release therapeutic agents as a response to defined stimuli [35]. MIPs can also be synthesized to develop materials that release molecules entrapped inside cavities with a particular kinetic profile [36]. This can be achieved by selecting the appropriate type and amount of functional monomers present in the MIPs' structure. Our group recently published a detailed investigation about using MIPs for prolonged drug delivery of doxorubicin [37] and paclitaxel [38].

MIPs are synthesized as monoliths that are ground to obtain particles of desired dimensions for numerous applications. Although this process is frequently used in analytical applications such as SPE, the development of DDS requires more rigorous control over polymer particles' size. New synthetic methods that allow obtaining MIPs as uniform nanoparticles or microparticles have been developed to solve this problem. As synthesized materials possess a significantly increased surface area, more cavities are closer to the surface, making them easily accessible. The additional benefit of small particle sizes is their ability to form stable dispersions in various solvents, which increases their attractiveness as potential materials to be used as DDS [39–41].

In this work, we report the synthesis and characterization of MIPs microparticles imprinted with 5-fluorouracil (5-FU). 5-FU is a cytotoxic drug with a broad activity spectrum against numerous tumors. It undergoes fast metabolism in the human body, and thus a therapy consisting of high doses is required, which can cause severe toxic effects in many patients [42]. As a result, new DDS for 5-FU are being developed [43]. The advantage of using microparticles for drug delivery is that they do not traverse into the interstitium over the size of 100 nm transported by the lymph, and thus they only act locally in a place of administration [44]. Methacrylate-based particulate carriers are widely applied for biomedical applications to obtain drug delivery systems. The therapeutic agent's release from their structure typically occurs in a biphasic way with an incomplete drug release. The release mechanism can be described using both Fickian and non-Fickian kinetic models. The improvement of drug release can be achieved by increasing polymer hydrophilicity by synthesizing functional microspheres with additional functional groups or formulating composites with hydrophilic polymers [45]. The microparticles were synthesized by using precipitation polymerization. Methacrylic acid (MAA) was used as a functional monomer, whereas 2-hydroxyethyl methacrylate (HEMA) was used as a hydrophilic monomer to improve the water compatibility of the final materials. The microparticles were prepared using two cross-linkers, particularly ethylene glycol dimethacrylate (EGDMA) and trimethylolpropane trimethacrylate (TRIM), to examine the influence of the polymer matrix structure on the final properties of microparticles. The interactions of MIPs with 5-FU were investigated by examining their adsorption properties and release profiles of 5-FU. Finally, the in vitro activity of 5-FU-loaded MIPs against HeLa, U87 MG, A-549, KB, and MCF-7 cancer cell lines was investigated. The results were compared with the measurements performed against human dermal fibroblasts (HDF) normal cell lines. To the best of our knowledge, this is the first research describing the synthesis of 5-FU-imprinted hydrophilic microparticles and presenting their in vitro activity against cancer and normal cell lines.

2. Materials and Methods

2.1. Materials and Chemicals

Ethylene glycol dimethacrylate, methacrylic acid, trimethylolpropane trimethacrylate, 2-hydroxyethyl methacrylate, 5-fluorouracil, 2,2'-azobisisobutyronitrile solution (AIBN; 0.2 M in toluene), and all solvents (HPLC grade) were obtained from Sigma-Aldrich (St. Louis, MO, USA).

2.2. Instruments

The FTIR measurements were performed using a IFS 66/s spectrometer (Bruker, Billerica, MA, USA). To obtain spectra, 1.5 mg of each material was mixed with ca. 200 mg of KBr, and the resulting powder was converted into tablets used in FTIR measurements. Thermogravimetric (TG) analysis was performed using a Setsys 1200 (Setaram, Caluire, France) apparatus. The analysis was performed in an air stream (50 mL min^{-1}) at a heating rate of 10 °C min^{-1}. Scanning electron microscopy (SEM) images were recorded using a Scanning Electron Microscope SU3500 (Hitachi, Tokyo, Japan). UV-Vis absorption spectra were obtained using a 8453 (Agilent, Santa Clara, CA, USA) spectrophotometer. The solutions' pH values were controlled with an CP-505 pH meter (Elmetron, Zabrze, Poland).

2.3. Synthesis of Microparticles

MIPs created using EGDMA cross-linker were synthesized as follows. The prepolymerization mixture consisting of 5-FU (0.9 mmol), MAA (1.8 mmol), HEMA (0.9 mmol), and methanol (90 mL) was prepared in a glass pressure tube. The mixture was degassed for 30 min using an ultrasound bath and purging with inert gas (nitrogen). Subsequently, EGDMA (8 mmol) and AIBN solutions (1 mL) were added, and the tube was degassed for an additional 10 min. Afterward, the tube was sealed and placed at 60 °C for 18 h. As-synthesized microparticles were filtered off and dried under a vacuum. The resulting drug-loaded MIPs were placed in a dialysis tubing to remove template molecules and were dialyzed against acidified methanol (9:1 methanol/acetic acid). This process was continued until 5-FU was no longer detected in a dialysis solution. The drug-unloaded MIPs cross-linked with EGDMA (denoted as MIP$_{EGDMA}$) were finally dried under a vacuum.

MIPs created using the TRIM cross-linker were prepared by the same setup and procedure as EGDMA-based MIPs, but with different amounts of reagents. The polymerization mixture consisted of 5-FU (2 mmol), MAA (4 mmol), HEMA (2 mmol), and methanol (80 mL). After degassing and sonication, TRIM (4 mmol) and AIBN solutions (1.4 mL) were added, and the tube was degassed for an additional 10 min. The remaining procedure is identical to EGDMA-based MIPs. Finally, the drug-unloaded MIPs cross-linked with TRIM were denoted as MIP$_{TRIM}$.

The corresponding non-imprinted polymers (NIPs) were synthesized using analogous procedures but without adding template molecules. The obtained microparticles were denoted as NIP$_{EGDMA}$ and NIP$_{TRIM}$. The structures of all used monomers and templates are presented in Figure S1.

2.4. Adsorption Studies

Adsorption isotherms were established using batch experiments, in which 10 mL of methanolic 5-FU solution at concentrations ranging from 0.39 to 50 mg L^{-1} was added to 10 mg of MIPs or NIPs. The obtained mixtures were equilibrated for 24 h at ambient conditions (25 °C). UV-vis absorption spectra were taken prior to and after adsorption solutions to establish the concentration of 5-FU. The amount of adsorbed 5-FU (q_{eq}, mg g^{-1}) was calculated as follows:

$$q_{eq} = \frac{(C_0 - C_{eq})V}{m} \quad (1)$$

where C_0 is the initial concentration of 5-FU solution (mg mL^{-1}), C_{eq} is the equilibrium concentration of 5-FU solution (mg mL^{-1}), m is the polymer (MIPs/NIPs) mass (g), and V is the volume of 5-FU solution (mL). The experiments were repeated three times, and mean values were used for calculation.

The adsorption kinetics studies were conducted when 50 mg of either MIPs or NIPs were placed in 50 mL of 5-FU solution in methanol with an initial concentration of 10 mg L^{-1}. The solution was stirred at ambient conditions, and the concentration of 5-FU was measured at defined time intervals by UV-vis absorption spectra. The q_t value (mg g^{-1}), which represents the amount of adsorbed 5-FU, was calculated as follows:

$$q_t = \frac{(C_0 - C_t)V}{m} \quad (2)$$

where C_t is the concentration of 5-FU after time t (h).

2.5. Release Experiments

The dissolution method was used to conduct the in vitro release studies. The experiments were conducted by dispersing 5-FU-loaded MIPs (20 mg) in buffer solutions (10 mL) at pH 2.2, 5.0, and 7.4. The solutions were stirred continuously for seven days at 37 °C, and at specified time intervals, the samples were collected, centrifuged, and their UV-vis absorption spectra were recorded. The obtained data allowed to calculate the amount of drug releases from studied MIPs. The experiments were repeated three times, and the obtained mean values were used for subsequent calculations.

The results obtained for the 5-FU release were fitted using various models that characterize the mechanism of the release process. The following models were applied: zero-order (Equation (3)), first-order (Equation (4)), simplified Higuchi (Equation (5)), Hixson–Crowell (Equation (6)), and Korsmeyer–Peppas (Equation (7)). The following equations mathematically represent these models:

$$F_t = k_0 t \quad (3)$$

$$F_t = 1 - e^{-k_1 t} \quad (4)$$

$$F_t = k_H \sqrt{t} \quad (5)$$

$$\sqrt[3]{F_0} - \sqrt[3]{F_t} = k_{HC} t \quad (6)$$

$$F_t = k_{KP} t^n \quad (7)$$

where F_t is the amount of 5-FU released at a specified time "t", F_0 is the initial amount of 5-FU in MIPs structure, k_0, k_1, k_H, k_{HC}, and k_{KP} are the release constants of corresponding equations, and n is the diffusion exponent.

2.6. Cytotoxicity Assays

The cytotoxicity assays were performed using the experimental procedure described in our previous research. Briefly, KB, HeLa, and MCF-7 cell lines were obtained from The European Collection of Cell Cultures (ECACC) supplied by Sigma-Aldrich (St. Louis, MO, USA). whereas A-549, U-87MG, and HDF cell lines were purchased from the American Type Cell Collection (ATCC) through LGC Standards. Approximately 0.1 mL of the diluted cell suspension (ca. 10,000 cells) was added to every well of the microtiter plate. A partial monolayer was formed after 24 h, and the supernatant was washed out. Then, 100 µL of 6 different 5-FU concentrations (0.1, 0.2, 1, 2, 10, and 20 µM) or number of MIPs that release the corresponding amount of 5-FU were added to the cells in microtiter plates. For NIPs blank experiments, the same mass of microparticles was used as for MIPs. Other experimental details were described in our previous work.

2.7. Statistical Analysis

One-way ANOVA with the post-hoc Tukey HSD test was used to test statistical significance. A *p*-value lower than 0.05 was considered as statistically significant.

3. Results and Discussion

The synthesized imprinted microparticles and corresponding non-imprinted microparticles present similar bands in the IR spectra (Figures S2 and S3), which indicates that the polymeric materials' main structure for all samples is similar. For all synthesized materials, the stretching and bending O-H vibrations of carboxyl groups originating from MAA can be observed at 3562 and 1389 cm^{-1}, respectively. These bands overlap with O-H stretching, and bending vibrations originated from hydroxyl groups of HEMA. The asymmetric stretching vibrations of CH_2 groups can be observed at 2955 cm^{-1} for EGDMA-based microparticles and 2974 for TRIM-based microparticles. The stretching vibrations of C=O bonds are observed at 1730 cm^{-1} for EGDMA-based microparticles and 1734 cm^{-1} for TRIM-based microparticles. The symmetric and asymmetric C-O stretching vibrations of ester groups can be observed at 1258 and 1161 cm^{-1} for EGDMA-based microparticles, respectively. For TRIM-based microparticles, these signals are observed at 1268 and 1156 cm^{-1}, respectively [46,47]. Drug-loaded MIP_{EGDMA} shows a band characteristic for 5-FU at 816 cm^{-1}, whereas for drug-loaded MIP_{TRIM}, these additional bands can be observed at 816, 553, and 471 cm^{-1}.

SEM images of EGDMA- and TRIM-based microparticles are presented in Figure 1. For drug-loaded and drug-unloaded MIPs microparticles, no differences in SEM images were observed. SEM images of NIP_{EGDMA} present microparticles of 0.7–1.3 µm in diameter, whereas MIP_{EGDMA} microparticles are much smaller, ranging between 300 and 600 nm. This result clearly indicates that the presence of 5-FU in the polymerization mixture results in the formation of smaller microparticles. The NIP_{TRIM} microparticles are also smaller than those synthesized using EGDMA as a cross-linker, and their size ranges between 400 and 800 nm. Contrary to the previous observation, the presence of 5-FU in the polymerization mixture has not influenced the size of MIP_{TRIM} microparticles. This result may indicate that the presence of three methacrylate groups in the cross-linker structure that undergo polymerization makes the process less prone to the presence of additional substances in the reaction mixture that can affect the reaction.

The TG results obtained for EGDMA-based microparticles are presented in Figure S4, whereas those obtained for TRIM-based microparticles are shown in Figure S5. All EGDMA-based microparticles demonstrate one major decomposition step, from ca. 200 to ca. 450 °C. This step refers to the polymer structure's decomposition and results in almost complete oxidation of organic material. For 5-FU-loaded MIP_{EGDMA} microparticles, an increased weight loss is observed in the initial decomposition step compared to drug-unloaded material. It probably results from the decomposition of 5-FU molecules. Compared to EGDMA-based microparticles, all TRIM-based microparticles are characterized by only one major decomposition step, that starts at around 300 °C and ends at around 475 °C, reflecting almost complete oxidation of the polymer material. For 5-FU-loaded MIP_{TRIM}, only a slight increase in the weight loss of the initial decomposition step is observed compared with drug-unloaded material, which is connected with the decomposition of 5-FU molecules. The increase in the weight loss occurs within the 260–320 °C range, which is in accordance with the decomposition temperature of 5-FU reported in the literature [48].

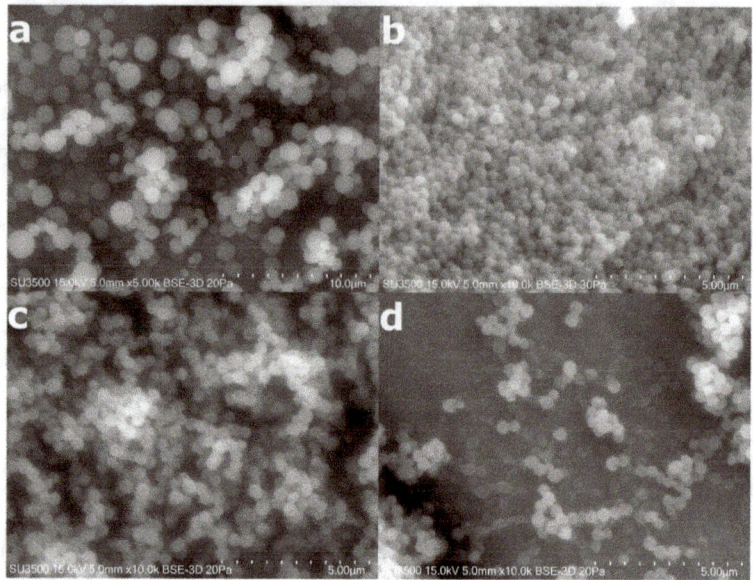

Figure 1. SEM images of (**a**) NIP$_{EGDMA}$, (**b**) MIP$_{EGDMA}$, (**c**) NIP$_{TRIM}$, and (**d**) MIP$_{TRIM}$.

3.1. Adsorption Isotherms

The adsorption process's characterization is achieved by plotting adsorption isotherms for experimental data obtained during adsorption of 5-FU by the microparticles at the equilibrium state. Figure 2 shows the relationship between the 5-FU equilibrium concentration and the amount of 5-FU adsorbed by 1 g of the appropriate adsorbent. To characterize the experimental data, Langmuir and Freundlich adsorption isotherm models were used for interpretation.

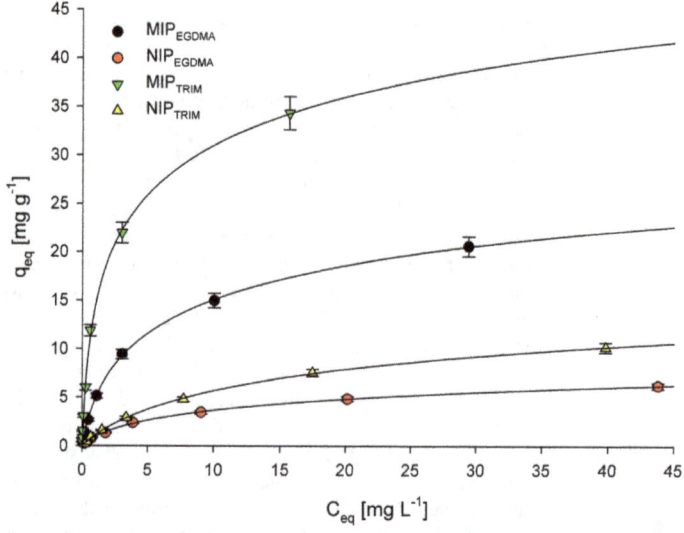

Figure 2. Adsorption isotherms of 5-FU onto MIP$_{EGDMA}$, NIP$_{EGDMA}$, MIP$_{TRIM}$, and NIP$_{TRIM}$.

The following equation represents the Langmuir adsorption isotherm:

$$\frac{C_{eq}}{q_{eq}} = \frac{C_{eq}}{q_m} + \frac{1}{Kq_m} \quad (8)$$

where K (L mg^{-1}) is the binding equilibrium constant, q_m (mg g^{-1}) is the maximum amount of bonded 5-FU, C_{eq} (mg L^{-1}) is the equilibrium concentration of 5-FU, and q_{eq} (mg g^{-1}) is the amount of 5-FU adsorbed at the equilibrium concentration. Table 1 summarizes the values of K, q_m, and R^2 (correlation coefficients). The R^2 values obtained for all synthesized microparticles are high (above 0.98), indicating that the Langmuir adsorption model nicely fits the experimental data. The confirmation of successful imprinting can be found in the high difference between the maximum adsorption capacity (q_m) between MIPs and corresponding to them NIPs. For both MIPs–NIPs pairs, the q_m value is around three times higher in favor of MIPs. There is also a significant difference in q_m values between the corresponding materials synthesized using different cross-linkers. The calculated data clearly show that the maximum adsorption capacity values are higher for MIP$_{TRIM}$ and NIP$_{TRIM}$ than for MIP$_{EGDMA}$ and NIP$_{EGDMA}$, respectively. As a result, it can be concluded that a TRIM cross-linker allows constructing a polymer network that has a higher affinity to 5-FU and allows to imprint it more effectively than a network synthesized using the EGDMA cross-linker.

Table 1. Parameters of 5-FU adsorption by MIPs and NIPs.

Polymer	Langmuir			Freundlich		
	q_m (mg g^{-1})	K (L mg^{-1})	R^2	K_f (mg g^{-1} (L mg^{-1})$^{1/n}$)	$1/n$	R^2
MIP$_{EGDMA}$	22.57 ± 1.96	0.285 ± 0.025	0.991	3.59 ± 0.32	0.62 ± 0.06	0.973
NIP$_{EGDMA}$	7.00 ± 0.64	0.142 ± 0.013	0.989	0.82 ± 0.08	0.60 ± 0.06	0.977
MIP$_{TRIM}$	37.30 ± 3.46	0.676 ± 0.064	0.997	9.72 ± 0.92	0.66 ± 0.06	0.937
NIP$_{TRIM}$	12.89 ± 1.63	0.088 ± 0.008	0.992	0.95 ± 0.09	0.73 ± 0.07	0.972

The following equations represent the Freundlich adsorption isotherm:

$$q_{eq} = K_f C_{eq}^{1/n} \quad (9)$$

$$\log q_{eq} = \log K_f + \frac{1}{n} \log C_{eq} \quad (10)$$

where K_f and n represent the Freundlich constants, C_{eq} (mg L^{-1}) is the equilibrium concentration of 5-FU, and q_{eq} (mg g^{-1}) is the amount of 5-FU adsorbed at the equilibrium concentration. Table 1 summarizes the values of K_f, $1/n$, and R^2 (correlation coefficients). The R^2 values obtained for MIP$_{EGDMA}$, NIP$_{EGDMA}$, and NIP$_{TRIM}$ are higher than 0.97, suggesting that the experimental data can be fitted using the Freundlich adsorption model. On the other hand, the R^2 values obtained for MIP$_{TRIM}$ are lower than 0.94, indicating that the Freundlich adsorption model should not be used to characterize the experimental results. The $1/n$ value calculated from the Freundlich adsorption model is considered a measure of adsorption intensity or surface heterogeneity. Basically, the closer the $1/n$ value to zero, the more heterogeneous the surface [49]. Moreover, if the value of $1/n$ is higher than one, then adsorption is considered cooperative [50]. The $1/n$ values obtained for all microparticles are similar and are in the 0.60–0.73 range. This result indicates that all materials' heterogeneity is similar and the adsorption of 5-FU on them follows a similar mechanism.

3.2. Adsorption Kinetics

Adsorption kinetics can be established by finding a relationship between the adsorption capacity of examined materials and the contact time with the adsorbate solution. For all synthesized microparticles, the plots of q_t versus t were obtained and are shown in Figure S6. The experimental data were fitted using two adsorption kinetics models, the pseudo-first-order model given by Langergren and Svenska and the pseudo-second-order model based on the equilibrium adsorption. The following equation represents the pseudo-first-order model:

$$\log(q_e - q_t) = \log q_e - \frac{k_1}{2.303}t \quad (11)$$

where k_1 (h^{-1}) is the pseudo-first-order rate constant, q_e (mg g^{-1}) is the amount of 5-FU adsorbed at the equilibrium concentration, and q_t (mg g^{-1}) is the amount of 5-FU adsorbed at time t (h). The k_1 and R^2 values are presented in Table 2. The calculated R^2 values range between 0.979 and 0.993, which means that this model could be applied to characterize the kinetics of the adsorption of 5-FU on synthesized microparticles. However, after comparing these results with much higher R^2 values calculated for the pseudo-second-order kinetic model, it becomes clear that the pseudo-first-order kinetic model only partially fits the experimental data obtained for all microparticles.

Table 2. Kinetic parameters calculated for pseudo-first-order and pseudo-second-order models.

Polymer	Pseudo-First-Order Kinetic Model		Pseudo-Second-Order Kinetic Model	
	k_1 (h^{-1})	R^2	k_2 (g mg^{-1} h^{-1})	R^2
MIP$_{EGDMA}$	3.46 ± 0.31	0.987	1.03 ± 0.09	0.998
NIP$_{EGDMA}$	1.50 ± 0.14	0.979	1.00 ± 0.09	0.997
MIP$_{TRIM}$	2.47 ± 0.22	0.993	0.52 ± 0.05	0.999
NIP$_{TRIM}$	1.74 ± 0.15	0.980	1.73 ± 0.14	0.994

The following equation represents the pseudo-second-order kinetic model:

$$\frac{t}{q_t} = \frac{1}{k_2 q_e^2} + \frac{1}{q_e}t \quad (12)$$

where k_2 (g mg^{-1} h^{-1}) is the pseudo-second-order rate constant. The k_2 and R^2 values are presented in Table 2. The R^2 values calculated using this model are much higher than those obtained for the pseudo-second-order kinetic model and are in the 0.994–0.999 range. This result indicates that the pseudo-second-order kinetic model should be used to characterize the adsorption of 5-FU by synthesized microparticles. For both MIPs and NIPs, the k_2 values are quite similar and are within a level of 0.52–1.73, indicating that the kinetics of 5-FU adsorption for all materials is similar.

3.3. In Vitro Release Studies

Three buffer solutions of pH 2.2 (simulated gastric fluid), pH 5.0 (simulated tumor interstitium of tumor cells) [51,52], and pH 7.4 (simulated intravenous conditions) were used as release media for in vitro release studies of 5-FU from the structures of drug-loaded imprinted materials. The release profiles of 5-FU from these materials (Figures 3 and 4) clearly show entirely different properties of both MIPs depending on the cross-linker used. For each material and each examined pH value, a significant difference in the cumulative 5-FU release after statistical analysis has been noted ($p < 0.05$). For EGDMA-based MIPs (5-FU loading equal to 11.3 mg g^{-1}), the highest cumulative release (ca. 40%) was observed for pH 2.2, whereas a much lower release, equal to 20% and 12%, was observed for pH 5.0 and 7.4, respectively. This result clearly shows that the higher the pH, the lower the cumulative release. The observed behavior is probably caused by the higher protonation of carboxylic groups present in the polymer structure at lower pH values. This process disrupts interactions between carboxylic groups and drug molecules, leading to increased drug release. The release profiles obtained for the examined pH values are different depending on the buffer pH. For pH 5.0 and 7.4, a very high initial burst release is observed, which means that after around 10 h, the release almost entirely reaches the final cumulative release values. On the other hand, the initial burst release at pH 2.2 is much lower, and after that, a steady release is observed, which lasts around 50 h, after which a final cumulative release value is reached. For TRIM-based MIPs (drug loading 24.2 mg g^{-1}), a completely different behavior is observed regarding the dependence of pH of the release medium on the cumulative 5-FU release. The highest cumulative release (ca. 86%) was observed for pH 7.4, whereas for pH 5.0 and 2.2, these values were lower, equal to 70% and 53%, respectively. This behavior, almost opposite to the results obtained for MIP$_{EGDMA}$, is presumably caused by the much higher hydrophobicity of the TRIM-based polymer network than the one observed for EGDMA-based polymers. As 5-FU is sparingly soluble in water, it possesses a relatively high affinity towards the hydrophilic environment, and therefore the hydrophobic TRIM-based polymer network does not form strong interactions with 5-FU drug molecules. As a result, at pH 7.4, the drug release from MIP$_{TRIM}$ has the highest value because, at this pH value, the polymer network's hydrophobicity is the highest. When the pH is lowered, the polymer's structure becomes protonated, increasing its hydrophilicity, leading to a lower 5-FU release from its structure. Moreover, a rapid release at pH 2.2 can be observed, which is probably connected with breaking interactions occurring between MIPs cavities and 5-FU. As a result, during release, the amount of 5-FU maintained in the MIP$_{TRIM}$ structure is physisorbed primarily on its structure due to the polymer's increased hydrophilicity. The higher hydrophobicity of the TRIM-based polymer network than the EGDMA-based network also explains the much higher cumulative percentage release of 5-FU for MIP$_{TRIM}$ than the one observed for MIP$_{EGDMA}$.

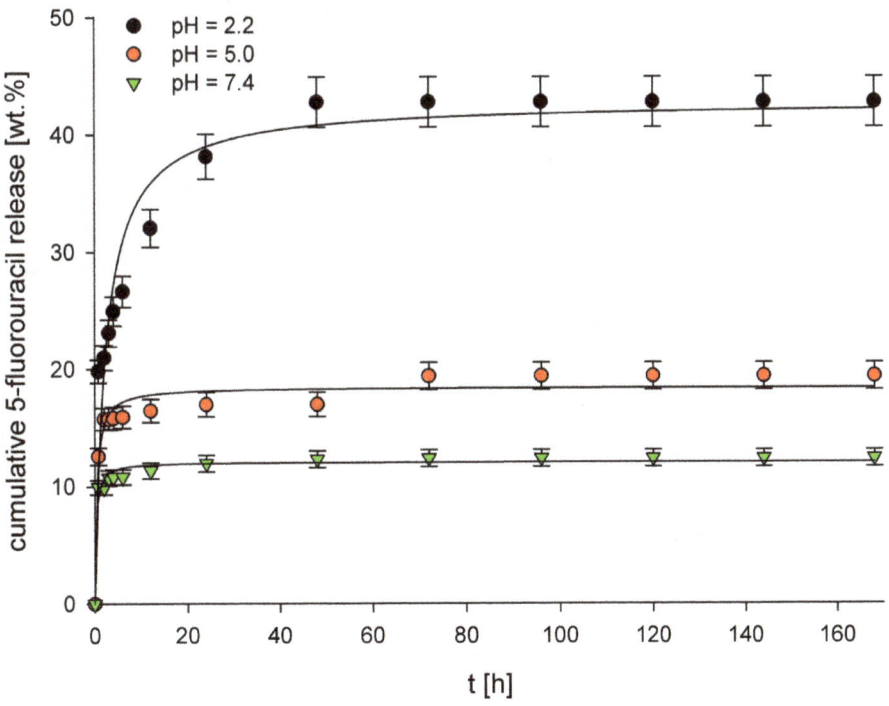

Figure 3. Release profiles of 5-FU from MIP$_{EGDMA}$ at pH 2.2, 5.0, and 7.4.

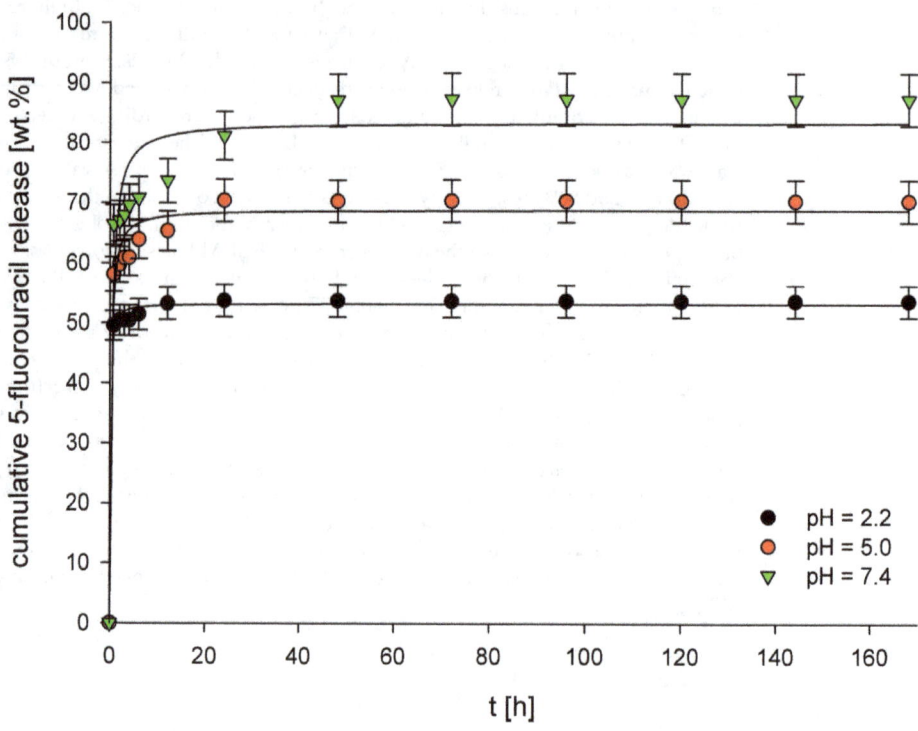

Figure 4. Release profiles of 5-FU from MIP$_{TRIM}$ at pH 2.2, 5.0, and 7.4.

Several mathematical models were used to fit the experimental data obtained from the 5-FU release from MIP$_{EGDMA}$ and MIP$_{TRIM}$. The release profiles were fitted with zero-order, first-order, Higuchi, Hixson–Crowell, and Korsmeyer–Peppas release models. The values of corresponding release constants (k), correlation coefficients (R^2), and the diffusion exponents (n) are presented in Table 3. For almost all pH values and both imprinted materials, the highest R^2 values were calculated for the Korsmeyer–Peppas release model, which indicates that the experimental data is best fitted with this model. The only exception is the release of 5-FU from MIP$_{TRIM}$ at pH 5.0, for which a higher R^2 value was obtained for the Higuchi model ($R^2 = 0.977$) compared to the Korsmeyer–Peppas model ($R^2 = 0.956$). However, the differences in the R^2 values are not significant; therefore, it can be concluded that the experimental data is nicely fitted with both of these models. These results indicate that the release mechanism of 5-FU from both MIPs at all examined pH values is similar. Moreover, for all MIPs at all examined pH values, the diffusion exponent value in the Korsmeyer–Peppas model was lower than 0.45, indicating a Fickian diffusion-controlled release mechanism.

Table 3. Release kinetic data of 5-FU from MIP$_{EGDMA}$ and MIP$_{TRIM}$.

Polymer	pH	Zero-Order		First-Order		Higuchi		Hixson–Crowell		Korsmeyer–Peppas		
		k_0 (h^{-1})	R^2	k_1 (h^{-1})	R^2	k_H (h$^{-1/2}$)	R^2	k_{HC} (h$^{-1/3}$)	R^2	k_{KP} (h^{-n})	n	R^2
MIP$_{EGDMA}$	2.2	0.106 ± 0.093	0.621	1.63 × 10^{-3} ± 0.14 × 10^{-3}	0.634	1.73 ± 0.15	0.791	2.19 × 10^{-3} ± 0.20 × 10^{-3}	0.630	20.7 ± 1.8	0.16 ± 0.2	0.900
	5.0	0.110 ± 0.009	0.405	1.30 × 10^{-3} ± 0.11 × 10^{-3}	0.413	0.77 ± 0.07	0.555	1.91 × 10^{-3} ± 0.17 × 10^{-3}	0.410	13.9 ± 1.2	0.07 ± 0.01	0.726
	7.4	0.047 ± 0.004	0.784	0.53 × 10^{-3} ± 0.05 × 10^{-3}	0.787	0.40 ± 0.04	0.910	0.79 × 10^{-3} ± 0.07 × 10^{-3}	0.786	9.9 ± 0.9	0.06 ± 0.01	0.944
MIP$_{TRIM}$	2.2	0.085 ± 0.008	0.670	1.77 × 10^{-3} ± 0.14 × 10^{-3}	0.674	0.70 ± 0.07	0.803	1.93 × 10^{-3} ± 0.16 × 10^{-3}	0.661	49.3 ± 4.7	0.03 ± 0.01	0.896
	5.0	0.494 ± 0.42	0.944	6.82 × 10^{-3} ± 0.61 × 10^{-3}	0.783	3.01 ± 0.28	0.977	1.46 × 10^{-2} ± 0.11 × 10^{-2}	0.966	56.5 ± 5.2	0.06 ± 0.01	0.956
	7.4	0.130 ± 0.011	0.705	6.40 × 10^{-3} ± 0.59 × 10^{-3}	0.737	1.96 ± 0.16	0.867	5.78 × 10^{-3} ± 0.53 × 10^{-3}	0.728	64.8 ± 6.2	0.06 ± 0.01	0.947

3.4. In Vitro Cell Viability

Cytotoxicity against U87 MG, HeLa, KB, A-549, and MCF-7 cancer cell lines, and HDF normal cell lines, was examined for drug-loaded and drug-unloaded MIPs and pure 5-FU. Table 4 summarizes the IC$_{50}$ values. Both drug-loaded MIPs present very high cytotoxicity against cancer cell lines, contrary to the result obtained for drug-unloaded MIPs. The results clearly show that drug-loaded MIP$_{TRIM}$ shows almost identical cytotoxicity against HeLa cancer cell lines as pure 5-FU. In contrast, its cytotoxicity against normal HDF cell lines is nearly three times lower than that measured for 5-FU. As a result, it can be concluded that MIP$_{TRIM}$ can be a good candidate as a DDS for 5-FU because it allows maintaining high cytotoxicity against HeLa cell lines while lowering the cytotoxicity against healthy cells.

Table 4. The IC$_{50}$ values (µg mL^{-1}) obtained for drug-loaded and unloaded MIPs and pure 5-FU. Standard deviations are presented in brackets.

Material	U87 MG	HeLa	KB	A-549	MCF-7	HDF
unloaded MIP$_{EGDMA}$	62.13 (0.11)	67.02 (0.49)	67.39 (0.07)	71.08 (0.73)	63.35 (0.19)	98.19 (0.51)
unloaded MIP$_{TRIM}$	81.12 (0.04)	69.02 (0.31)	81.01 (1.04)	77.82 (0.59)	81.93 (0.94)	102.02 (0.87)
loaded MIP$_{EGDMA}$	0.38 (0.07)	1.02 (0.11)	0.41 (0.16)	0.31 (0.91)	0.19 (0.01)	0.44 (0.49)
loaded MIP$_{TRIM}$	0.50 (0.04)	0.16 (0.01)	0.21 (0.17)	0.26 (0.19)	0.22 (0.01)	1.94 (0.55)
5-FU	0.07 (0.01)	0.16 (0.13)	0.03 (0.01)	0.08 (0.03)	0.08 (0.07)	0.71 (0.05)

4. Conclusions

We have shown that microparticles imprinted with 5-FU can be applied for the prolonged release of this drug. The microparticles synthesized using the TRIM cross-linker showed higher adsorption properties towards 5-FU than those synthesized using EGDMA. The experiments proved that prolonged drug release could last up to 50 h when an EGDMA cross-linker is used. The calculated highest cumulative release was highly dependent on the cross-linker applied during the synthesis. For EGDMA-based MIPs, the highest cumulative release was observed at pH 2.2, lower at pH 5.0, and the lowest at pH 7.4. Opposite results were obtained for TRIM-based MIPs, as the highest cumulative release was obtained for pH 7.4 and the lowest for pH 2.2. Moreover, the overall cumulative release was much higher for TRIM-based than EGDMA-based MIPs. The 5-FU release from all examined materials at all pH values was well fitted with the Korsmeyer–Peppas model. The IC$_{50}$ values proved that drug-loaded MIP$_{TRIM}$ possesses high cytotoxic activity against cancer cell lines, while lowering the toxicity towards normal HDF cell lines. Therefore, it has been proven that the selection of a cross-linker has a significant impact on the final properties of microparticles and has to be considered during the design of materials used for drug delivery.

Supplementary Materials: The following supporting information can be downloaded at: https://www.mdpi.com/article/10.3390/polym14051027/s1, Figure S1: Chemical structures of 5- fluorouracil, functional monomers, and cross-linkers. Figure S2: FT-IR spectra of loaded MIP$_{EGDMA}$, unloaded MIP$_{EGDMA}$, and NIP$_{EGDMA}$. Figure S3: FT-IR spectra of loaded MIP$_{TRIM}$, unloaded

MIP$_{TRIM}$, and NIP$_{TRIM}$. Figure S4: Results of thermogravimetric analysis for NIP$_{EGDMA}$, unloaded MIP$_{EGDMA}$, and loaded MIP$_{EGDMA}$. Figure S5: Results of thermogravimetric analysis for NIP$_{TRIM}$, unloaded MIP$_{TRIM}$, and loaded MIP$_{TRIM}$. Figure S6: The relationship between q$_t$ and t obtained for MIP$_{EGDMA}$, NIP$_{EGDMA}$, MIP$_{TRIM}$, and NIP$_{TRIM}$ during adsorption kinetics experiments.

Author Contributions: Conceptualization, M.C.; materials' synthesis and characterization, M.C. and A.L.; adsorption experiments, M.C. and T.N.; release experiments, J.K.; interpretation of the adsorption and release data, M.C. and J.K.; cytotoxicity assays and data interpretation, P.R.; writing—original draft preparation, M.C.; writing—review and editing, M.C.; supervision, M.C.; funding acquisition, M.C. All authors have read and agreed to the published version of the manuscript.

Funding: This work was supported by the National Science Centre, Poland, under grant number 2020/37/B/ST5/01938.

Institutional Review Board Statement: Not applicable.

Informed Consent Statement: Not applicable.

Data Availability Statement: The additional data that support the findings of this study are available from the corresponding author upon request.

Conflicts of Interest: The authors declare no conflict of interest. The funders had no role in the design of the study; in the collection, analyses, or interpretation of data; in the writing of the manuscript, or in the decision to publish the results.

References

1. Allen, T.M.; Cullis, P.R. Drug Delivery Systems: Entering the Mainstream. *Science* **2004**, *303*, 1818–1822. [CrossRef]
2. Mura, S.; Nicolas, J.; Couvreur, P. Stimuli-responsive nanocarriers for drug delivery. *Nat. Mater.* **2013**, *12*, 991–1003. [CrossRef] [PubMed]
3. Kataoka, K.; Harada, A.; Nagasaki, Y. Block copolymer micelles for drug delivery: Design, characterization and biological significance. *Adv. Drug Deliv. Rev.* **2001**, *47*, 113–131. [CrossRef]
4. Abdou, E.M.; Fayed, M.A.A.; Helal, D.; Ahmed, K.A. Assessment of the hepatoprotective effect of developed lipid-polymer hybrid nanoparticles (LPHNPs) encapsulating naturally extracted β-Sitosterol against CCl$_4$ induced hepatotoxicity in rats. *Sci. Rep.* **2019**, *9*, 19779. [CrossRef] [PubMed]
5. Iachetta, G.; Falanga, A.; Molino, Y.; Masse, M.; Jabès, F.; Mechioukhi, Y.; Laforgia, V.; Khrestchatisky, M.; Galdiero, S.; Valiante, S. gH625-liposomes as tool for pituitary adenylate cyclase-activating polypeptide brain delivery. *Sci. Rep.* **2019**, *9*, 9183. [CrossRef]
6. Wei, Y.; Huang, Y.-H.; Cheng, K.-C.; Song, Y.-L. Investigations of the Influences of Processing Conditions on the Properties of Spray Dried Chitosan-Tripolyphosphate Particles loaded with Theophylline. *Sci. Rep.* **2020**, *10*, 1155. [CrossRef]
7. Babu, A.; Amreddy, N.; Muralidharan, R.; Pathuri, G.; Gali, H.; Chen, A.; Zhao, Y.D.; Munshi, A.; Ramesh, R. Chemodrug delivery using integrin-targeted PLGA-Chitosan nanoparticle for lung cancer therapy. *Sci. Rep.* **2017**, *7*, 14674. [CrossRef]
8. Zhang, D.-X.; Yoshikawa, C.; Welch, N.G.; Pasic, P.; Thissen, H.; Voelcker, N.H. Spatially Controlled Surface Modification of Porous Silicon for Sustained Drug Delivery Applications. *Sci. Rep.* **2019**, *9*, 1367. [CrossRef]
9. Cai, D.; Liu, L.; Han, C.; Ma, X.; Qian, J.; Zhou, J.; Zhu, W. Cancer cell membrane-coated mesoporous silica loaded with superparamagnetic ferroferric oxide and Paclitaxel for the combination of Chemo/Magnetocaloric therapy on MDA-MB-231 cells. *Sci. Rep.* **2019**, *9*, 14475. [CrossRef]
10. Amolegbe, S.A.; Hirano, Y.; Adebayo, J.O.; Ademowo, O.G.; Balogun, E.A.; Obaleye, J.A.; Krettli, A.U.; Yu, C.; Hayami, S. Mesoporous silica nanocarriers encapsulated antimalarials with high therapeutic performance. *Sci. Rep.* **2018**, *8*, 3078. [CrossRef]
11. Cheng, C.; Gao, Y.; Song, W.; Zhao, Q.; Zhang, H.; Zhang, H. Halloysite nanotube-based H2O2-responsive drug delivery system with a turn on effect on fluorescence for real-time monitoring. *Chem. Eng. J.* **2020**, *380*, 122474. [CrossRef]
12. Sharif, S.; Abbas, G.; Hanif, M.; Bernkop-Schnürch, A.; Jalil, A.; Yaqoob, M. Mucoadhesive micro-composites: Chitosan coated halloysite nanotubes for sustained drug delivery. *Colloids Surf. B* **2019**, *184*, 110527. [CrossRef]
13. Nagiah, N.; Murdock, C.J.; Bhattacharjee, M.; Nair, L.; Laurencin, C.T. Development of Tripolymeric Triaxial Electrospun Fibrous Matrices for Dual Drug Delivery Applications. *Sci. Rep.* **2020**, *10*, 609. [CrossRef]
14. Fortuni, B.; Inose, T.; Ricci, M.; Fujita, Y.; Van Zundert, I.; Masuhara, A.; Fron, E.; Mizuno, H.; Latterini, L.; Rocha, S.; et al. Polymeric Engineering of Nanoparticles for Highly Efficient Multifunctional Drug Delivery Systems. *Sci. Rep.* **2019**, *9*, 2666. [CrossRef]
15. Kharaghani, D.; Gitigard, P.; Ohtani, H.; Kim, K.O.; Ullah, S.; Saito, Y.; Khan, M.Q.; Kim, I.S. Design and characterization of dual drug delivery based on in-situ assembled PVA/PAN core-shell nanofibers for wound dressing application. *Sci. Rep.* **2019**, *9*, 12640. [CrossRef]
16. Gao, J.; Dutta, K.; Zhuang, J.; Thayumanavan, S. Cellular- and Subcellular-Targeted Delivery Using a Simple All-in-One Polymeric Nanoassembly. *Angew. Chem. Int. Ed.* **2020**, *59*, 23466–23470. [CrossRef]

17. Li, Y.; Maciel, D.; Rodrigues, J.; Shi, X.; Tomás, H. Biodegradable Polymer Nanogels for Drug/Nucleic Acid Delivery. *Chem. Rev.* **2015**, *115*, 8564–8608. [CrossRef]
18. Gao, J.; Wu, P.; Fernandez, A.; Zhuang, J.; Thayumanavan, S. Cellular AND Gates: Synergistic Recognition to Boost Selective Uptake of Polymeric Nanoassemblies. *Angew. Chem. Int. Ed.* **2020**, *59*, 10456–10460. [CrossRef]
19. Molina, M.; Asadian-Birjand, M.; Balach, J.; Bergueiro, J.; Miceli, E.; Calderón, M. Stimuli-responsive nanogel composites and their application in nanomedicine. *Chem. Soc. Rev.* **2015**, *44*, 6161–6186. [CrossRef]
20. Kokil, G.R.; Veedu, R.N.; Le, B.T.; Ramm, G.A.; Parekh, H.S. Self-assembling asymmetric peptide-dendrimer micelles—a platform for effective and versatile in vitro nucleic acid delivery. *Sci. Rep.* **2018**, *8*, 4832. [CrossRef]
21. Zhang, H.; Ma, Y.; Xie, Y.; An, Y.; Huang, Y.; Zhu, Z.; Yang, C.J. A Controllable Aptamer-Based Self-Assembled DNA Dendrimer for High Affinity Targeting, Bioimaging and Drug Delivery. *Sci. Rep.* **2015**, *5*, 10099. [CrossRef]
22. Smoluch, M.; Cegłowski, M.; Kurczewska, J.; Babij, M.; Gotszalk, T.; Silberring, J.; Schroeder, G. Molecular Scavengers as Carriers of Analytes for Mass Spectrometry Identification. *Anal. Chem.* **2014**, *86*, 11226–11229. [CrossRef]
23. Narkiewicz, U.; Pełech, I.; Podsiadły, M.; Cegłowski, M.; Schroeder, G.; Kurczewska, J. Preparation and characterization of magnetic carbon nanomaterials bearing APTS–silica on their surface. *J. Mater. Sci.* **2010**, *45*, 1100–1106. [CrossRef]
24. Sellergren, B.; Allender, C.J. Molecularly imprinted polymers: A bridge to advanced drug delivery. *Adv. Drug Deliv. Rev.* **2005**, *57*, 1733–1741. [CrossRef]
25. Mokhtari, P.; Ghaedi, M. Water compatible molecularly imprinted polymer for controlled release of riboflavin as drug delivery system. *Eur. Polym. J.* **2019**, *118*, 614–618. [CrossRef]
26. Tuwahatu, C.A.; Yeung, C.C.; Lam, Y.W.; Roy, V.A.L. The molecularly imprinted polymer essentials: Curation of anticancer, ophthalmic, and projected gene therapy drug delivery systems. *J. Control. Release* **2018**, *287*, 24–34. [CrossRef]
27. Zhang, Z.; Cao, X.; Zhang, Z.; Yin, J.; Wang, D.; Xu, Y.; Zheng, W.; Li, X.; Zhang, Q.; Liu, L. Synthesis of dummy-template molecularly imprinted polymer adsorbents for solid phase extraction of aminoglycosides antibiotics from environmental water samples. *Talanta* **2020**, *208*, 120385. [CrossRef]
28. Tamayo, F.G.; Turiel, E.; Martín-Esteban, A. Molecularly imprinted polymers for solid-phase extraction and solid-phase microextraction: Recent developments and future trends. *J. Chromatogr. A* **2007**, *1152*, 32–40. [CrossRef]
29. Andersson, L.I. Molecular imprinting for drug bioanalysis: A review on the application of imprinted polymers to solid-phase extraction and binding assay. *J. Chromatogr. B* **2000**, *739*, 163–173. [CrossRef]
30. Whitcombe, M.J.; Chianella, I.; Larcombe, L.; Piletsky, S.A.; Noble, J.; Porter, R.; Horgan, A. The rational development of molecularly imprinted polymer-based sensors for protein detection. *Chem. Soc. Rev.* **2011**, *40*, 1547–1571. [CrossRef]
31. Kidakova, A.; Boroznjak, R.; Reut, J.; Öpik, A.; Saarma, M.; Syritski, V. Molecularly imprinted polymer-based SAW sensor for label-free detection of cerebral dopamine neurotrophic factor protein. *Sens. Actuat. B-Chem.* **2020**, *308*, 127708. [CrossRef]
32. Zhao, X.; He, Y.; Wang, Y.; Wang, S.; Wang, J. Hollow molecularly imprinted polymer based quartz crystal microbalance sensor for rapid detection of methimazole in food samples. *Food Chem.* **2020**, *309*, 125787. [CrossRef] [PubMed]
33. Hashemi-Moghaddam, H.; Kazemi-Bagsangani, S.; Jamili, M.; Zavareh, S. Evaluation of magnetic nanoparticles coated by 5-fluorouracil imprinted polymer for controlled drug delivery in mouse breast cancer model. *Int. J. Pharm.* **2016**, *497*, 228–238. [CrossRef] [PubMed]
34. Li, L.; Chen, L.; Zhang, H.; Yang, Y.; Liu, X.; Chen, Y. Temperature and magnetism bi-responsive molecularly imprinted polymers: Preparation, adsorption mechanism and properties as drug delivery system for sustained release of 5-fluorouracil. *Mater. Sci. Eng. C* **2016**, *61*, 158–168. [CrossRef]
35. Parisi, O.I.; Morelli, C.; Puoci, F.; Saturnino, C.; Caruso, A.; Sisci, D.; Trombino, G.E.; Picci, N.; Sinicropi, M.S. Magnetic molecularly imprinted polymers (MMIPs) for carbazole derivative release in targeted cancer therapy. *J. Mater. Chem. B* **2014**, *2*, 6619–6625. [CrossRef]
36. Norell, M.C.; Andersson, H.S.; Nicholls, I.A. Theophylline molecularly imprinted polymer dissociation kinetics: A novel sustained release drug dosage mechanism. *J. Mol. Recognit.* **1998**, *11*, 98–102. [CrossRef]
37. Cegłowski, M.; Kurczewska, J.; Ruszkowski, P.; Liberska, J.; Schroeder, G. The influence of cross-linking agent onto adsorption properties, release behavior and cytotoxicity of doxorubicin-imprinted microparticles. *Colloids Surf. B* **2019**, *182*, 110379. [CrossRef]
38. Cegłowski, M.; Kurczewska, J.; Ruszkowski, P.; Schroeder, G. Application of paclitaxel-imprinted microparticles obtained using two different cross-linkers for prolonged drug delivery. *Eur. Polym. J.* **2019**, *118*, 328–336. [CrossRef]
39. Jia, C.; Zhang, M.; Zhang, Y.; Ma, Z.-B.; Xiao, N.-N.; He, X.-W.; Li, W.-Y.; Zhang, Y.-K. Preparation of Dual-Template Epitope Imprinted Polymers for Targeted Fluorescence Imaging and Targeted Drug Delivery to Pancreatic Cancer BxPC-3 Cells. *ACS Appl. Mater. Interfaces* **2019**, *11*, 32431–32440. [CrossRef]
40. Hashemi-Moghaddam, H.; Zavareh, S.; Karimpour, S.; Madanchi, H. Evaluation of molecularly imprinted polymer based on HER2 epitope for targeted drug delivery in ovarian cancer mouse model. *React. Funct. Polym.* **2017**, *121*, 82–90. [CrossRef]
41. Marcelo, G.; Ferreira, I.C.; Viveiros, R.; Casimiro, T. Development of itaconic acid-based molecular imprinted polymers using supercritical fluid technology for pH-triggered drug delivery. *Int. J. Pharm.* **2018**, *542*, 125–131. [CrossRef]
42. Arias, J.L. Novel strategies to improve the anticancer action of 5-fluorouracil by using drug delivery systems. *Molecules* **2008**, *13*, 2340–2369. [CrossRef]
43. Cegłowski, M.; Jerca, V.V.; Jerca, F.A.; Hoogenboom, R. Reduction-Responsive Molecularly Imprinted Poly(2-isopropenyl-2-oxazoline) for Controlled Release of Anticancer Agents. *Pharmaceutics* **2020**, *12*, 506. [CrossRef]

44. Lengyel, M.; Kállai-Szabó, N.; Antal, V.; Laki, A.J.; Antal, I. Microparticles, Microspheres, and Microcapsules for Advanced Drug Delivery. *Sci. Pharm.* **2019**, *87*, 20. [CrossRef]
45. Bettencourt, A.; Almeida, A.J. Poly(methyl methacrylate) particulate carriers in drug delivery. *J. Microencapsul.* **2012**, *29*, 353–367. [CrossRef]
46. Farzaneh, S.; Asadi, E.; Abdouss, M.; Barghi-Lish, A.; Azodi-Deilami, S.; Khonakdar, H.A.; Gharghabi, M. Molecularly imprinted polymer nanoparticles for olanzapine recognition: Application for solid phase extraction and sustained release. *RSC Adv.* **2015**, *5*, 9154–9166. [CrossRef]
47. Javanbakht, M.; Attaran, A.M.; Namjumanesh, M.H.; Esfandyari-Manesh, M.; Akbari-adergani, B. Solid-phase extraction of tramadol from plasma and urine samples using a novel water-compatible molecularly imprinted polymer. *J. Chromatogr. B* **2010** *878*, 1700–1706. [CrossRef]
48. Gupta, A.; Tiwari, G.; Tiwari, R.; Srivastava, R.; Rai, A.K. Enteric coated HPMC capsules plugged with 5-FU loaded microsponges A potential approach for treatment of colon cancer. *Braz. J. Pharm. Sci.* **2015**, *51*, 591–606. [CrossRef]
49. Haghseresht, F.; Lu, G.Q. Adsorption Characteristics of Phenolic Compounds onto Coal-Reject-Derived Adsorbents. *Energy Fuels* **1998**, *12*, 1100–1107. [CrossRef]
50. Fytianos, K.; Voudrias, E.; Kokkalis, E. Sorption–desorption behaviour of 2,4-dichlorophenol by marine sediments. *Chemosphere* **2000**, *40*, 3–6. [CrossRef]
51. Dand, N.; Patel, P.; Ayre, A.; Kadam, V. Polymeric micelles as a drug carrier for tumor targeting. *Chron. Young Sci.* **2013**, *4*, 94–101. [CrossRef]
52. Bai, J.; Zhang, Y.; Chen, L.; Yan, H.; Zhang, C.; Liu, L.; Xu, X. Synthesis and characterization of paclitaxel-imprinted microparticles for controlled release of an anticancer drug. *Mater. Sci. Eng. C* **2018**, *92*, 338–348. [CrossRef]

Article

Application of the Remote Interaction Effect and Molecular Imprinting in Sorption of Target Ions of Rare Earth Metals

Talkybek Jumadilov [1], Ruslan Kondaurov [1,2,*] and Aldan Imangazy [1]

1. Laboratory of Synthesis and Physicochemistry of Polymers, JSC "Institute of Chemical Sciences after A.B. Bekturov", Sh. Valikhanov St. 106, Almaty 050010, Kazakhstan; jumadilov_kz@mail.ru (T.J.); imangazy.aldan@mail.ru (A.I.)
2. Department of Chemistry and Technology of Organic Substances, Natural Compounds and Polymers, Al-Farabi Kazakh National University, Al-Farabi Ave. 71, Almaty 050040, Kazakhstan
* Correspondence: r-kondaurov@mail.ru

Abstract: The goal of the present work is a comparative study of the effectiveness of the application of intergel systems and molecularly imprinted polymers for the selective sorption and separation of neodymium and scandium ions. The following physico-chemical methods of analysis were used in this study: colorimetry and atomic-emission spectroscopy. The functional polymers of polyacrylic acid (hPAA) and poly-4-vinylpyridine (hP4VP) in the intergel system undergo significant changes in the initial sorption properties. The remote interaction of the polymers in the intergel system hPAA–hP4VP provides mutual activation of these macromolecules, with subsequent transfer into a highly ionized state. The maximum sorption of neodymium and scandium ions is observed at molar ratios of 83%hPAA:17%hP4VP and 50%hPAA:50%hP4VP. Molecularly imprinted polymers MIP(Nd) and MIP(Sc) show good results in the sorption of Nd and Sc ions. Based on both these types of these macromolecular structures, principally new sorption methods have been developed. The method based on the application of the intergel system is cheaper and easier in application, but there is some accompanying sorption (about 10%) of another metal from the model solution during selective sorption and separation. Another method, based on the application of molecularly imprinted polymers, is more expensive and the sorption properties are higher, with the simultaneous sorption of the accompanying metal from the model solution.

Keywords: rare earth metals; sorption; separation; remote interaction of macromolecules; molecular imprinting

1. Introduction

Nowadays, one of the elements highly in demand for industrial use is rare earth metals (REMs). REMs can be named as a critical component in almost all important technologies, which, in turn, drive the modern industrial development across the world. REMs are represented by 15 elements of the lanthanoid group (from La to Lu) plus two elements: Sc and Y; these metals can be considered as a unique row of elements that are used in many areas: lanthanum (battery alloys, metal alloys, auto catalysts, petroleum refining, polishing powders, glass additives, phosphors, ceramics, and optics); cerium (battery alloys, metal alloys, auto catalysts (emissions control), petroleum refining, polishing powders, glass additives, phosphors, and ceramics); praseodymium (battery alloys, metal alloys, auto catalysts, polishing powders, glass additives, and coloring ceramics); neodymium (permanent magnets, battery alloys, metal alloys, auto catalysts, glass additives, and ceramics); promethium (watches, pacemakers, and research; promethium is a radioactive metal, which has no stable isotopes; it is present in the crust of Earth in low quantities); samarium (magnets, ceramics, and medical radiation treatment (cancer diseases)); europium (phosphors); gadolinium (ceramics, nuclear energy, and medicine (magnetic resonance imaging and X-rays)); terbium (fluorescent lamp phosphors and magnets (especially for high temperatures and defense); dysprosium (permanent magnets); holmium (permanent magnets, nuclear energy, and microwave equipment); erbium (nuclear energy, fiber optic communications, and glass coloring); thulium (X-rays (medical) and lasers); ytterbium (cancer treatment and stainless steel); lutetium (age determination and petroleum refining); yttrium (battery alloys, phosphors, and ceramics); and scandium (high-strength, low-weight aluminum–scandium alloys). The increased importance of REMs over the last 100 years is due to their unique properties, which have no existing analogues. This fact can be seen from the rate of annual growth of 13.7% (expected growth between 2017 and 2021) for the global REM market [1–5].

REMs not only replace each other in the structure of the minerals but also occur within different minerals' structures in the same deposit [6]. As mentioned above, most of the REMs are lanthanoids and their chemical properties are similar. These metals occur together within minerals in varying quantities [7]. The similarity of chemical properties of the REMs creates serious difficulties in their separation after extraction from the minerals where they are found [8,9]. The process of separation of one targeted REM from the amount is difficult, environmentally challenging, and expensive, wherein over 1% of the REM is recycled owing to the many challenges of collecting various end products and separating the REM from other metals/contaminants [10,11]. The main focus for investments in recycling is applications of REMs, such as in magnets, in which the economies of scale allow it. The REM market is relatively small and can be easily disrupted. The main factors that can impact the market are the increase of REM production from existing mines, development of mine prospects advanced during price spikes, research and development efforts focused on improving REM recoveries, recycling, substitution, alternate sources of REMs, and governmental policies [12–14].

Hydrometallurgical solutions in various branches of industry have complex chemical composition, which is the limiting stage for the efficient sorption of the target REM ions by ion-exchange resins [15–18]. Existing ion exchangers have selectivity to a certain metal ion, the first problem showing that the sorption of each ion of REM requires a certain ion-exchange resin. Another drawback of this type of macromolecular structure is its regeneration process, which assumes continuous washing of the ion exchangers, first with acids and then with distilled water for renewing the exchange capacity to initial values after each cycle of sorption of the target metal ions [19–23].

One of the proposed analogues to existing sorption technologies is using the remote interaction effect for the selective sorption of targeted REM ions [24,25]. Functional polymers in intergel systems undergo mutual activation, with further transfer into a highly ionized state, resulting in a significant increase in the initial sorption properties. The use of remote interaction of functional macromolecules has some advantages over the existing sorption methods:

(1) Each intergel system can be efficiently used for the selective sorption of several REM ions by varying polymer molar ratios in it.
(2) The mutually activated macromolecules can be used as independent sorbents.
(3) The high ionization degree of the components in intergel systems leads to significant growth in sorption properties of the initial macromolecules.

Another possible variant of selective sorption of the targeted REM ions can be implemented by using the molecular imprinting technique [26–31]. A molecular-imprinted polymer (MIP) is a polymer treated by using a special molecular imprinting technique, which results in the appearance of cavities in the polymer matrix with an affinity for the selected molecular "template" [32,33]. This process typically involves initiating the polymerization of the monomers in the presence of a template molecule, which is subsequently removed, thus leaving complementary cavities. Molecular imprinting is, in fact, an artificial tiny "lock" for a particular molecule, which serves as a miniature "key". Molecular imprinting is a fairly effective technique for incorporating specific pattern recognition of the analyzed object into polymers [34–37]. Molecular recognition characteristics of these polymers directly depend on the complementary size and shape of the binding objects, imparted to the polymers by template molecules [38–40]. The concept of complementarity includes the correspondence of an imprint to a template both in size and shape and in the presence of complementary functional groups in the imprint that are capable of interacting with the functional groups of the template molecule [41–44].

In this regard, the goal of the present work is a comparative study of selective sorption of targeted REM ions (on neodymium and scandium) by intergel systems and molecularly imprinted polymers.

2. Materials and Methods

2.1. Materials

Monomers: acrylic acid (AA), methacrylic acid (MAA), and 4-vinylpyridine (4VP) were purchased from "LaborPharma" Ltd. (Almaty, Kazakhstan); linear polymer: poly-4-vinylpyridine (P4VP); cross-linking agents: N,N'-methylenebis(acrylamide) (MBAA), epichlorohydrin (ECH), and ethyleneglycol dimethacrylate (EGDMA); initiator: azobisisobutyronitrile (AIBN); redox system: $K_2S_2O_8$–$Na_2S_2O_3$; solvent: dimethylformamide (DMFA); porogen: toluene; stabilizer: hydroxyethylcellulose (HEC). the rest of mentioned chemicals were purchased from Sigma-Aldrich company (Saint-Louis, MO, USA). Deionized water was used in all experiments (χ = 11 µS/cm; pH = 6.97) was obtained in the DV-1 deionizer (Technokom, Ekaterinburg, Russia).

The monomers of AA and MAA initially undergo vacuum distillation at the following conditions: temperature 75 °C (for AA) and 105 °C (for MAA); rate 2 drops per 4–5 s. The necessity of vacuum distillation is to purify the AA monomers from the inhibitor monomethyl ether of hydroquinone (MeHQ).

2.2. Methods

2.2.1. Synthesis of Polymer Hydrogels

Synthesis of Polyacrylic Acid Hydrogels

Polymerization of rare-cross-linked AA hydrogels was conducted in the following order: AA monomers were polymerized in the MBAA cross-linking agent in the presence of the redox system $K_2S_2O_8$–$Na_2S_2O_3$ in a reactor. The polymerization reaction occurred in the following order: 12 mL of the monomer was put into the volumetric flask (100 mL); after that, 2 drops of the cross-linking agent dissolved in water was added; after that, 1 mL of the initiator was added. After that, the polymerizate was poured into special ampoules (special cylinders without bottom and top) and placed in an oven for 5 h at a temperature of 45–50 °C. During the drying procedure, the ampoules were periodically taken and weighted to the constant weight on analytical scales (importantly, the ampoules were closed when being taken out of the oven to exclude the intake of moisture from the air). The swelling degree of the synthetized hydrogels of PAA was 30.33 g/g.

Synthesis of Poly-4-Vinylpyridine Hydrogels

Hydrogels of P4VP were obtained in the following order: initially, linear polymers of P4VP were dissolved in the medium of DMFA and cross-linked by an ECH cross-linking agent under permanent stirring. The polymerization reaction occurred in the following order: a weighed portion of a linear polymer (5 g) was filled with 20 mL of DMFA until the polymer sample was completely dissolved in the solvent; moreover, before the dissolution of the linear polymer, there was a stage of swelling of the polymer in the solvent. After that, 2.5 mL of ECH was added dropwise to the solution, with constant stirring, at a temperature of 60 °C. The swelling degree of the synthetized hydrogels of P4VP was 3.72 g/g.

After the synthesis procedure, the obtained rare-cross-linked hydrogels of PAA and P4VP were purified from unreacted products and soluble polymer fractions by long-term washing (for 14 days) under stationary conditions. The synthesized hydrogels were washed in wash columns. The water was changed 2–3 times a day. The control of the degree of purification was carried out by determining the specific electrical conductivity and pH of water after gel purification. After 14 days, the values of specific electrical conductivity and pH of the wash water remained constant. This indicated that the process of purification of the polymer hydrogels from unreacted products was over. Then the obtained samples of hydrogels were subjected to dispersion by grinding in an analytical mill. Subsequently, the samples of the hydrogels were filtered through a sieve by fractions. For subsequent experiments, samples of PAA and P4VP hydrogels were selected the granule sizes of which were higher than 120 µm and lower than 180 µm.

2.2.2. Preparation of the Intergel System

Previous studies have shown that the intergel system should at least contain two components [45,46]. In the present study, the acidic and basic components of the system were PAA and P4VP hydrogels, respectively. The obtained dispersions of PAA and P4VP hydrogels were put in special polypropylene cells with pores, the cells separated from each other by no more than 100 µm (one cell contained a dispersion of PAA and another a dispersion of P4VP). These cells were impermeable for the macromolecular dispersion but permeable for low-molecular ions. Subsequently, the cells were put in the common solution in a glass; the distance between them was 2 cm. The total amount of the dispersion was 100 mol.% either for the presence of individual polymers or for the presence of intergel pairs. Molar ratios of the polymer hydrogels in intergel systems were taken for convenience. The concentration of the PAA hydrogel decreased from 1.67 mmol/L to 0.28 mmol/L with an increase in the P4VP hydrogel share in the intergel system (molar ratios hPAA:hP4VP 100%:0%–17%:83%), while the concentration of hP4VP increased from 0.28 mmol/L to 1.67 mmol/L with a decrease in the hPAA share.

2.2.3. Synthesis of Molecularly Imprinted Polymers

Molecularly imprinted polymers (MIPs) were synthesized by the suspension polymerization technique. Neodymium and scandium nitrates were chosen as templates. MAA and 4VP were chosen as functional monomers, EGDMA was used as a cross-linking agent, AIBN was used as an initiator, HEC was chosen as stabilizer, and toluene was chosen as a porogen. Polymerization of the MIPs was carried out in deionized water. The composition of the reaction mixture was as follows: template ion:MAA:4VP:EGDMA = 1:2:2:8. The stirring speed was 250 rpm. The reaction was carried out for 15 min at room temperature, then for 6 h at 70 °C in a stream of nitrogen. After polymerization, the resulting MIP particles were thoroughly washed with deionized water and acetone to remove impurities and residues of unreacted monomers. The resulting granules were vacuum-dried for 24 h. To control the selectivity of the MIPs, control samples of cMIPs were synthesized, differing in that no metal template was added during their synthesis. To remove the template from the MIPs, 1 M nitric acid was used, with stirring for 1 h. To completely remove the metals, the washing cycle was repeated 30 times, after which the MIPs were washed with deionized water and dried in vacuum for 24 h.

2.2.4. Sorption Experiments

For the present study, model salt solutions were made: neodymium sulfate hydrate and scandium sulfate hydrate (the concentration was 100 mg/L for each REM salt solution). The macromolecular dispersion (individual polymer hydrogel, intergel system, or molecularly imprinted polymer) was put into the salt solution for 48 h; at specific time intervals (0.5, 1, 2, 6, 24, and 48 h after the beginning of sorption), the aliquots were taken for further analysis of the residual concentration of neodymium/scandium ions.

2.2.5. Laboratory Experiments on Selective Sorption and Separation of Nd and Sc Ions

Figure 1 shows the scheme of the selective sorption and separation of Nd and Sc ions using the developed unit and the hPAA–hP4VP intergel system. As seen, the scheme involves the following stages:

(1) A solution containing Nd and Sc ions is pumped into the first laboratory unit. The unit is filled with the intergel system 83%hPAA:17%hP4VP for the selective sorption of

neodymium ions. One cartridge of the unit contains a dispersion of PAA hydrogel and another a dispersion of P4VP hydrogel. Here, the solution is stored for 48 h. Aliquots are taken at a specific time for further measurement of the residual concentration of the metals' ions. This stage can be called "sorption of neodymium."

(2) At this stage, the solution is pumped into the second laboratory unit, which contains a 50%hPAA:50%hP4VP intergel system for the selective sorption of scandium ions for 48 h. The placement of the cartridges inside the unit is similar to that in the first unit (neodymium sorption); one cartridge contains the dispersion of PAA hydrogel and another a P4VP dispersion. Aliquots are taken at the same time for control of the REMs' residual concentration.

(3) After the sorption of Nd and Sc is fully complete, the cartridges are removed from the units. The new cartridges (four cartridges, two of which contain the intergel system 83%hPAA:17%hP4VP and two of which contain the intergel system 50%hPAA:50%hP4VP) are put into the units, and all is ready for a new cycle of the selective sorption of neodymium and scandium ions.

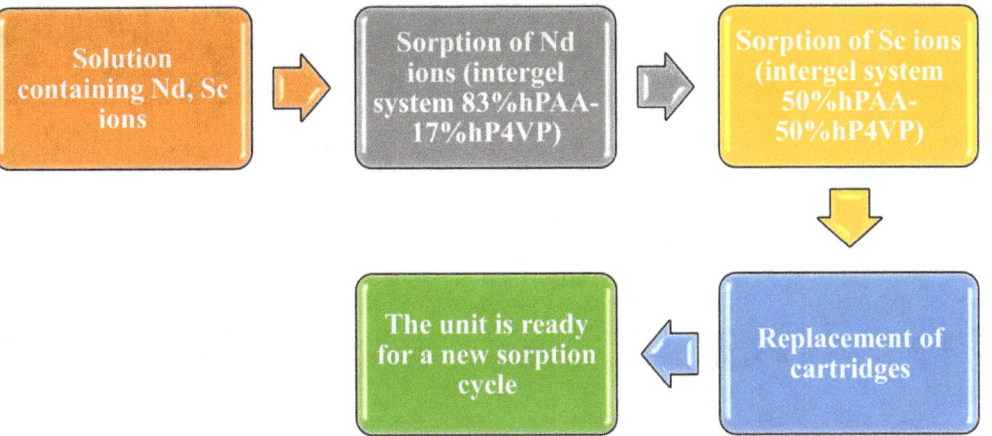

Figure 1. Scheme of Nd and Sc ion sorption and separation using the intergel system hPAA–hP4VP.

Figure 2 presents a scheme for the sequential selective extraction and separation of Nd and Sc ions using the developed laboratory unit and imprinted structures MIP1(Nd) and MIP1(Sc) as sorbents. This scheme involves two identical developed units placed one after another. The fill of the cartridges in this scheme is quite similar to the case when sorption in the unit is based on intergel systems (Figure 1), but there are some differences: each unit contains only one macromolecular structure: the first unit contains MIP(Nd), and the second unit contains MIP(Sc). From the figure, it can be seen that there are three main stages in the scheme:

(1) The solution containing Nd and Sc ions is pumped into the first unit; one cartridge of the unit is filled with imprinted structure MIP(Nd). In this unit, the solution is stored for 48 h. Aliquots are taken at a specific time for further measurement of the residual concentration of the metals' ions. The sorption of neodymium occurs at this stage.

(2) Further occurrence of the sorption and separation process proposes that the model solution is pumped to the second unit, one cartridge of which contains imprinted structure MIP(Sc). Aliquots are taken at the same time for control of the REMs' residual concentration.

(3) After the sorption process of Nd and Sc, the cartridges with MIP(Nd) and MIP(Sc) are removed from the units. Two new cartridges with MIP(Nd) and MIP(Sc) are put into the units, and all is ready for a new cycle of the selective sorption of Nd and Sc ions.

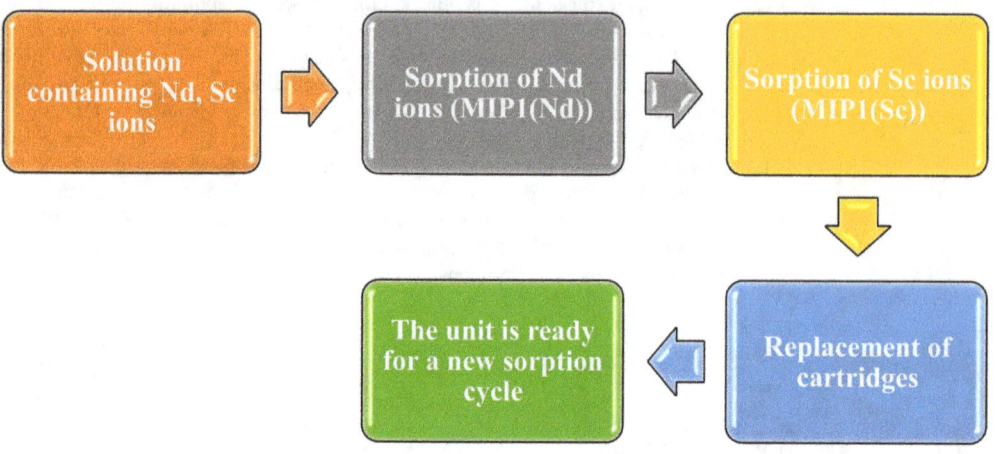

Figure 2. Scheme of Nd and Sc ion sorption and separation using the molecularly imprinted polymers MIP(Nd) and MIP(Sc).

2.2.6. Determination of Initial Electrochemical Properties

For determination of the electric conductivity, the conductometer Expert 002 (Econics-expert, Moscow, Russian, Federation) was used. Measurement of pH values was performed on pH-meter 780 Metrohm (Metrohm, Herizau, Switzerland). These electrochemical properties were controlled during purification of the synthetized polymers (polymer hydrogels, MIPs). Analytical scales Shimadzu AY120 (Shimadzu, Kyoto, Japan) were used for the measurement of the weight of the synthetized polymer hydrogels and further determination of the swelling degree.

2.2.7. Determination of the Residual Concentration of the REM's Ions

The photocolorimeter KFK-3KM (Unico Sys, Saint-Petersburg, Russian, Federation) was used for determination of the optical density of the REM's salt solutions. It should be noted that the concentration was determined on the ICP-OES spectrometer 8300 ICP-OES (Perkin Elmer, Waltham, MA, USA).

2.3. Calculation of Parameters

The swelling degree of the synthetized rare-cross-linked polymer hydrogels of PAA and P4VP was calculated in accordance with the equation

$$\alpha \frac{m_2 - m_1}{m_1} \tag{1}$$

where m_1 is the mass (g) of the dry polymer hydrogel and m_2 is the mass (g) of the swollen polymer hydrogel.

Based on the residual concentration after the sorption of Nd and Sc ions, the following sorption parameters were calculated:

(1) Sorption degrees of Nd^{3+} or Sc^{3+} ions:

$$\eta = \frac{C_0 - C_e}{C_0} \times 100\% \tag{2}$$

where C_0 is the initial concentration (mg/L) of the REM's ions and C_e is the initial equilibrium concentration (mg/L) of the REM's ions.

(2) Dynamic exchange capacity of the polymer structures:

$$Q = \frac{m_{sorbed}}{m_{sorbent}} \quad (3)$$

where m_{sorbed} is the mass (mg) of the sorbed REM's ions and m is the macromolecule's portion (g). If there are two macromolecules in the corresponding salt solution (presence of the intergel system), this value is determined as the sum of total weight of each macromolecule.

(3) Growth in the sorption parameters (sorption degree/dynamic exchange capacity):

$$\omega_i = \frac{P_i}{P_0} \times 100\% - 100\% \quad (4)$$

where P_i is the sorption parameter (sorption degree or dynamic exchange capacity) of the intergel system or MIPs at a specific time and P_0 is the sorption parameter (sorption degree or dynamic exchange capacity) of the PAA or P4VP hydrogel at the same time.

(4) Mean growth in the sorption parameters:

$$\varpi = \frac{\omega_\eta + \omega_Q}{2} \quad (5)$$

where ω_η is the growth in the sorption degree (%) at a specific time and ω_Q is the growth in the sorption capacity (%) at the same time.

3. Results and Discussion

The remote interaction effect leads to significant changes in the initial properties of the macromolecules in the intergel systems due to changes in the structures of the polymers, wherein direct contact between interacting polymers is absent. Remote interaction of rare-cross-linked polymer hydrogels is accompanied by the following reactions:

(1) Dissociation of -COOH groups of the PAA hydrogel (Figure 3):

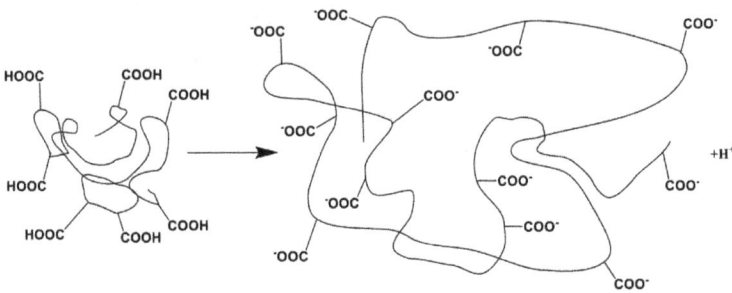

Figure 3. Dissociation of the polyacrylic acid hydrogel.

Initially, ionization occurs along with formation of ionic pairs; subsequently, the ionic pairs partially dissociate on separate charged particles.

(2) Ionization and partial dissociation of the heteroatom of the pyridine ring (nitrogen atom):

$$\equiv N + H_2O \rightarrow \equiv NH^+ \ldots OH^- \rightarrow \equiv NH^+ + OH^-$$

(3) Further interaction provides binding of protons cleaved from carboxyl groups by the heteroatoms of the pyridine ring (Figure 4):

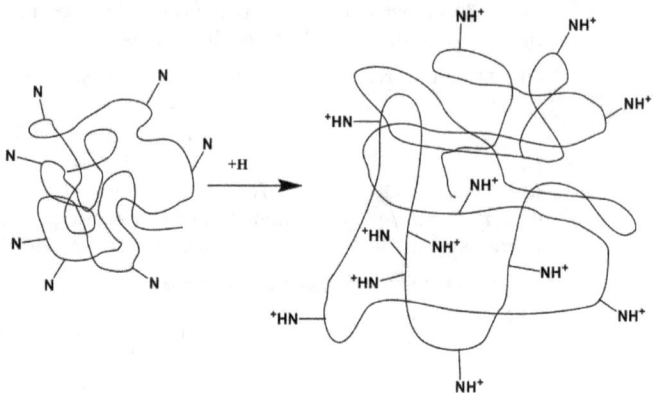

Figure 4. Association of protons by the heteroatoms of the P4VP hydrogel.

(4) Formation of water molecules due to the interaction of H^+ and OH^- ions (right for equimolar concentrations of protons and hydroxyl ions):

$$H^+ + OH^- \rightarrow H_2O$$

As can be seen from these reactions, the dissociation of carboxyl groups with the consequent binding of the cleaved protons leads to a decrease in the amount of H^+ ions in the solution. In turn, there is additional dissociation of other (undissociated) carboxyl groups (owing to Le-Chatelier principle). Such interactions lead to the formation of uncompensated charged groups on the internode links of each hydrogel, which undergo repulsion, leading to the unfolding of the polymer globe. Such transition into the ionized state of each initial macromolecule in the intergel system is called mutual activation. Stages of mutual activation of PAA and P4VP hydrogels are presented on Figure 5.

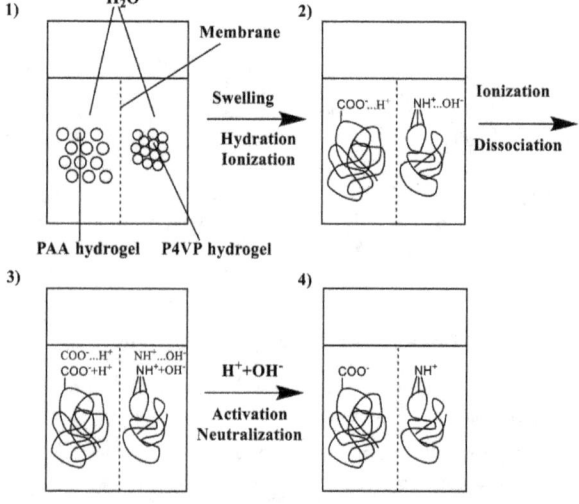

Figure 5. Stages of mutual activation of PAA and P4VP hydrogels.

The mutual activation results in significant changes in the initial electrochemical, conformational, and sorption properties.

3.1. Sorption of Nd^{3+} and Sc^{3+} Ions Based on Remote Interaction

Interaction of individual initial hydrogels of PAA and P4VP and the intergel system on their basis (hPAA–hP4VP) with neodymium and scandium nitrates provides sorption of the REMs.

Figure 6 presents the dependence of the sorption degrees of neodymium (a) and scandium (b) ions from molar ratios of PAA and P4VP hydrogels over time. The sorption of neodymium ions by individual polymers hPAA and hP4VP does not have a strict intense character. As seen from the figure, the sorption degree increases slightly (a sharp increase observed only during the first 6 h of interaction) with the time of the interaction of the macromolecules with the REM's salt solution. During the time intervals of 0.5, 1, 2, and 6 h, the increase in the sorption degree is 9.81%, 15.46%, 25.35%, and 37.18% for hPAA and 4.86%, 8.49%, 15.87%, and 27.58% for hP4VP, respectively. After that, it can be said that the studied polymer hydrogels, which interact with the salt solution, are close to the equilibrium state: at 24 h of interaction, the sorption degree is 56.27% for hPAA and 48.69% for hP4VP; at 48 h, it is 61.22% and 54.15%, respectively. Higher values of the sorption degree (comparatively with individual macromolecules) in the presence of the intergel system hPAA–hP4VP point to the high ionization of the initial rare-cross-linked polymer hydrogels PAA and P4VP in the intergel pairs. Strong sorption of neodymium ions by the intergel system hPAA–hP4VP occurs during 6 h of remote interaction at the following molar ratios: 83%hPAA:17%hP4VP and 50%hPAA:50%hP4VP; these ratios are areas of maximum sorption of neodymium ions, wherein the highest values of the sorption degree are observed at the 83%hPAA:17%hP4VP ratio (the sorption degree is 93.44%). At the end of the sorption time (48 h) also, high values of the sorption degree are observed at the ratio 67%hPAA:33%hP4VP (86.57%). The maximum amount of scandium ions is sorbed by the intergel system hPAA–hP4VP at ratios 50%hPAA:50%hP4VP and 33%hPAA:67%hP4VP, wherein an overwhelming majority of scandium (more than 70%) is sorbed after 6 h of remote interaction of PAA and P4VP hydrogels in these intergel pairs; it is more than half of all the sorbed scandium; at 6 h, the sorption degree is 76.57% and 73.64%, respectively. During the same time of interaction, individual hydrogels PAA and P4VP sorb about 40% of the scandium (the sorption degree is 39.30% for hPAA and 29.70% for hP4VP). The highest values of the sorption parameter at 48 h are observed at ratios 67%hPAA:33%hP4VP (89.50%), 50%hPAA:50%hP4VP (94.24%), and 33%hPAA:67%hP4VP (92.73%).

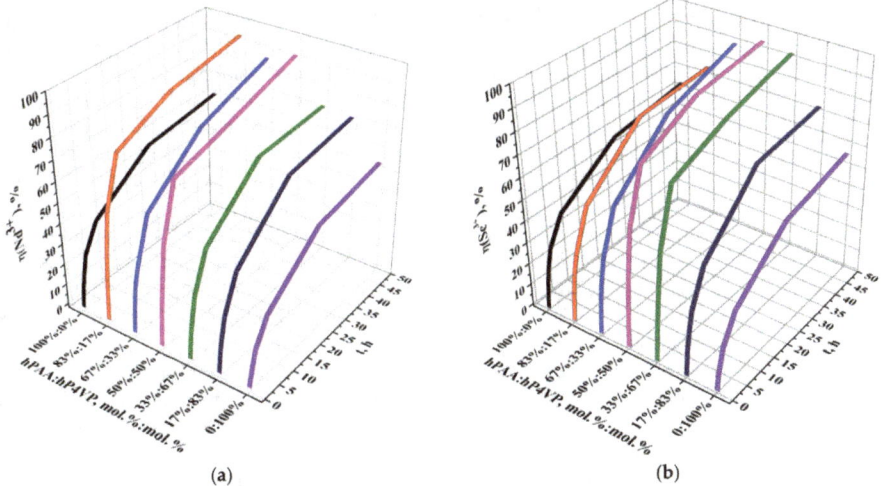

Figure 6. Sorption degrees of Nd^{3+} (a) and Sc^{3+} (b) by the intergel system hPAA–hP4VP.

The values of the sorption degrees of neodymium and scandium ions by the intergel system hPAA–hP4VP are presented in Tables 1 and 2, respectively.

Table 1. Sorption degree of Nd^{3+} ions by the intergel system hPAA–hP4VP.

t, h	$\eta(Nd^{3+})$, % hPAA:hP4VP, mol.%:mol.%						
	100%:0%	83%:17%	67%:33%	50%:50%	33%:67%	17%:83%	0%:100%
0	0	0	0	0	0	0	0
0.5	9.81	19.71	12.13	15.97	11.73	10.51	4.86
1	15.46	33.14	20.92	26.37	17.58	15.87	8.49
2	24.35	50.61	29.00	45.66	28.19	25.06	15.87
6	37.18	74.04	51.12	72.23	46.17	41.52	27.58
24	56.27	86.67	73.94	80.10	69.80	66.47	48.69
48	61.22	93.44	86.57	91.62	72.02	71.11	54.15

Table 2. Sorption degree of Sc^{3+} ions by the intergel system hPAA–hP4VP.

T, h	$\eta(Sc^{3+})$, % hPAA:hP4VP, mol.%:mol.%						
	100%:0%	83%:17%	67%:33%	50%:50%	33%:67%	17%:83%	0%:100%
0	0	0	0	0	0	0	0
0.5	12.33	12.63	14.35	21.52	17.89	11.93	7.28
1	18.09	20.92	22.63	34.96	29.10	18.49	10.31
2	26.57	29.40	33.14	52.93	47.78	27.99	17.79
6	39.30	48.29	53.14	76.57	73.64	43.94	29.70
24	57.98	72.02	77.48	89.60	84.34	69.50	49.60
48	63.34	74.65	89.50	94.24	92.73	73.23	56.47

The dynamic exchange capacity (in relation to Nd^{3+} (a) and Sc^{3+} (b) ions) of the intergel system hPAA–hP4VP is shown on Figure 7. The sorption of both metals is accompanied by a significant increase in the dynamic exchange capacity of the intergel system. Strong sorption of neodymium ions is observed at the molar ratios 83%hPAA:17%hP4VP and 50%hPAA:50%hP4VP during 6 h of remote interaction, while the highest sorption values for the scandium sorption process are observed at ratios 50%hPAA:50%hP4VP and 33%hPAA:67%hP4VP at the same time interval. During this interaction time, the dynamic exchange capacity (in relation to neodymium ions) increases in the following order: 0.5 h, 821.04 mg/g; 1 h, 1380.75 mg/g; 2 h, 2108.79 mg/g; and 6 h, 3085.13 mg/g for the ratio 83%hPAA:17%hP4VP; for the ratio 50%hPAA:50%hP4VP, the parameters are 665.33, 1098.79, 1902.58, and 3009.38 mg/g, respectively. The exchange capacity (in relation to scandium ions) increases as follows: 0.5 h, 896.67 mg/g; 1 h, 1456.67 mg/g; 2 h, 2205.42 mg/g; and 6 h, 3190.42 mg/g for the molar ratio 50%hPAA:50%hP4VP. The ratio 33%hPAA:67%hP4VP has lower values of exchange capacity: 497.08, 770.42, 1166.25, and 1830.83 mg/g, respectively. The individual polymers PAA and P4VP have the following values of capacity at the same time of interaction (0.5–6 h): in relation to Nd ions, for hPAA, 408.63, 644.29, 1014.63, and 1549.08 mg/g, and for hP4VP, 202.42, 353.92, 661.13, and 1149.29 mg/g; in relation to Sc ions, for hPAA, 513.75, 753.75, 1107.08, and 1637.50 mg/g, and for hP4VP, 303.33, 429.58, 741.25, and 1237.50 mg/g. Such strong differences in the values of the sorption parameter between the intergel system and individual hydrogels is due to the high ionization of the polymers in the intergel pairs. Further interaction of the intergel system with the corresponding neodymium and scandium salt solutions leads to the consequent sorption of these metals and increase in the exchange capacity. The highest values of the exchange capacity (in relation to Nd ions) are observed at 48 h at molar ratios 83%hPAA:17%hP4VP (3893.13 mg/g) and 50%hPAA:50%hP4VP (3817.38 mg/g). The maximum values of the dynamic exchange capacity (in relation to Sc ions) at 48 h of remote interaction at molar ratios 50%hPAA:50%hP4VP and 33%hPAA:67%hP4VP are 3926.67 and 3863.75 mg/g, respectively. The sorption parameters are 2550.67 mg/g for hPAA and 2256.08 mg/g for hP4VP at Nd sorption and 2639.17 mg/g for hPAA and 2352.92 mg/g for hP4VP at Sc sorption.

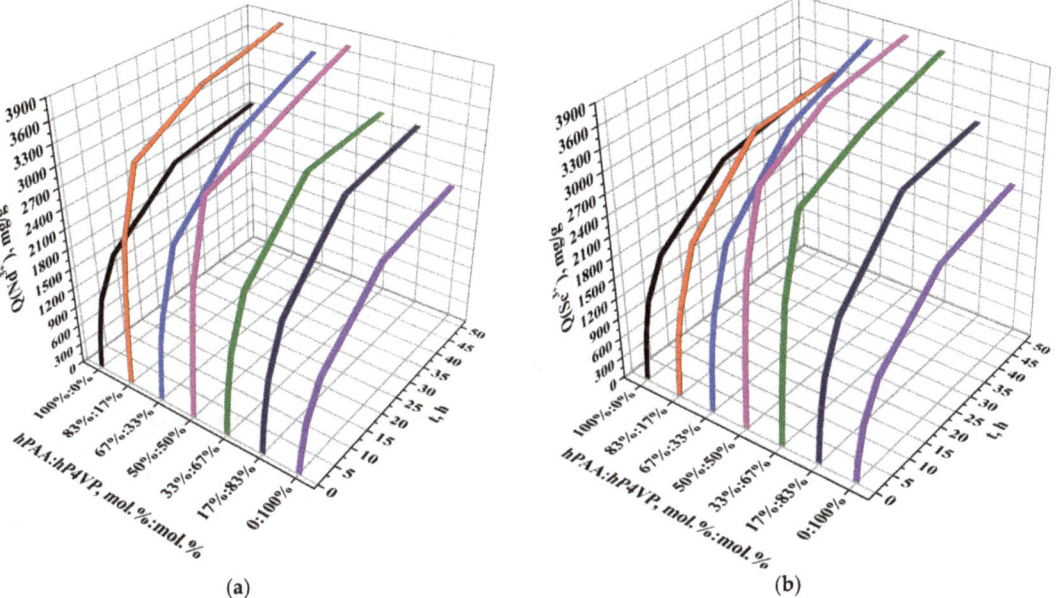

Figure 7. Dynamic exchange capacity (in relation to Nd^{3+} (**a**) and Sc^{3+} (**b**) ions) of the intergel system hPAA–hP4VP.

Values of the dynamic exchange capacity of neodymium and scandium ions of the intergel system hPAA–hP4VP are presented in Tables 3 and 4, respectively.

Table 3. Dynamic exchange capacity (in relation to Nd^{3+} ions) of the intergel system hPAA–hP4VP.

t, h	$Q(Nd^{3+})$, mg/g						
	hPAA:hP4VP, mol.%:mol.%						
	100%:0%	83%:17%	67%:33%	50%:50%	33%:67%	17%:83%	0%:100%
0	0	0	0	0	0	0	0
0.5	408.63	821.04	505.42	665.33	488.58	438.08	202.42
1	644.29	1380.75	871.54	1098.79	732.67	661.13	353.92
2	1014.63	2108.79	1208.21	1902.58	1174.54	1044.08	661.13
6	1549.08	3085.13	2129.83	3009.38	1923.63	1730.04	1149.29
24	2344.46	3611.17	3080.92	3337.63	2908.38	2769.50	2028.83
48	2550.67	3893.13	3606.96	3817.38	3000.96	2963.08	2256.08

Table 4. Dynamic exchange capacity (in relation to Sc^{3+} ions) of the intergel system hPAA–hP4VP.

t, h	$Q(Sc^{3+})$, mg/g						
	hPAA:hP4VP, mol.%:mol.%						
	100%:0%	83%:17%	67%:33%	50%:50%	33%:67%	17%:83%	0%:100%
0	0	0	0	0	0	0	0
0.5	513.75	526.25	597.92	896.67	745.42	497.08	303.33
1	753.75	871.67	942.92	1456.67	1212.50	770.42	429.58
2	1107.08	1225.00	1380.83	2205.42	1990.83	1166.25	741.25
6	1637.50	2012.08	2214.17	3190.42	3068.33	1830.83	1237.50
24	2415.83	3000.83	3228.33	3733.33	3514.17	2895.83	2066.67
48	2639.17	3110.42	3729.17	3926.67	3863.75	3051.25	2352.92

As seen from the obtained data, a significant increase in the initial sorption properties (sorption degree and dynamic exchange capacity) in the intergel system occurs due to the formation of optimal conformation for sorption of neodymium and scandium ions at certain molar ratios. Mutual activation of the rare-cross-linked polymer hydrogels during their remote interaction enables their transition into a highly ionized state.

3.2. Sorption of Nd^{3+} and Sc^{3+} Ions Based on Molecular Imprinting

The sorption properties (sorption degree and dynamic exchange capacity) of synthetized structures MIP(Nd) and MIP(Sc) and the control sample cMIPs are presented in Tables 5 and 6. These data show that the sorption ability of these synthetized MIPs is sufficiently high. Strong sorption of Nd and Sc occurs due to the formation of complementary to these REM cavities in the structure of molecularly imprinted polymers during the synthesis procedure. As seen from Table 5, the sorption degree increases with time for both neodymium and scandium sorption. During 6 h of interaction, the sorption degree increases in the following order: for Nd sorption 0.5 h, 17.79%; 1 h, 29.80%; 2 h, 40.61%; and 6 h, 61.32%; for Sc sorption 0.5 h, 20.51%; 1 h, 32.94%; 2 h, 44.15%; and 6 h, 62.12%. The overwhelming majority of neodymium and scandium is sorbed during 24 h of interaction: the sorption degree of MIP(Nd) is 83.64%; the sorption degree of MIP(Sc) is 85.66%. The remaining neodymium is sorbed during the last 24 h (up to 48 h of interaction), wherein the sorption that occurs is not intense, evidenced by the fact that the sorption degree increases by 4–5%. It can be said that the system MIP–REM salt solution reaches an equilibrium state. As seen from Table 6, the dynamic exchange capacity of MIP(Nd) increases from 741.25 to 2555.00 mg/g (0.5 h, 741.25 mg/g; 1 h, 1241.67 mg/g; 2 h, 1692.08 mg/g; and 6 h, 2555.00 mg/g) during 6 h for neodymium sorption. At the same time, the increase in the capacity of MIP(Sc) during scandium sorption is 0.5 h, 854.58 mg/g; 1 h, 1372.50 mg/g; 2 h, 1839.58 mg/g; and 6 h, 2588.33 mg/g. The almost final values of the parameter for MIP(Nd), 3485.00 mg/g, and for MIP(Sc), 3569.17 mg/g, are observed at 24 h of interaction with the corresponding REM's salt solution. Any further increase in the dynamic exchange capacity is insignificant in the case of Nd sorption by MIP(Nd); the parameter increases up to 3699.58 mg/g; in the case of Sc sorption, the capacity increases up to 3783.75 mg/g. Non-imprinted sample cMIPs do not participate in the sorption of either Nd or Sc due to the absence of complementary cavities to Nd or Sc ions.

Table 5. Sorption degrees of Nd^{3+} and Sc^{3+} ions by the synthetized molecular-imprinted polymers.

t, h	$\eta(Nd^{3+})$, %		$\eta(Sc^{3+})$, %	
	MIP(Nd)	cMIP	MIP(Sc)	cMIP
0	0	0	0	0
0.5	17.79	0	20.51	0
1	29.80	0	32.94	0
2	40.61	0	44.15	0
6	61.32	0	62.12	0
24	83.64	0	85.66	0
48	88.79	0	90.81	0

Table 6. Dynamic exchange capacity (in relation to Nd^{3+} ions) of the synthetized molecular-imprinted polymers.

t, h	$Q(Nd^{3+})$, mg/g		$Q(Sc^{3+})$, mg/g	
	MIP(Nd)	cMIP	MIP(Sc)	cMIP
0	0	0	0	0
0.5	741.25	0	854.58	0
1	1241.67	0	1372.50	0
2	1692.08	0	1839.58	0
6	2555.00	0	2588.33	0
24	3485.00	0	3569.17	0
48	3699.58	0	3783.75	0

The decrease in neodymium (a) and scandium (b) concentrations during their sorption by individual PAA and P4VP hydrogels and the intergel system and MIPs is shown in Figure 8. The figure provides comparative characteristics on the sorption efficiency of Nd or Sc ions (based on the REM concentration decrease) of the polymer structures: PAA and P4VP hydrogels, MIP(Nd)- and MIP(Sc)-imprinted polymers, and the intergel systems 83%hPAA:17%hP4VP and 50%hPAA:50%hP4VP. The character of the Nd ion concentration decrease is different for these macromolecules: the sorption of Nd ions is accompanied by the following decrease: hP4VP, 100 mg/L-95.14 mg/L-91.51 mg/L-84.13 mg/L-72.42 mg/L-51.31 mg/L-45.85 mg/L; hPAA, 100 mg/L-90.19 mg/L-84.54 mg/L-75.65 mg/L-62.83 mg/L-43.73 mg/L-38.78 mg/L; MIP(Nd), 100 mg/L-82.21 mg/L-70.20 mg/L-59.39 mg/L-38.68mg/L-16.36mg/L-11.21mg/L; 83%hPAA-17%hP4VP, 100 mg/L-80.30 mg/L-66.86 mg/L-49.39 mg/L-25.96 mg/L-13.33 mg/L-5.57 mg/L, respectively, for interaction time 0, 0.5, 1, 2, 6, 24, and 48 h. The decrease in the Sc ion concentration during sorption by these macromolecules occurs as follows: hP4VP, 100 mg/L-92.72 mg/L-89.69 mg/L-82.21 mg/L-70.30 mg/L-50.40 mg/L-43.53 mg/L; hPAA, 100 mg/L-87.67 mg/L-81.91 mg/L-73.43 mg/L-60.70 mg/L-42.02 mg/L-36.66 mg/L; MIP(Sc), 100 mg/L-79.49 mg/L-67.06 mg/L-55.85 mg/L-37.88 mg/L-14.34 mg/L-9.19 mg/L; 50%hPAA-50%hP4VP, 100 mg/L-78.48 mg/L-65.04 mg/L-47.07 mg/L-23.43 mg/L-10.40 mg/L-5.76 mg/L, respectively, for interaction time 0, 0.5, 1, 2, 6, 24, and 48 h. From these results, it can be seen that concentration decrease occurs more intensely when sorption is carried out by the intergel systems 83%hPAA:17%hP4VP and 50%hPAA:50%hP4VP and MIP(Nd)- and MIP(Sc)-imprinted polymers. Such difference in sorption intensities between these macromolecular structures and individual rare-cross-linked polymer hydrogels PAA and P4VP is due to high ionization, with the consequent formation of optimal conformation for the sorption of neodymium and scandium in these intergel systems and due to the formation of complementary cavities in the structure of the molecularly imprinted polymers during the synthesis procedure.

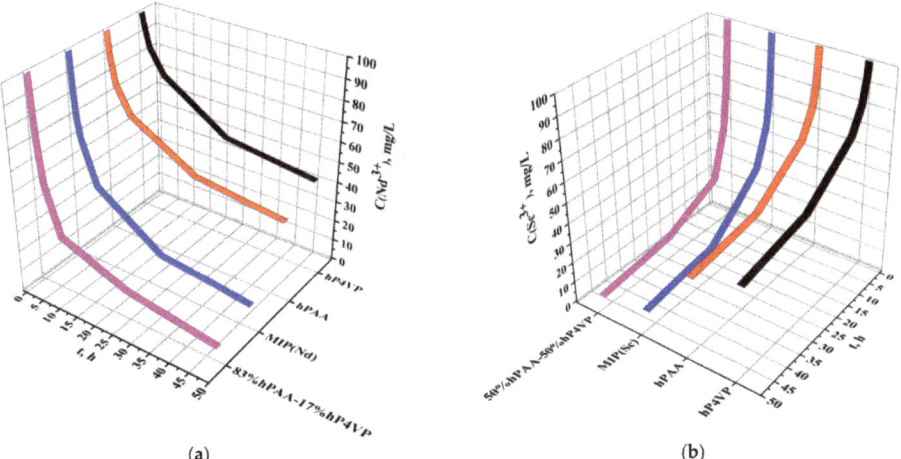

Figure 8. Neodymium (a) and scandium (b) concentration decrease during sorption by macromolecular sorbents.

Figure 9 presents the average growth (mean value of ω(η) and ω(Q) for the sorption of Nd and Sc ions, respectively) in the sorption properties of intergel systems 83%hPAA:17%hP4VP and 50%hPAA:50%hP4VP and MIP(Nd) and MIP(Sc) compared to those of PAA (Figure 9a,c) and P4VP (Figure 9b,d) during Nd^{3+} and Sc^{3+} ion sorption. The most significant growth in the sorption properties (in comparison with PAA hydrogel at Nd sorption) in Figure 9a occurs during 6 h for the intergel system 83%hPAA:17%hP4VP; the growth occurs as follows: 105.98% (0.5 h)-120.03% (1 h)-113.23% (2 h)-104.12% (6 h); in the same time interval, the growth in the sorption properties for MIP(Nd) occurs in the following order: 85.47% (0.5 h)-97.36% (1 h)-70.11 (2 h)-68.19 (6 h). The subsequent growth in the sorption properties is not intense: for the intergel system, it is 56.73% at 24 h and 55.26% at 48 h of remote interaction, while the growth for MIP(Nd) is 51.08% at 24 h and 47.30% at 48 h. Higher growth in sorption properties for both macromolecular structures are observed in comparison with the P4VP hydrogel (Figure 9b). The growth in the sorption properties for the intergel system 83%hPAA:17%hP4VP occurs as follows: 320.90% (0.5 h)-304.64% (1 h)-229.92% (2 h)-176.86% (6 h); at the same time, the growth for the MIP(Nd) is as follows: 279.51% (0.5 h)-263.38% (1 h)-163.74% (2 h)-128.43% (6 h); after this time, there is a growth decrease for both polymer structures: 81.89% at 24 h and 76.19% at 48 h for 83%hPAA:17%hP4VP and 75.36% (24 h) and 67.18% (48 h) for MIP(Nd). Obtained data on sorption properties' growth in relation to scandium ions compared with that of PAA hydrogel (Figure 9c) indicate that maximum growth values are observed at 2 h of remote interaction in the presence of the intergel system 50%hPAA:50%hP4VP, wherein the growth character changes in the following order: 78.26% (0.5 h)-97.92% (1 h)-104.17% (2 h)-99.58% (6 h)-57.27% (24 h)-51.22% (48 h); the growth values for MIP(Sc) are 69.66% (0.5 h)-86.20 (1 h)-69.47% (2 h)-60.97% (6 h)-50.13% (24 h)-45.54% (48 h). The mean growth in the sorption properties compared with that of P4VP hydrogel (Figure 9d) for the intergel system 50%hPAA:50%hP4VP is 205.38% (0.5 h)-251.05% (1 h)-207.41% (2 h)-165.70% (6 h)-84.68% (24 h)-70.23 (48 h); for the imprinted polymer MIP(Sc), the growth is 190.82% (0.5 h)-230.48% (1 h)-155.58% (2 h)-114.62% (6 h)-76.34% (24 h)-63.85% (48 h). As seen from Figure 9, the growth in the sorption properties significantly increases in most cases during the first 2 h of interaction. After that, it slightly decreases up to 6 h, with a sharp decrease at 24 h. After this time, the decrease in the growth is insignificant (over 10%).

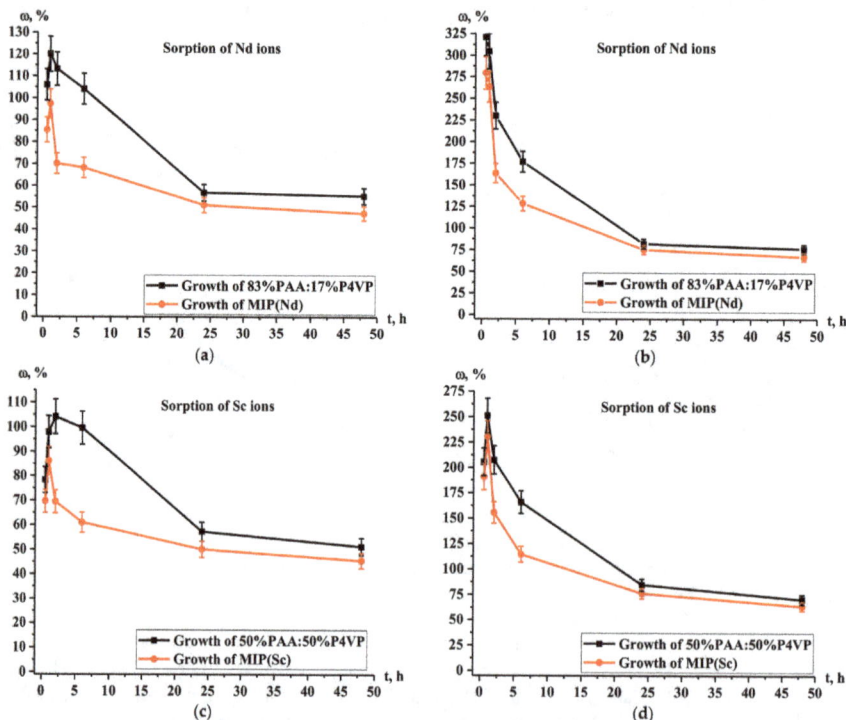

Figure 9. Average growth in sorption properties of intergel systems 83%hPAA:17%hP4VP and 50%hPAA:50%hP4VP and MIP(Nd) and MIP(Sc) compared to that of PAA (**a,c**) and P4VP (**b,d**) during Nd^{3+} and Sc^{3+} ion sorption.

3.3. Laboratory Tests on Selective Sorption of Nd^{3+} and Sc^{3+} Ions

To carry out laboratory tests on the selective sorption of neodymium and scandium ions, a prototype of a laboratory unit was created. The creation of a unit prototype for the selective extraction of neodymium and scandium ions presupposes the initial development of technical requirements for the structure of the unit itself. The following basic requirements were developed:

(1) The design and elements of the unit must be resistant to aggressive environments, since testing involves interaction with strongly acidic solutions (pH = 3.5–4.5) containing ions of rare earth and rare metals.
(2) The material from which the unit is made should not enter into chemical reactions with product solutions.
(3) The unit should provide the ability to quickly change the filling polymer structures with sorbed ions of neodymium and scandium to unused ones (ready for sorption). The laboratory unit should be relatively easy to reconstruct for the use of various highly selective polymer structures: highly selective intergel systems, interpenetrating polymer networks, and molecularly imprinted polymers.

Work on the design of a unit prototype installation for the selective sorption of Nd and Sc ions from industrial solutions of hydrometallurgy was carried out. Figure 10 is a photograph of an assembled unit. The laboratory unit is a structure made of plexiglass, glued with dichloroethane, containing two cartridges. The placement of the cartridges inside the unit is shown in Figure 11. Operational removal of the cartridges is shown in Figure 12 (the cartridges move along special skids back and forth). Each cartridge is covered with a special polymer membrane (the material is a polypropylene analogue of the one used for the laboratory studies). A pair of cartridges enables remote interaction of polymer structures (polyacids are placed in one cartridge and polybases in the other) in the case of using intergel systems for the sorption of Nd and Sc ions. The dimensions of the unit (outer contour) are 300 × 200 × 500 mm^3. The size of the cartridges is 280 × 18 × 480 mm^3. The cartridges involve the loading of functional polyacids and polybases in intergel pairs (certain molar ratios) of hPAA–hP4VP that have the maximum sorption properties relatively to neodymium and scandium ions. In addition to the above intergel systems, in the developed laboratory unit, it is possible to use molecularly imprinted polymers.

Figure 10. Unit for selective sorption of neodymium and scandium ions.

Figure 11. Location of cartridges inside the unit.

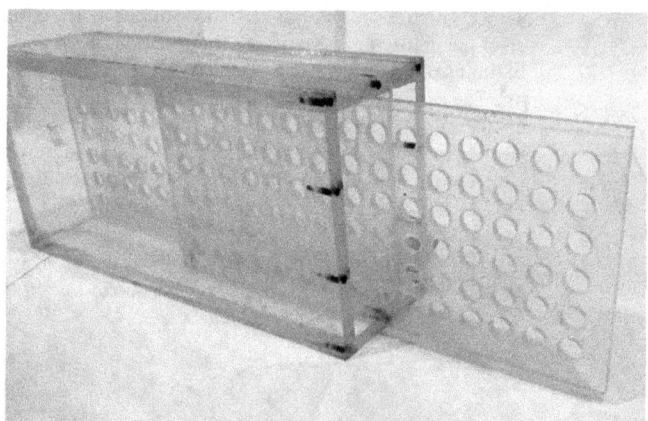

Figure 12. Way to quickly remove the cartridges.

The initial results of laboratory studies showed that the maximum sorption of neodymium and scandium occurs at molar ratios 83%hPAA:17%hP4VP and 50%hPAA:50%hP4VP, respectively. High values of the sorption properties are observed in the sorption of these metals by the molecularly imprinted structures MIP(Nd) and MIP(Sc). Two pieces of the developed laboratory unit were used (one for Nd ion sorption and one for Sc ion sorption) one after another for the selective sorption and separation of neodymium and scandium ions from each other from the model solution (the solution containing neodymium nitrate hydrate and scandium nitrate hydrate, each salt at a concentration of 100 mg/L).

3.3.1. Selective Sorption and Separation Based on the Intergel Systems

Tables 7 and 8 contain values of the sorption properties (sorption degree and dynamic exchange capacity) in relation to Nd and Sc ions during the selective sorption of these ions.

As can be seen from these results, the selective sorption of Nd ions by the intergel system 83%hPAA:17%hP4VP (first laboratory unit) is hampered by the accompanying sorption of scandium (Table 7). The values of the sorption degree in relation to Nd ions are as follows: 15.84% (0.5 h)-26.37% (1 h)-39.61% (2 h)-51.22% (6 h)-73.67% (24 h)-81.43% (48 h); the sorption degree of the accompanying Sc ions is 1.17% (0.5 h)-1.76% (1 h)-2.33% (2 h)-4.09% (6 h)-7.06% (24 h)-10.58% (48 h). The selective sorption of Nd ions leads to the following changes in the dynamic exchange capacity: 660.00 mg/g (0.5 h)-1098.75 mg/g (1 h)-1650.42 mg/g (2 h)-2134.17 mg/g (6 h)-3069.58 mg/g (24 h)-3392.92 mg/g (48 h); this parameter in relation to Sc ions is 48.75 mg/g (0.5 h)-73.33 mg/g (1 h)-97.08 mg/g (2 h)-170.42 mg/g (6 h)-294.17 mg/g (24 h)-440.83 mg/g (48 h). If the sorbed amount of scandium is considered to be a share of the sorbed neodymium, the mean values of the sorption properties in relation to scandium ions are 7.39% (0.5 h)-6.67% (1 h)-5.88% (2 h)-7.99% (6 h)-9.58% (24 h)-12.99% (48 h), wherein the average share of the interfering scandium during the entire time of the selective sorption of neodymium is 8.42%.

Similarly, the selective sorption of Sc ions by the intergel system 50%hPAA:50%hP4VP (second laboratory unit) is hampered by the remaining part of the sorbed neodymium (Table 8). The values of the sorption degree of scandium ions are 19.87% (0.5 h)-32.45% (1 h)-50.58% (2 h)-60.20% (6 h)-71.23% (24 h)-83.57% (48 h); the parameters in relation to neodymium ions are 2.15% (0.5 h)-2.65% (1 h)-4.33% (2 h)-7.08% (6 h)-10.01% (24 h)-13.22% (48 h). The average values of sorption properties (if the sorbed amount of scandium is considered as a share of the sorbed neodymium) in relation to neodymium ions that interfere with the selective sorption of scandium ions are as follows: 10.82% (0.5 h)-8.17% (1 h)-8.56% (2 h)-11.76% (6 h)-14.05% (24 h)-15.82% (48 h), wherein the average share of interfering neodymium during the entire time of the selective sorption of neodymium is 11.53%.

Table 7. Sorption properties of the intergel system hPAA–hP4VP during neodymium ion sorption.

t, h	Nd Selective Sorption 83%hPAA:17%hP4VP			
	η(Nd), %	Q(Nd), mg/g	η(Sc), %	Q(Sc), mg/g
0	0	0	0	0
0.5	15.84	660.00	1.17	48.75
1	26.37	1098.75	1.76	73.33
2	39.61	1650.42	2.33	97.08
6	51.22	2134.17	4.09	170.42
24	73.67	3069.58	7.06	294.17
48	81.43	3392.92	10.58	440.83

Table 8. Sorption properties of the intergel system hPAA–hP4VP during scandium ion sorption.

t, h	Sc Selective Sorption 50%hPAA:50%hP4VP			
	η(Nd), %	Q(Nd), mg/g	η(Sc), %	Q(Sc), mg/g
0	0	0	0	0
0.5	2.15	89.58	19.87	827.92
1	2.65	110.42	32.45	1352.08
2	4.33	180.42	50.58	2107.50
6	7.08	295.00	60.20	2508.33
24	10.01	417.08	71.23	2967.92
48	13.22	550.83	83.57	3482.08

3.3.2. Selective Sorption and Separation Based on Molecularly Imprinted Polymers

The sorption properties (sorption degree and dynamic exchange capacity) of the macromolecular structures MIP(Nd) and MIP(Sc) during the selective sorption of Nd and

Sc ions are presented in Tables 9 and 10. The selective sorption of neodymium ions occurs intensively during 24 h. The remaining 24 h provide an insignificant increase in the sorption properties; the sorption degree of neodymium ions increases as follows: 18.10% (0.5 h)-29.95% (1 h)-41.13% (2 h)-61.60% (6 h)-83.45% (24 h)-88.78% (48 h). The increase in the exchange capacity is 754.17 mg/g (0.5 h)-1247.92 mg/g (1 h)-1713.75 mg/g (2 h)-2566.67 mg/g (6 h)-3477.08 mg/g (24 h)-3699.17 mg/g (48 h). The selective sorption of scandium from the model solution (after neodymium sorption) points to the increase in the sorption properties; the sorption degree increases with time as follows: 21.30% (0.5 h)-33.56% (1 h)-44.69% (2 h)-62.43% (6 h)-85.78% (24 h)-90.81% (48 h). The dynamic exchange capacity increases as follows: 887.50 mg/g (0.5 h)-1398.33 mg/g (1 h)-1862.08 mg/g (2 h)-2601.25 mg/g (6 h)-3574.17 mg/g (24 h)-3783.75 mg/g (48 h).

Table 9. Sorption properties of the imprinted structures MIP(Nd) and MIP(Sc) during neodymium ion sorption.

t, h	Nd Selective Sorption			
	MIP(Nd)		MIP(Sc)	
	η(Nd), %	Q(Nd), mg/g	η(Nd), %	Q(Nd), mg/g
0	0	0	0	0
0.5	18.10	754.17	0	0
1	29.95	1247.92	0	0
2	41.13	1713.75	0	0
6	61.60	2566.67	0	0
24	83.45	3477.08	0	0
48	88.78	3699.17	0	0

Table 10. Sorption properties of the imprinted structures MIP(Nd) and MIP(Sc) during scandium ion orption.

t, h	Sc Selective Sorption			
	MIP(Nd)		MIP(Sc)	
	η(Sc), %	Q(Sc), mg/g	η(Sc), %	Q(Sc), mg/g
0	0	0	0	0
0.5	0	0	21.30	887.50
1	0	0	33.56	1398.33
2	0	0	44.69	1862.08
6	0	0	62.43	2601.25
24	0	0	85.78	3574.17
48	0	0	90.81	3783.75

The obtained results in the laboratory tests on selective sorption and separation showed the advantages and disadvantages of the proposed methods. The main advantage of the sorption method based on the intergel system hPAA–hP4VP is the possibility to "change" the selectivity of the system by changing molar ratios in the intergel pairs (as mentioned earlier, the maximum sorption of neodymium occurs at the ratio 83%hPAA:17%hP4VP and the maximum sorption of scandium occurs at the ratio 50%hPAA:50%hP4VP); in other words, one intergel system can be successfully applied for the selective sorption and separation of Nd and Sc ions. Nevertheless, the drawback is the accompanying sorption of Nd and Sc ions (average share of the sorbed accompanying REMs is over 10%), which, in turn, decreases the sorption properties during selective sorption and decreases the efficiency of the separation process. The advantage of the sorption method based on the imprinted polymers MIP(Nd) and MIP(Sc) is the higher values of the sorption properties (compared to that of the intergel system) and the absence of the simultaneous sorption of the accompanying metal; a drawback is that the developed MIPs are focused on the selective extraction of only one REM (each metal requires a specific MIP).

The universality of the intergel system (despite the accompanying slight sorption of the other metals) and the high selectivity of the molecularly imprinted polymers (despite the high cost and complex procedure of synthesis) are the advantages of the molecular imprinting technique that make them preferrable in the development of new-generation sorption technologies. The principle of application of molecularly imprinted polymers lies in the cost of the sorbed rare earth metal; in the case of the extraction of the most expensive rare earth metals, it would be cost effective.

4. Conclusions

The two developed methods for the selective sorption of Nd and Sc ions and their separation from each other showed sufficiently good results to be successfully applied in the upgrade of the existing sorption technologies. From the point of view of economical efficiency, it is more appropriate to use intergel systems for selective sorption and further separation of Nd and Sc ions despite the lower values of the sorption properties in comparison with molecularly imprinted polymers. However, there is no doubt about the absolute maximum efficiency of sorption and separation of Nd and Sc ions from each other by the application of molecularly imprinted polymers despite their complicated synthesis procedure and the high initial cost of their production.

Author Contributions: Conceptualization, T.J. and R.K.; data curation, T.J. and R.K.; formal analysis, R.K. and A.I.; funding acquisition, R.K.; investigation, R.K. and A.I.; methodology, T.J. and R.K.; project administration, R.K.; resources, A.I.; supervision, R.K.; validation, A.I.; visualization, R.K. and A.I.; writing—original draft, R.K. and A.I.; writing—review and editing, T.J. and R.K. All authors have read and agreed to the published version of the manuscript.

Funding: This research was funded by the Ministry of Education and Science of the Republic of Kazakhstan, grant numbers AP09563137 and AP08856668.

Institutional Review Board Statement: Not applicable.

Informed Consent Statement: Not applicable.

Data Availability Statement: The data presented in this study are available upon request from the corresponding author.

Conflicts of Interest: The authors declare no conflict of interest.

References

1. Hedrick, J.B. *Rare Earths: U.S. Geological Survey in Metals and Minerals in Minerals Yearbook*; U.S. Geological Survey: Reston, VA, USA, 1999; Volume 1, pp. 61.1–61.12.
2. Cotton, S. *Lanthanides and Actinides*; Oxford University Press: Oxford, UK, 1991; 192p.
3. Norman, A.; Zou, X.; Barnett, J. *Critical Minerals: Rare Earths and the U.S. Economy*; backgrounder #175; NCPA: Washington, DC, USA, 2014; 16p.
4. Kryukov, V.A.; Yatsenko, V.A.; Kryukov, Y.V. Rare-earth industry—To realize available opportunities. *Min. Ind.* **2020**, *5*, 68–84.
5. Kilbourn, B.T. *A Lanthanide Lanthology*; Molycorp, Inc.: New York, NY, USA, 1993; pp. 1–61.
6. Jones, A.P.; Wall, F.; Williams, C.T. *Rare Earth Minerals; Chemistry, Origin, and Ore Deposits*; Chapman and Hall: London, UK, 1996; 372p.
7. Lipin, B.R.; McKay, G.A. *Geochemistry and Mineralogy of the Rare Earth Elements: Reviews in Mineralogy*; De Gruyter: Berlin, Germany, 1989; Volume 21, 348p.
8. Bautista, R.G. Separation Chemistry. In *Handbook on the Physics and Chemistry of Rare Earths*; Elsevier B.V.: Amsterdam, The Netherlands, 1995; Chapter 139; pp. 1–27. [CrossRef]
9. Alguacil, F.J.; Rodriguez, F. Separation processes in rare earths. *Rev. Metal.* **1997**, *33*, 187–196. [CrossRef]
10. Larsson, K.; Binnemans, K. Separation of rare earths by split-anion extraction. *Hydrometallurgy* **2015**, *115*, 206–214. [CrossRef]
11. Innocenzi, V.; De Michelis, I.; Ferella, F.; Veglio, F. Rare earths from secondary sources: Profitability study. *Adv. Environ. Res.-Int. J.* **2016**, *5*, 125–140. [CrossRef]
12. Maestro, P.; Huguenin, D. Industrial applications of rare earths: Which way for the end of the century. *J. Alloys Compd.* **1995**, *225*, 520–528. [CrossRef]
13. Ryan, N.E. High-Temperature Corrosion Protection. In *Handbook on the Physics and Chemistry of Rare Earths*; Elsevier B.V.: Amsterdam, The Netherlands, 1995; Chapter 141; Volume 2, pp. 93–132. [CrossRef]

14. Omodara, L.; Pitkäaho, S.; Turpeinen, E.-M.; Saavalainen, P.; Oravisjärvi, K.; Keiski, R.L. Recycling and substitution of light rare earth elements, cerium, lanthanum, neodymium, and praseodymium from end-of-life applications—A review. *J. Clean. Prod.* **2019**, *236*, 117573. [CrossRef]
15. Jelinek, L.; Wei, Y.Z.; Arai, T.; Kumagai, M. Study on separation of Eu(II) from trivalent rare earths via electro-reduction and ion exchange. *J. Alloys Compd.* **2008**, *451*, 341–343. [CrossRef]
16. Moldoveanu, G.; Papangelakis, V. Chelation-Assisted Ion-Exchange Leaching of Rare Earths from Clay Minerals. *Metals* **2021**, *11*, 1265. [CrossRef]
17. Rozelle, P.L.; Khadilkar, A.B.; Pulati, N.; Soundarrajan, N.; Klima, M.S.; Mosser, M.M.; Miller, C.E.; Pisupati, S.V. A Study on Removal of Rare Earth Elements from U.S. Coal Byproducts by Ion Exchange. *Metall. Mater. Trans. e-Mater. Energy Syst.* **2016**, *3*, 6–17. [CrossRef]
18. Miller, D.D.; Siriwardane, R.; Mcintyre, D. Anion structural effects on interaction of rare earth element ions with Dowex 50W X8 cation exchange resin. *J. Rare Earths* **2018**, *36*, 879–890. [CrossRef]
19. Chandrasekara, N.P.G.N.; Pashley, R.M. Study of a new process for the efficient regeneration of ion exchange resins. *Desalination* **2015**, *357*, 131–139. [CrossRef]
20. Greenleaf, J.E.; SenGupta, A.K. Carbon dioxide regeneration of ion exchange resins and fibers: A review. *Solvent Extr. Ion Exch.* **2012**, *30*, 350–371. [CrossRef]
21. Zhang, J.; Amini, A.; O'Neal, J.A.; Boyer, T.H.; Zhang, Q. Development and validation of a novel modeling framework integrating ion exchange and resin regeneration for water treatment. *Water Res.* **2015**, *84*, 255–265. [CrossRef]
22. Melnikov, Y.A.; Ergozhin, E.E.; Chalov, T.K.; Nikitina, A.I. The anion exchange resin based on diglycidyl benzylamine and polyethylenimine to extract perrenate ions. *Int. J. Chem. Sci.* **2015**, *13*, 990–996.
23. Melnikov, Y.A.; Ergozhin, E.E.; Chalov, T.K.; Nikitina, A.I. Sorption of chromium (VI) ions by anionites based on epoxidized derivatives of aniline and benzylamine. *Life Sci. J.* **2014**, *11*, 252–254.
24. Jumadilov, T.K.; Kondaurov, R.G.; Abilov, Z.A.; Grazulevicius, J.V.; Akimov, A.A. Influence of polyacrylic acid and poly-4-vinylpyridine hydrogels mutual activation in intergel system on their sorption properties in relation to lanthanum (III) ions. *Polym. Bull.* **2017**, *74*, 4701–4713. [CrossRef]
25. Jumadilov, T.; Abilov, Z.; Grazulevicius, J.; Zhunusbekova, N.; Kondaurov, R.; Agibayeva, L.; Akimov, A. Mutual activation and sorption ability of rare cross-linked networks in intergel system based on polymethacrylic acid and poly-4-vinylpyridine hydrogels in relation to lanthanum ions. *Chem. Chem. Technol.* **2017**, *11*, 188–194. [CrossRef]
26. Nicholls, I.A.; Adbo, K.; Andersson, H.S.; Andersson, P.O.; Ankarloo, J.; Hedin-Dahlstrom, J.; Jokela, P.; Karlsson, J.G.; Olofsson, L.; Rosengren, J.; et al. Can we rationally design molecularly imprinted polymers? *Anal. Chim. Acta* **2001**, *435*, 9–18. [CrossRef]
27. Toth, B.; Pap, T.; Horvath, V.; Horvai, G. Which molecularly imprinted polymer is better? *Anal. Chim. Acta* **2001**, *591*, 7–21. [CrossRef]
28. Ying, T.L.; Gao, M.J.; Zhang, X.L. Highly selective technique-molecular imprinting. *Chin. J. Anal. Chem.* **2001**, *29*, 99–102.
29. Lai, J.P.; He, X.W.; Guo, H.S.; Liang, H. A review on molecular imprinting technique. *Chin. J. Anal. Chem.* **2001**, *29*, 836–844.
30. Haupt, K.; Linares, A.V.; Bompart, M.; Bernadette, T.S.B. Molecularly Imprinted Polymers. *Mol. Impr.* **2012**, *325*, 1–28. [CrossRef]
31. Komiyama, M.; Mori, T.; Ariga, K. Molecular Imprinting: Matrials Nanoarchitectonics with Molecular Information. *Bull. Chem. Soc. Jpn.* **2018**, *91*, 1075–1111. [CrossRef]
32. Chen, W.; Liu, F.; Xu, Y.T.; Li, K.A.; Tong, S.Y. Molecular recognition of procainamide-imprinted polymer. *Anal. Chim. Acta* **2001**, *432*, 277–282. [CrossRef]
33. Byrne, M.E.; Park, K.; Peppas, N.A. Molecular imprinting within hydrogels. *Adv. Drug Deliv. Rev.* **2002**, *54*, 149–161. [CrossRef]
34. Yano, K.; Karube, I. Molecularly imprinted polymers for biosensor applications. *TrAC—Trends Anal. Chem.* **1999**, *18*, 199–204. [CrossRef]
35. Allender, C.J.; Richardson, C.; Woodhouse, B.; Heard, C.M.; Brain, K.R. Pharmaceutical applications for molecularly imprinted polymers. *Int. J. Pharm.* **2000**, *195*, 39–43. [CrossRef]
36. Cai, W.S.; Gupta, R.B. Molecularly-imprinted polymers selective for tetracycline binding. *Sep. Purif. Technol.* **2004**, *35*, 215–221. [CrossRef]
37. Saylan, Y.; Yilmaz, F.; Ozgur, E.; Derazshamshir, A.; Yavuz, H.; Denizli, A. Molecular Imprinting of Macromolecules for Sensor Applications. *Sensors* **2017**, *17*, 898. [CrossRef]
38. Qiao, F.; Sun, H.; Yan, H.; Row, K.H. Molecularly imprinted polymers for solid phase extraction. *Chromatographia* **2006**, *64*, 625–634. [CrossRef]
39. Szatkowska, P.; Koba, M.; Koslinski, P.; Szablewski, M. Molecularly Imprinted Polymers' Applications: A Short Review. *Mini-Rev. Org. Chem.* **2013**, *10*, 400–408. [CrossRef]
40. Pohanka, M. Sensors Based on Molecularly Imprinted Polymers. *Int. J. Electrochem. Sci.* **2017**, *12*, 8082–8094. [CrossRef]
41. Cormack, P.A.G.; Elorza, A.Z. Molecularly imprinted polymers: Synthesis and characterisation. *J. Chromatogr. B—Anal. Technol. Biomed. Life Sci.* **2004**, *804*, 173–182. [CrossRef]
42. Yemis, F.; Alkan, P.; Yenigul, B.; Yenigul, M. Molecularly Imprinted Polymers and Their Synthesis by Different Methods. *Polym. Polym. Compos.* **2013**, *21*, 145–150. [CrossRef]
43. Ye, L. Synthetic Strategies in Molecular Imprinting. In *Molecularly Imprinted Polymers in Biotechnology*; Springer: Cham, Switzerland, 2015; Volume 150, pp. 1–24. [CrossRef]

44. Balamurugan, S.; Spivak, D.A. Molecular imprinting in monolayer surfaces. *J. Mol. Recognit.* **2011**, *24*, 915–929. [CrossRef]
45. Jumadilov, T.K.; Kondaurov, R.G.; Imangazy, A.M.; Myrzakhmetova, N.O.; Saparbekova, I. Phenomenon of remote interaction and sorption ability of rare cross–linked hydrogels of polymethacrylic acid and poly-4-vinylpyridine in relation to erbium ions. *Chem. Chem. Technol.* **2019**, *13*, 451–458. [CrossRef]
46. Jumadilov, T.K.; Kondaurov, R.G.; Imangazy, A.M. Features of sorption of rare-earth metals of cerium group by intergel systems based on polyacrylic acid, polymethacrylic acid and poly-4-vinylpyridine hydrogels. *Bull. Karaganda Univ. Chem. Ser.* **2020**, *98*, 58–67. [CrossRef]

Article

Upconversion Nanoparticles Encapsulated with Molecularly Imprinted Amphiphilic Copolymer as a Fluorescent Probe for Specific Biorecognition

Hsiu-Wen Chien [1,*], Chien-Hsin Yang [2], Yan-Tai Shih [2] and Tzong-Liu Wang [2,*]

1 Department of Chemical and Materials Engineering, National Kaohsiung University of Science and Technology, Kaohsiung 807, Taiwan
2 Department of Chemical and Materials Engineering, National University of Kaohsiung, Kaohsiung 811, Taiwan; yangch@nuk.edu.tw (C.-H.Y.); yan.tai.shih@gmail.com (Y.-T.S.)
* Correspondence: hsiu-wen.chien@nkust.edu.tw (H.-W.C.); tlwang@nuk.edu.tw (T.-L.W.)

Abstract: A fluorescent probe for specific biorecognition was prepared by a facile method in which amphiphilic random copolymers were encapsulated with hydrophobic upconversion nanoparticles (UCNPs). This method quickly converted the hydrophobic UCNPs to hydrophilic UNCPs. Moreover, the self-folding ability of the amphiphilic copolymers allowed the formation of molecular imprinting polymers with template-shaped cavities. LiYF$_4$:Yb^{3+}/Tm^{3+}@LiYF$_4$:Yb^{3+} UCNP with upconversion emission in the visible light region was prepared; this step was followed by the synthesis of an amphiphilic random copolymer, poly(methacrylate acid-co-octadecene) (poly(MAA-co-OD)). Combining the UCNPs and poly(MAA-co-OD) with the templates afforded a micelle-like structure. After removing the templates, UCNPs encapsulated with the molecularly imprinted polymer (MIP) (UCNPs@MIP) were obtained. The adsorption capacities of UCNPs@MIP bound with albumin and hemoglobin, respectively, were compared. The results showed that albumin was more easily bound to UCNPs@MIP than to hemoglobin because of the effect of protein conformation. The feasibility of using UCNPs@MIP as a fluorescent probe was also studied. The results showed that the fluorescence was quenched when hemoglobin was adsorbed on UCNPs@MIP; however, this was not observed for albumin. This fluorescence quenching is attributed to Förster resonance energy transfer (FRET) and overlap of the absorption spectrum of hemoglobin with the fluorescence spectrum of UCNPs@MIP. To our knowledge, the encapsulation approach for fabricating the UCNPs@MIP nanocomposite, which was further used as a fluorescent probe, might be the first report on specific biorecognition.

Keywords: upconversion nanoparticles; molecularly imprinted polymers; amphiphilic random copolymer; fluorescence probes

1. Introduction

Upconversion nanoparticles (UCNPs) are trivalent lanthanide (Ln^{3+})-doped nanoparticles, which can up-convert two or more lower-energy photons into one high-energy photon [1–4]. Because Ln^{3+} undergoes f-f transitions within the 4f shell, when Ln^{3+} is embedded in an insulating host lattice, the energies of the excited states will generate a series of states with many closely spaced energy levels, which creates an excited state with a longer lifetime and a sharper optical line shape [2]. Therefore, UCNPs are a type of photostable nanocrystal with a high signal-to-noise ratio, and can be used for photodynamic therapy, light-induced drug delivery, (targeted) cell imaging, and immunoassay sensors [3,4].

The most common strategies for synthesizing small, monodisperse, and bright UCNPs are coprecipitation, thermal decomposition, and solvothermal syntheses [1,5]. Among these strategies, using oleic acid to cap the UCNP is the most common approach. For example, synthesizing hexagonal-phase NaYF$_4$ nanoparticles doped with Ln^{3+} involves heating rare earth chlorides in a mixture of octadecene and oleic acid [6,7]. First, oleate salts are generated as in situ precursors. Subsequently, ammonium fluoride and sodium hydroxide are added, and the temperature of the reaction is increased to 300 °C. These processes generate highly monodisperse and oleate-capped UCNPs, which are hydrophobic and can only be dispersed in nonpolar solvents [8].

Nanoparticles for biomedical applications should be water-soluble; to meet this criterion, surficial modifications are usually required [8]. For example, ligand exchange is a well-known method for binding hydrophilic heads to the surface of nanoparticles [9]. Other methods that do not use ligand exchange for nanoparticle solubilization include the formation of a stable silica shell [10] or the use of amphiphilic surfactants or block copolymers to form a bilayer, which occurs via hydrophobic interactions between oleic acid and oleate ions [11,12]. However, if nanoparticles are to be further used as biomarkers, the surfaces of these nanoparticles also need to have distinctive functional groups. Generally, nanoparticles are functionalized after being transferred into water. The common method involves a carbodiimide reaction, which allows amine-containing biomolecules to couple with COOH groups on the surface of the nanoparticles [13,14]. However, 1-ethyl-3-(3-dimethylaminopropyl) carbodiimide hydrochloride (EDC), which is required in this method, is expensive and has a low conversion rate. These shortcomings have therefore stimulated the exploration of other methods of functionalization.

Recently, molecular imprinting for biological recognition has attracted significant attention [15–17]. This approach uses the 'lock and key' mechanism; therefore, the molecule can identify a cavity with a complementary shape to achieve specific binding. Molecularly imprinted polymers (MIPs) are synthesized by the in-situ polymerization of functional monomers and crosslinkers in the presence of target templates [16]. The monomer can interact with the template through covalent or non-covalent bonds. After copolymerization, chemical decomposition or solvent extraction is used to remove the templates from the polymer network. The resulting cavities retain the corresponding steric and chemical memories of the templates. Therefore, the target species can selectively rebind to MIPs through specific interactions with these imprinted sites [16]. In recent years, some studies have combined MIP with fluorescent nanoparticles to measure fluorescence. The combined system can detect a lower content of the target analytes when compared to MIP only, and the fluorescent nanoparticles provide a readout signal, which improves the sensitivity [18–20]. For example, Yu et al. detected acetamiprid with fabricated $NaYF_4$:Yb UCNPs encapsulated in a molecularly imprinted polymer [20]. According to their procedure, the UCNPs required surficial modification via a sol–gel method to generate methacrylate groups on the surface of UCNPs. The typical polymerization of MIPs was then applied to UCNPs [20]; however, the procedure is cumbersome.

Folding amphiphilic polymers to form MIPs have been developed recently [21]. In this method, an amphiphilic random copolymer was first synthesized. Next, the amphiphilic random copolymer was mixed and interacted with the templates. The polymer could self-fold to form micelles and encapsulate the templates inside the polymer. After the templates were removed, an imprinted polymer with specific template cavities was formed [21]. Compared with the traditional in-situ polymerization for MIP preparation, this method is easier to remove templates because of the physical interaction between the polymer and the templates. Although a few literature used amphiphilic random copolymers to modify the surface of UCNPs [22], most of them mainly rely on the functional groups of the copolymers graft to UCNPs via the ligand exchange process. In addition, there was no literature that used the combination of amphiphilic random copolymers and UCNP for preparing MIPs. In this study, we propose a strategy that utilizes an amphiphilic random copolymer to encapsulate hydrophobic UCNPs, which does not interfere with the original ligands and can self-fold into water-soluble imprint polymers within the cavities of the template. A oleate-capped $LiYF_4$:Yb^{3+}/Tm^{3+}@$LiYF_4$:Yb^{3+} core/shell UCNP with upconversion emission in the visible-light region and amphiphilic poly(methacrylic acid-co-1-octadecene) (poly(MAA-co-OD)) were first synthesized. MIPs are then formed from the surface of the

UCNPs modified with the as-prepared amphiphilic poly(MAA-co-OD). This approach not only confers hydrophilicity to the UCNPs, but also integrates specific templates into the polymer shell. Therefore, after dissolution in water, the UCNPs@MIP could be used as a biosensor without further functionalization. When the prepared UCNPs@MIP rebind with specific target analytes within absorption bands of the visible-light region, the fluorescence is quenched, verifying the feasibility of the prepared UCNPs@MIP as fluorescent probes. To the best of our knowledge, this is the first report of molecularly imprinted UCNPs prepared via the folding of amphiphilic random copolymers.

2. Materials and Methods

2.1. Materials

Lithium carbonate (Li_2CO_3), yttrium(III) oxide (Y_2O_3), ytterbium(III) oxide (Yb_2O_3), thulium(III) oxide (Tm_2O_3), trifluoroacetic acid (TFA), oleic acid, and 1-octadecene (OD) were purchased from Alfa Aesar (Heysham, Lancashire, United Kingdom). Methacrylic acid (MAA) was purchased from SHOWA. Azobisisobutyronitrile (AIBN) was obtained from Aencore (Surrey Hills, Australia). Albumin (Alb), hemoglobin (Hb), and all other chemicals were purchased from Sigma-Aldrich (Saint Louis, MO, USA).

2.2. Synthesis of $LiYF_4$: Yb^{3+}/Tm^{3+}@$LiYF_4$:Yb^{3+} Core/Shell UCNPs

The $LiYF_4$:Yb^{3+}/Tm^{3+} core was synthesized by thermal decomposition of lanthanide and lithium trifluoroacetate precursors in the presence of oleic acid, coordinating ligands, and noncoordinating 1-octadecene molecules [23,24]. Briefly, Li_2CO_3 (1.44 mmol), Y_2O_3 (0.72 mmol), Yb_2O_3 (0.25 mmol), and Tm_2O_3 (0.01 mmol) were dissolved in 10 mL of aqueous TFA (50%) and stirred at 90 °C until the solution became transparent. Oleic acid (15 mL) and OD (15 mL) were then added to the solution. The resultant solution was then heated to 120 °C under nitrogen gas to remove water and oxygen until the solution turned light yellow. The solution was then heated to 300 °C at a rate of approximately 30 °C/min, and reacted at 300 °C under vigorous stirring for 1 h. The mixture was cooled to room temperature and precipitated with ethanol. The solid was collected by centrifugation at 8000 rpm for 10 min. The solid was then dispersed in n-hexane and re-precipitated with ethanol. The above steps were repeated twice to obtain oleate-capped $LiYF_4$:Yb^{3+}/Tm^{3+} core nanoparticles (Yields: 88–90%).

For shell growth, Li_2CO_3 (1.54 mmol), Y_2O_3 (0.77 mmol), and Yb_2O_3 (0.2 mmol) were dissolved in 10 mL of aqueous TFA (50%) at 90 °C until the solution became transparent. Oleic acid (15 mL) and OD (15 mL) were then added to the solution. The resultant solution was heated to 120 °C at a rate of approximately 2 °C/min for 30 min under argon gas. The $LiYF_4$:Yb^{3+}/Tm^{3+} core nanoparticles were then added to the flask. After the solution turned light yellow, it was heated to 300 °C at a rate of approximately 30 °C/min and allowed to react for 1 h. The mixture was cooled to room temperature and precipitated with ethanol. The solid was collected by centrifugation at 8000 rpm for 10 min. The solid was then dispersed and re-precipitated twice with n-hexane and ethanol to obtain the oleate-capped $LiYF_4$:Yb^{3+}/Tm^{3+}@$LiYF_4$:Yb^{3+} core/shell UCNPs (Yields: 88–90%).

2.3. Synthesis of Amphiphilic Random Copolymer Poly(MAA-co-OD)

A free radical reaction was used to synthesize amphiphilic poly(MAA-co-OD) with a hydrophilic monomer, MAA, and a hydrophobic monomer, OD. Briefly, MAA (0.1 mol), OD (0.1 mol), and AIBN (0.6 mmol) were added to ethanol (100 mL) under nitrogen protection. The polymerization was initiated by heating the mixture at 65 °C for 12 h under constant stirring. After polymerization, the product was purified via precipitation in diethyl ether and centrifugation. After executing this procedure three times, the product was dried under vacuum. The product was further soluble in chloroform for gel permeation chromatography (GPC, YL9100 GPC System, Young Lin Instrument Co., Ltd., Anyang, South Korea) analysis via a polystyrene (PS) standard calibration yielded M_n = 22,894, and a polydispersity of M_w/M_n = 1.507 (Figure S1).

2.4. Preparation of UCNPs@MIP

UCNPs@MIP were prepared via the encapsulation of amphiphilic random copolymers. Briefly, the poly(MAA-co-OD) (14 mg/mL), as-synthesized hydrophobic UCNPs (8.4 mg/mL), and templates (6 mg/mL) were separately dissolved or dispersed in methanol. The three solutions were mixed at a volume ratio of 1:1:1, and then sonicated for 2 h. The mixture in methanol (2 mL) was then injected into DI water (12 mL), which triggered the self-assembly of the poly(MAA-co-OD) and UCNPs. After the methanol was evaporated, the obtained mixture was centrifuged at 10,000 rpm for 30 min. The precipitate was collected and re-dispersed three times in water to extract the templates, which were Alb and Hb. At the end of this process, a stable colloidal dispersion in water was obtained. For the control experiment, the same procedure without the templates was applied to prepare the non-imprinted polymer (NIP) UCNPs, and the obtained sample is termed UCNPs@NIP.

2.5. Characterization of UCNPs and UCNPs@MIP

Wide-angle X-ray diffractograms (WAXD) were obtained with a Bruker D8 ADVANCE diffractometer (Karlsruhe, Germany), using Cu-Kα radiation with a step size of 0.05° and a scanning speed of 4°/min. A JEOL JEM1230 transmission electron microscope (Tokyo, Japan) was used to obtain transmission electron microscopy (TEM) images. A PerkinElmer Lambda 35 UV–vis spectrophotometer (Waltham, MA, USA) was used to perform ultraviolet–visible (UV–vis) spectroscopic analysis. A Hitachi F-7000 fluorescence spectrophotometer (Tokyo, Japan) was used to record the photoluminescence (PL) spectra. A SDL980-LM-5000T laser diode (980 nm, 3 W/cm^2) from Shanghai Dream Lasers Technology Co., Ltd. (Shanghai, China) was used to obtain the emission spectra of the nanocrystals, after NIR excitation at 980 nm.

2.6. Application in Biorecognition

The kinetic adsorption test in this experiment was performed in a tube, where the UCNPs@MIP (50 mg) were mixed with 10 mL of the corresponding template solution (50 mg/mL). The mixed solution was shaken for 0, 15, 30, 45, 60, and 75 min at room temperature. The templates were rebound to the UCNPs@MIP, after which the mixture was centrifuged at 10,000 rpm for 30 min. The supernatant was then collected and analyzed using a UV–vis spectrophotometer. A standard curve was used to calculate the concentration of each sample. The UCNPs@NIP was similarly treated for comparison. The mass balance equation was used to calculate the adsorption capacity (Q) [25,26]

$$Q\ (\text{mg/mg}) = \frac{(C_i - C_r)V}{m}$$

where C_i (mg/mL) is the initial concentration of the template in the aqueous solution, C_r (mg/mL) is the concentration of the template in the supernatant after adsorption, m (mg) is the mass of the adsorbent, and V (mL) is the volume of the solution.

For the fluorescence measurements, the templates and UCNPs@MIP were rebound for 45 min, and the mixture was centrifuged at 10,000 rpm for 30 min. The precipitate was collected and re-dispersed in water, and the PL spectra were acquired. The quenching efficiency is expressed as $(F_0 - F)/F_0$, where F_0 and F are the fluorescence intensities at 450 nm, without or with the addition of the template solution, respectively [27,28].

3. Results and Discussion

3.1. Characterization of the UCNPs

The LiYF$_4$:Yb^{3+}/Tm^{3+}@LiYF$_4$:Yb^{3+} UCNPs were synthesized via thermal decomposition. Tetragonal crystals were readily obtained when the reaction temperature was high [29]; hence, the reaction temperature was set to 300 °C. Based on the TEM images, both the core and core–shell nanoparticles were indeed tetragonal with an octahedral morphology (Figure 1a). The average dimensions of the core nanoparticles, which were calculated from the TEM images, were approximately 75 nm along the long axis and 45 nm along the short axis. After coating the LiYF$_4$:Yb^{3+}/Tm^{3+} cores with the LiYF$_4$:Yb^{3+} shell, the length of the short axis of the core/shell nanoparticles increased to 50 nm. Wide-angle X-ray diffraction (WAXD) was further used to investigate the phase structures of the nanoparticles; peaks were observed at 2θ values of 18°, 29°, 31°, 33°, 34°, 40°, 42°, 46°, 47°, 49°, 50°, 54°, 58°, 59°, 62°, 63°, and 66°, respectively attributed to the (101), (112), (103), (004), (200), (202), (211), (114), (105), (123), (204), (220), (301), (116), (132), (224), and (206) planes of the tetragonal LiYF$_4$ crystal (Figure 1b). The relative intensities and positions of all the diffraction peaks in the WAXD pattern were consistent with the Joint Committee on Powder Diffraction Standards (JCPDS) file no. 17e0874, indicating the absence of impurity phases in the synthesized crystals. After coating with the LiYF$_4$: Yb^{3+} shell, the XRD patterns of the core–shell nanoparticles were almost the same as those of the core nanoparticles, which indicates that the shell coating did not change the core structure of the crystal. The similarity of the patterns also shows that the shell was likely very thin.

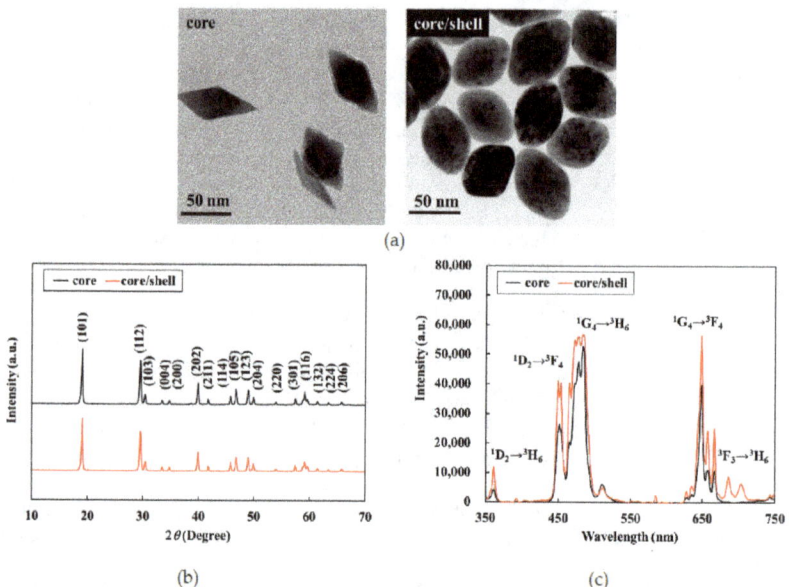

Figure 1. (a) TEM images, (b) XRD patterns, and (c) photoluminescence spectra of LiYF$_4$: Yb^{3+}/Tm^{3+} core and LiYF$_4$: Yb^{3+}/Tm^{3+}@LiYF$_4$:Yb^{3+} core/shell nanoparticles.

Figure 1c shows that upon excitation at 980 nm, the emission spectra of the LiYF$_4$: Yb^{3+}/Tm^{3+} core nanoparticles had characteristic emission peaks with electronic transitions at 360 nm ($^1D_2 \rightarrow ^3H_6$), 450 nm ($^1D_2 \rightarrow ^3F_4$), 475 nm ($^1G_4 \rightarrow ^3H_6$), and 650 nm ($^1G_4 \rightarrow ^3F_4$). This finding confirmed that LiYF$_4$:Yb^{3+}/Tm^{3+} had upconversion emissions in the region of visible light. After coating with the LiYF$_4$: Yb^{3+} shell, the emissions of the core/shell nanoparticles were apparently stronger than the corresponding emissions of the core nanoparticles. During energy transfer, the photoexcited dopants located on or near the surface can be directly deactivated by neighboring quenching centers [30,31]. Moreover, it is plausible that the energy contained in the photoexcited dopants located in the center of the nanophosphors migrated randomly and traveled a long distance on or near the surface of the dopant, or directly to the surficial quenching sites [32,33]. Therefore, the active shell structure plausibly prevented energy quenching, which allowed for high upconversion of the photoluminescence efficiency. Therefore, the core/shell UCNPs could be used in a follow-up study for the encapsulation of MIP.

3.2. Hydrophilicity of UCNPs@MIP

Amphiphilic random copolymers of poly(MAA-co-OD) were used to encapsulate the UCNPs, which allowed the poly(MAA-co-OD) to form enclose the UCNPs. Generally, when amphiphilic random copolymers are dissolved in water, the hydrophobic groups self-aggregate via hydrophobic-hydrophobic interactions, while the hydrophilic groups surround the hydrophobic domain; therefore, micelles can be formed and dispersed in water [34]. When the amphiphilic poly(MAA-co-OD) and the hydrophobic UCNPs were mixed, hydrophobic-hydrophobic interactions incorporate the hydrophobic groups in the copolymer with the oleic acid in UCNPs. Consequently, the hydrophilic groups were exposed on the periphery of the hydrophobic aggregates [22]. The water-solubility of the UCNPs, before and after the encapsulation of poly(MAA-co-OD) was visually observed. Figure 2 shows that the hydrophobic UCNPs floated on the water, while the colloidal UCNPs@NIP settled to the bottom of the vial (Figure S2), which indicates that the obtained UCNPs@NIP were water-soluble. Similarly, UCNPs@MIP was also water-soluble, which means that poly(MAA-co-OD) dominated the conversion of hydrophobic UCNPs into hydrophilic, while templates were not a factor to affect the character. TEM was further used to examine the morphology of the prepared UCNPs@MIP (Figure 3). Regardless of whether Ab or Hb was used as a template to prepare the UCNPs@MIP, both types of UCNPs@MIPs were generally composed of several UCNPs, which were aggregated in a polymeric matrix to form quasi-microspheres. The average size of the spheres was approximately 300–500 nm. The results show that the poly(MAA-co-OD) could indeed convert the hydrophobic UCNPs into hydrophilic UCNPs that were capable of forming a micelle-like structure.

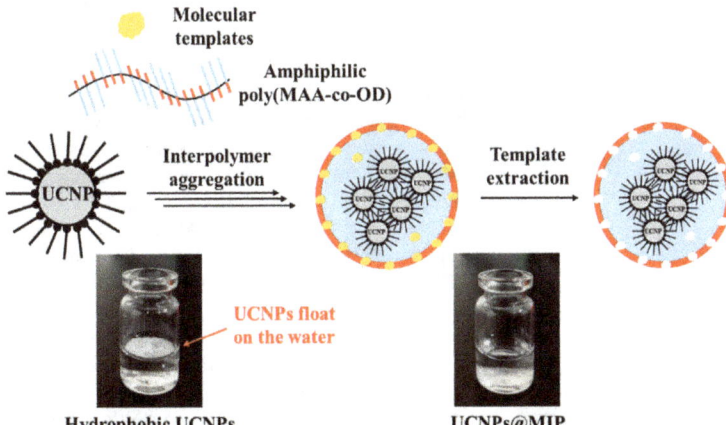

Figure 2. Schematic representation of preparation of UCNPs@MIP.

Traditional in situ polymerization, which was used to prepare the MIP, involves mixing monomers, crosslinkers, and templates, and then polymerizing and removing the templates to form imprinted polymers with specific template cavities. Recently, folding amphiphilic polymers to form MIPs has attracted attention [21]. In this method, an amphiphilic random copolymer was first synthesized. As the templates interacted with the amphiphilic random copolymer, the polymer self-folded to form micelles and the template structure was imprinted into the polymer. After the templates were removed, an imprinted polymer with specific template cavities was formed [21]. This strategy for folding amphiphilic random copolymers is advantageous. For example, the interaction between the template and the polymer occurs via physical interaction; therefore, the template is easier to remove. This method can use either a hydrophilic or hydrophobic template. Because of the water insolubility of the hydrophobic template, the hydrophobic template becomes completely entrapped inside the micelles. Consequently, the loading capacity of the template and the binding capacity of the MIPs increase [21]. Although there are only a few studies, the superiority of the encapsulation method has been demonstrated. In the present study, the hydrophobic UCNPs were encapsulated with an amphiphilic poly(MAA-co-OD). This approach not only ensured that hydrophilicity was conferred to the UCNPs, but also enabled the integration of specific templates into the polymer shell, which facilitated the formation of MIPs on the surface of the UCNPs. The feasibility of the prepared UCNPs@MIP for molecular recognition is evaluated in the subsequent sections.

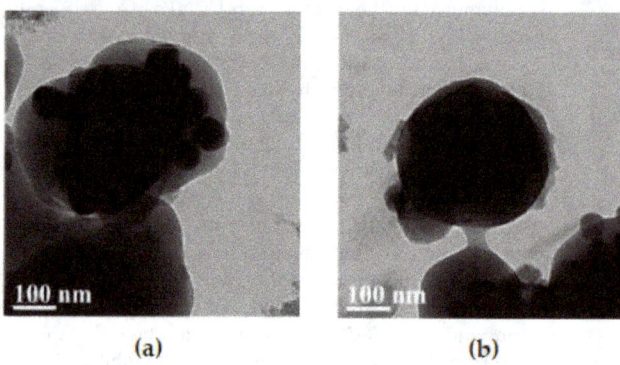

Figure 3. TEM images of UCNPs@MIP. Albumin (**a**) and hemoglobin (**b**) as templates.

3.3. Equilibrium Binding of UCNPs@MIP and UCNPs@NIP

UV–vis spectroscopy was used to analyze the kinetics of binding of the template molecules with the UCNPs@MIP and UCNPs@NIP. The binding kinetics were used to confirm that the UCNPs@MIPs had the ability to recognize corresponding template molecules. During the recognition of Alb, the first 15 min was the zone of linear increase, and the next 40 min was the saturation zone, regardless of the adsorption curve (UCNPs@MIP or UCNPs@NIP) (Figure 4a). The adsorption capacity of the UCNPs@MIP was much larger than that of the UCNPs@NIPs. This result is noteworthy because it indicates that the UCNPs@MIP had the cavities of Alb, which promoted mass transfer and enhanced rebinding of the template molecules. During the recognition of Hb, the adsorption capacity of UCNPs@NIP increased with the adsorption time (Figure 4b). However, the adsorption capacity was consistently very low; even when the time increased to 45 min, the adsorption capacity was only 0.012 mg/mg. In contrast, the adsorption curve of the UCNPs@MIP showed a linear increase within 40 min, whereas after 40 min, there was a tendency towards saturated adsorption. After 40 min, the adsorption capacity of the UCNPs@MIP was approximately 20 times that of the UCNPs@NIP. These results provide excellent evidence that the imprinting of the UCNPs@MIPs was efficient.

The adsorption capacity of the UCNPs@MIP for Alb was about 0.23 mg/mg at 45 min, compared to 0.18 mg/mg for Hb. This notable result shows that the adsorption capacity of the UCNPs@MIP for Alb was slightly better than that for Hb. In contrast, the adsorption capacity of the UCNPs@NIP for Hb was much smaller than that of Alb. In general, the efficiency of imprinting depended strongly on factors such as the strength of the interactions between the template and the polymeric matrix [26], or the shape of the template. Generally, the hydrophobic residues of most proteins are sequestered in the core of the native structure, while the polar residues are present on the surface [35]. In the structure of Hb, its 'arm' is a negatively charged propionate group, which faces the surface of the protein, and the hydrophobic part is buried in the hydrophobic amino acid of the protein. Comparatively, Alb is negatively charged when it is ionized in water at pH 7.4 [36]. Because this experiment used negatively charged MAA as the hydrophilic end of MIP, and Alb and Hb are also negatively charged, the polymer and protein may undergo repulsive interactions, which rationalizes the very low adsorption of Hb on the UCNPs@NIP.

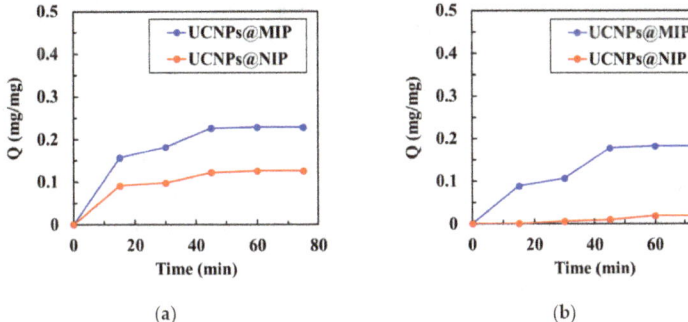

Figure 4. Kinetics of (**a**) albumin and (**b**) hemoglobin adsorption in UCNPs@MIP and UCNPs@NIP.

Alb has nominal dimensions of 7.5 × 6.5 × 4.0 nm^3 and a molecular mass of 66.4 kDa, while Hb has nominal dimensions of 6.0 × 5.0 × 5.0 nm^3 and a molecular mass of 64 kDa [37]. Although both Alb and Hb are globular proteins [37], Alb is not really spherical, but has a "V" shape [38]. An early study evaluating the adsorption of Alb on silica spheres showed when the V-shaped Alb approached the silica spheres in a direction parallel to the surface of the sphere, it was strongly rejected. However, because albumin has a very small curvature at its V tip, when the V-shaped Alb approached the silica sphere by its tip, it likely penetrated the repulsion field, and was easily adsorbed onto the silica sphere [38]. It is proposed herein that the V-shaped Alb penetrated the repulsive field of UCNPs@MIP or UCNPs@NIP by its tip, which allowed the Alb to enter the polymer matrix. Therefore, Alb adsorption on the UCNPs@NIP was detected. The same rationale likely explains why UCNPs@MIP had a greater adsorption capacity for Alb than for Hb.

3.4. Effect of Template Molecules on Fluorescence Quenching

The PL spectra of the UCNPs, UCNPs@MIP, and UCNPs@MIP bound to Alb and Hb are shown in Figure 5a. As aforementioned, the emission bands of the as-prepared UCNPs occurred at 360, 450, 475, and 650 nm. The results show that the fluorescence intensity of the encapsulated MIP was similar to that of the original UCNPs, which indicates that the encapsulated poly(MAA-co-OD) did not significantly affect the luminescence of the UCNPs. Furthermore, the fluorescence intensity of the UCNPs@MIP bound to Alb did not change. However, the fluorescence intensity at 360 and 450 nm was evidently reduced when Hb was bound to UCNPs@MIP. The fluorescence quenching may be attributed to Förster resonance energy transfer (FRET) [39,40]. According to Förster's non-radiative energy transfer theory, energy transfer occurs under the following conditions: (i) the donor can produce fluorescence; (ii) the fluorescence emission spectrum of the donor and the ultraviolet absorption spectrum of the acceptor overlap; (iii) the distance between the donor and the acceptor is <8 nm [41]. As shown in Figure 6b, Hb had six absorption peaks at 220, 280, 340, 400, 550, and 570 nm. The absorption band at 220 nm resulted from the absorption of light between the peptide bonds and amino acid residues. In the UV range of 250–300 nm, light was absorbed by the delocalized electrons of the aromatic side chains. In the wavelength range of 300–700 nm, the absorption of light was associated with the excitation of the porphyrin structure. [42]. Because the absorption spectrum of Hb overlapped with the fluorescence emission spectrum of the UCNPs@MIP, the UCNP emission was quenched by Hb. In contrast, there was no spectral overlap between the absorption spectra of Alb and the emission spectrum of UCNPs@MIP (Figure 6a). Upon binding of the UCNPs to Alb and Hb, the efficiency of fluorescence reduction was approximately 2.1% and 30%, respectively (Figure 5b). The data indicate that the prepared UCNPs@MIP has great potential as a fluorescent probe for detecting biomolecules.

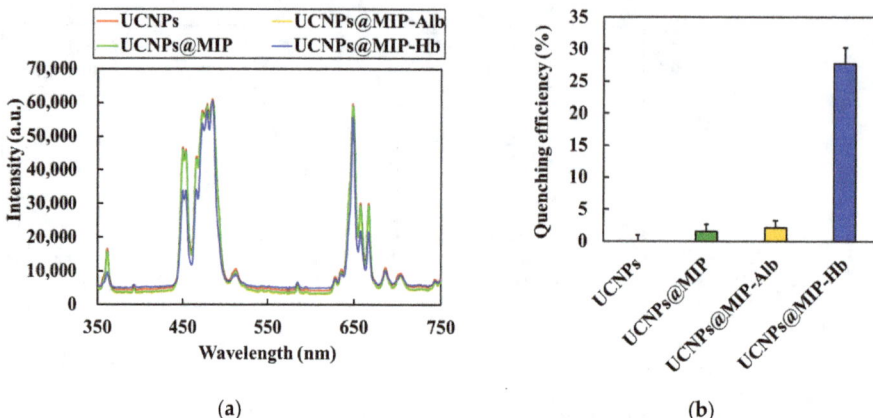

Figure 5. (a) Photoluminescence spectra and (b) efficiency of fluorescence reduction for the UCNPs, UCNPs@MIP, and UCNPs@MIP bound to albumin and hemoglobin. The quenching efficiency is expressed as $(F_o - F)/F_o$, where F_o and F are the fluorescence intensities at 450 nm.

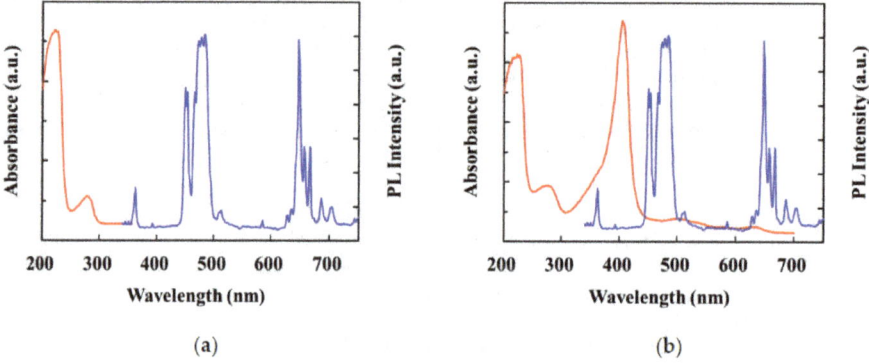

Figure 6. Spectral overlap between the UV–vis absorption spectrum of templates (red line) and the fluorescence spectra of the UCNPs@MIP (blue line). Albumin (a) and hemoglobin (b) as templates.

4. Conclusions

A facile method for encapsulating hydrophobic UCNPs with amphiphilic random copolymers was proposed. This method could convert the hydrophobic UCNPs to hydrophilic UCNPs, which self-folded to form imprinted polymers with template cavities. The UCNPs with up-conversion emission in the visible-light region of 360 nm and 450–500 nm were first synthesized; this step was followed by the synthesis of an amphiphilic random copolymer of poly(MAA-co-OD). When the UCNPs and poly(MAA-co-OD) were mixed with the templates, a micelle-like structure was formed. After removal of the templates, UCNPs@MIP with a particle size of approximately 500 nm was obtained. The UCNPs@MIP adsorbed more of the template than UCNPs@NIP because the former had template cavities. In addition, UCNPs@MIP had a higher capacity for Alb adsorption than for Hb adsorption because the V-shaped Alb was able to penetrate the repelling field, and was absorbed. Comparatively, because the absorption spectrum of Hb overlapped with the fluorescence spectrum of the UCNPs@MIP, the fluorescence was quenched at wavelengths of 360 nm and 450 nm when Hb was bonded to UCNPs@MIP. The data indicate that the prepared UCNPs@MIP could be used as a fluorescent probe.

Supplementary Materials: The following are available online at https://www.mdpi.com/article/10.3390/polym13203522/s1, Figure S1: Gel permeation chromatography (GPC) analyses on the synthesized poly(MAA-co-OD). Figure S2: Schematic representation of preparation of UCNPs@NIP.

Author Contributions: Conceptualization, H.-W.C., C.-H.Y., and T.-L.W.; Data curation, H.-W.C., C.-H.Y., and Y.-T.S.; Formal analysis, Y.-T.S.; Funding acquisition, T.-L.W.; Supervision, T.-L.W.; Validation, C.-H.Y.; Writing—original draft, H.-W.C.; Writing—review and editing, T.-L.W. All authors have read and agreed to the published version of the manuscript.

Funding: Authors are grateful to the support of the Ministry of Science and Technology in Taiwan through grant MOST 107-2221-E-390-004.

Conflicts of Interest: The authors declare no conflict of interest.

References

1. Zheng, K.; Loh, K.Y.; Wang, Y.; Chen, Q.; Fan, J.; Jung, T.; Nam, S.H.; Suh, Y.D.; Liu, X. Recent advances in upconversion nanocrystals: Expanding the kaleidoscopic toolbox for emerging applications. *Nano Today* **2019**, *29*, 100797. [CrossRef]
2. Bünzli, J.C.G. Lanthanide photonics: Shaping the nanoworld. *Trends Chem.* **2019**, *1*, 751. [CrossRef]
3. Luther, D.C.; Huang, R.; Jeon, T.; Zhang, X.; Lee, Y.W.; Nagaraj, H.; Rotello, V.M. Delivery of drugs, proteins, and nucleic acids using inorganic nanoparticles. *Adv. Drug Deliv. Rev.* **2020**, *156*, 188. [CrossRef]
4. Yi, Z.; Luo, Z.; Qin, X.; Chen, Q.; Liu, X. Lanthanide-activated nanoparticles: A toolbox for bioimaging, therapeutics, and neuromodulation. *Acc. Chem. Res.* **2020**, *53*, 2692. [CrossRef] [PubMed]
5. Zhen, X.; Kankala, R.K.; Liu, C.G.; Wang, S.B.; Chen, A.Z.; Zhang, Y. Lanthanides-doped near-infrared active upconversion nanocrystals: Upconversion mechanisms and synthesis. *Coord. Chem. Rev.* **2021**, *438*, 213870. [CrossRef]
6. Suter II, J.D.; Pekas, N.J.; Berry, M.T.; May, P.S. Real-time-monitoring of the synthesis of β-NaYF$_4$:17% Yb,3% Er nanocrystals using NIR-to-visible upconversion luminescence. *J. Phys. Chem. C* **2014**, *118*, 13238. [CrossRef]
7. Grzyb, T.; Kamiński, P.; Przybylska, D.; Tymiński, A.; Sanz-Rodríguezb, F.; Gonzalez, P.H. Manipulation of up-conversion emission in NaYF$_4$ core@shell nanoparticles doped by Er^{3+}, Tm^{3+}, or Yb^{3+} ions by excitation wavelength—Three ions—Plenty of possibilities. *Nanoscale* **2021**, *13*, 7322. [CrossRef]
8. Muhr, V.; Wilhelm, S.; Hirsch, T.; Wolfbeis, O.S. Upconversion nanoparticles: From hydrophobic to hydrophilic surfaces. *Acc. Chem. Res.* **2014**, *47*, 3481. [CrossRef]
9. Chen, Y.; D'Amario, C.; Gee, A.; Duong, H.T.T.; Shimoni, O.; Valenzuela, S.M. Dispersion stability and biocompatibility of four ligand-exchanged NaYF$_4$: Yb, Er upconversion nanoparticles. *Acta Biomater.* **2020**, *102*, 384. [CrossRef] [PubMed]
10. Reddy, K.L.; Sharma, P.K.; Singh, A.; Kumar, A.; Shankar, K.R.; Singh, Y.; Garg, N.; Krishnan, V. Amine-functionalized, porous silica-coated NaYF$_4$:Yb/Er upconversion nanophosphors for efficient delivery of doxorubicin and curcumin. *Mater. Sci. Eng. C* **2019**, *96*, 86. [CrossRef]
11. Wu, L.; Yu, J.; Chen, L.; Yang, D.; Zhang, S.; Han, L.; Ban, M.; He, L.; Xu, Y.; Zhang, Q. A general and facile approach to disperse hydrophobic nanocrystals in water with enhanced long-term stability. *J. Mater. Chem. C* **2017**, *5*, 3065. [CrossRef]
12. Su, Q.; Zhou, M.T.; Zhou, M.Z.; Sun, Q.; Ai, T.; Su, Y. Microscale self-assembly of upconversion nanoparticles driven by block copolymer. *Front. Chem.* **2020**, *8*, 836. [CrossRef]
13. Han, G.M.; Li, H.; Huang, X.X.; Kong, D.M. Simple synthesis of carboxyl-functionalized upconversion nanoparticles for biosensing and bioimaging applications. *Talanta* **2016**, *147*, 207. [CrossRef]
14. Vijayan, A.N.; Liu, Z.; Zhao, H.; Zhang, P. Nicking enzyme-assisted signal-amplifiable Hg^{2+} detection using upconversion nanoparticles. *Anal. Chim. Acta* **2019**, *1072*, 75. [CrossRef] [PubMed]
15. Fernando, P.U.A.I.; Glasscott, M.W.; Pokrzywinski, K.; Fernando, B.M.; Kosgei, G.K.; Moores, L.C. Analytical methods incorporating molecularly imprinted polymers (MIPs) for the quantification of microcystins: A mini-review. *Crit. Rev. Anal. Chem.* **2021**, *11*, 1. [CrossRef] [PubMed]
16. Batista, A.D.; Silva, W.R.; Mizaikoff, B. Molecularly imprinted materials for biomedical sensing. *Med. Devices Sens.* **2021**, *4*, e10166. [CrossRef]
17. Gao, M.; Gao, Y.; Chen, G.; Huang, X.; Xu, X.; Lv, J.; Wang, J.; Xu, D.; Liu, G. Recent advances and future trends in the detection of contaminants by molecularly imprinted polymers in food samples. *Front. Chem.* **2020**, *8*, 616326. [CrossRef]
18. Hu, X.; Cao, Y.; Tian, Y.; Qi, Y.; Fang, G.; Wang, S. A molecularly imprinted fluorescence nanosensor based on upconversion metal-organic frameworks for alpha-cypermethrin specific recognition. *Microchim. Acta* **2020**, *187*, 632. [CrossRef]
19. Elmasry, M.R.; Tawfik, S.M.; Kattaev, N.; Lee, Y.I. Ultrasensitive detection and removal of carbamazepine in wastewater using UCNPs functionalized with thin-shell MIPs. *Microchem. J.* **2021**, *170*, 106674. [CrossRef]
20. Yu, Q.; He, C.; Li, Q.; Zhou, Y.; Duan, N.; Wu, S. Fluorometric determination of acetamiprid using molecularly imprinted upconversion nanoparticles. *Mikrochim. Acta* **2020**, *187*, 222. [CrossRef]
21. Nagao, C.; Sawamoto, M.; Terashima, T. Molecular imprinting on amphiphilic folded polymers for selective molecular recognition in water. *J. Polym. Sci.* **2020**, *58*, 215. [CrossRef]

22. Liras, M.; González-Béjar, M.; Peinado, E.; Francés-Soriano, L.; Pérez-Prieto, J.; Quijada-Garrido, I.; García, O. Thin amphiphilic polymer-capped upconversion nanoparticles: Enhanced emission and thermoresponsive properties. *Chem. Mater.* **2014**, *26*, 4014 [CrossRef]
23. Chien, H.W.; Yang, C.H.; Tsai, M.T.; Wang, T.L. Photoswitchable spiropyran-capped hybrid nanoparticles based on UV-emissive and dual-emissive upconverting nanocrystals for bioimaging. *J. Photochem. Photobiol. A* **2020**, *392*, 112303. [CrossRef]
24. Chien, H.W.; Wu, C.H.; Yang, C.H.; Wang, T.L. Multiple doping effect of LiYF$_4$:Yb^{3+}/Er^{3+}/Ho^{3+}/Tm^{3+}@LiYF$_4$:Yb^{3+} core/shell nanoparticles and its application in Hg^{2+} sensing detection. *J. Alloys Compd.* **2019**, *806*, 272–282. [CrossRef]
25. EL-Sharif, H.F.; Yapati, H.; Kalluru, S.; Reddy, S.M. Highly selective BSA imprinted polyacrylamide hydrogels facilitated by a metal-coding MIP approach. *Acta Biomater.* **2015**, *28*, 121–127. [CrossRef] [PubMed]
26. Chien, H.W.; Tsai, M.T.; Yang, C.H.; Lee, R.H.; Wang, T.L. Interaction of LiYF$_4$:Yb^{3+}/Er^{3+}/Ho^{3+}/Tm^{3+}@LiYF$_4$:Yb^{3+} upconversion nanoparticles, molecularly imprinted polymers, and templates. *RSC Adv.* **2020**, *10*, 35600. [CrossRef]
27. Wu, N.; Wei, Y.; Pan, L.; Yang, X.; Qi, H.; Gao, Q.; Zhang, C. Lateral flow immunostrips for the sensitive and rapid determination of 8-hydroxy-2'-deoxyguanosine using upconversion nanoparticles. *Microchim. Acta* **2020**, *187*, 377. [CrossRef] [PubMed]
28. Tang, Y.; Li, M.; Gao, Z.; Liu, X.; Gao, X.; Ma, T.; Lu, X.; Li, J. Upconversion nanoparticles capped with molecularly imprinted polymer as fluorescence probe for the determination of ractopamine in water and pork. *Food Anal. Methods* **2017**, *10*, 2964. [CrossRef]
29. Zhang, Q.; Yan, B. Hydrothermal synthesis and characterization of LiREF$_4$ (RE = Y, Tb−Lu) nanocrystals and their core−shell nanostructures. *Inorg. Chem.* **2010**, *49*, 6834. [CrossRef]
30. Wen, S.; Zhou, J.; Schuck, P.J.; Suh, Y.D.; Schmidt, T.W.; Jin, D. Future and challenges for hybrid upconversion nanosystems. *Nat. Photonics* **2019**, *13*, 828. [CrossRef]
31. Tessitore, G.; Mandl, G.A.; Brik, M.G.; Park, W.; Capobianco, J.A. Recent insights into upconverting nanoparticles: Spectroscopy, modeling, and routes to improved luminescence. *Nanoscale* **2019**, *11*, 12015. [CrossRef]
32. Fang, Y.; Liu, L.; Zhang, F. Exploiting lanthanide-doped upconversion nanoparticles with core/shell structures. *Nano Today* **2019**, *25*, 68.
33. Shi, L.; Hu, J.; Wu, X.; Zhan, S.; Hu, S.; Tang, Z.; Chen, M.; Liu, Y. Upconversion core/shell nanoparticles with lowered surface quenching for fluorescence detection of Hg^{2+} ions. *Dalton Trans.* **2018**, *47*, 16445. [CrossRef] [PubMed]
34. Ohshio, M.; Ishihara, K.; Yusa, S. Self-association behavior of cell membrane-inspired amphiphilic random copolymers in water. *Polymers* **2019**, *11*, 327. [CrossRef] [PubMed]
35. Turner, N.W.; Jeans, C.W.; Brain, K.R.; Allender, C.J.; Hlady, V.; Britt, D.W. From 3D to 2D: A review of the molecular imprinting of proteins. *Biotechnol. Prog.* **2006**, *22*, 1474. [CrossRef]
36. Baler, K.; Martin, O.A.; Carignano, M.A.; Ameer, A.G.; Vila, J.A.; Szleifer, I. Electrostatic unfolding and interactions of albumin driven by pH Changes: A molecular dynamics study. *J. Phys. Chem. B* **2014**, *118*, 921. [CrossRef]
37. Erickson, H.P. Size and shape of protein molecules at the nanometer level determined by sedimentation, gel filtration, and electron microscopy. *Biol. Proced. Online* **2009**, *11*, 32. [CrossRef]
38. van Oss, C.J. Macroscopic and microscopic aspects of repulsion versus attraction in adsorption and adhesion in water. *Interface Sci. Technol.* **2008**, *16*, 167.
39. Sapsford, K.E.; Berti, L.; Medintz, I.L. Materials for fluorescence resonance energy transfer analysis: Beyond traditional donor–acceptor combinations. *Angew. Chem. Int. Ed. Engl.* **2006**, *45*, 4562–4589. [CrossRef]
40. Bagheri, A.; Arandiyan, H.; Boyer, C.; Lim, M. Lanthanide-doped upconversion nanoparticles: Emerging intelligent light-activated drug delivery systems. *Adv. Sci.* **2016**, *3*, 1500437. [CrossRef]
41. Suryawanshi, V.D.; Walekar, L.S.; Gore, A.H.; Anbhule, P.V.; Kolekar, G.B. Spectroscopic analysis on the binding interaction of biologically active pyrimidine derivative with bovine serum albumin. *J. Pharm. Anal.* **2016**, *6*, 56. [CrossRef] [PubMed]
42. Mahmoud, S.S.; Amal, E.I. Analysis of retinal b-wave by fourier transformation due to ammonia exposure and the role of blood erythrocytes. *Rom. J. Biophys.* **2010**, *20*, 269.

Article

Synergetic Effect of Dual Functional Monomers in Molecularly Imprinted Polymer Preparation for Selective Solid Phase Extraction of Ciprofloxacin

Ut Dong Thach [1,2,*], Hong Hanh Nguyen Thi [3], Tuan Dung Pham [3], Hong Dao Mai [3] and Tran-Thi Nhu-Trang [4,*]

1 Department of Polymer Chemistry, Graduate University of Science and Technology, Vietnam Academy of Science and Technology, Ha Noi 100000, Vietnam
2 Faculty of Pharmacy, Ton Duc Thang University, Ho Chi Minh City 700000, Vietnam
3 Faculty of Chemistry, University of Science, Vietnam National University, Ho Chi Minh City 700000, Vietnam; nthhanh2397@gmail.com (H.H.N.T.); t.dung597@gmail.com (T.D.P.); hongdao4567@gmail.com (H.D.M.)
4 Faculty of Environmental and Food Engineering, Nguyen Tat Thanh University (NTTU), Ho Chi Minh City 700000, Vietnam
* Correspondence: thachutdong@tdtu.edu.vn (U.D.T.); ttntrang@ntt.edu.vn (T.-T.N.-T.); Tel.: +84-028-37-761-043 (U.D.T.); +84-028-39-404-759 (T.-T.N.-T.)

Abstract: Background: Ciprofloxacin (CIP), an important broad-spectrum fluoroquinolone antibiotic, was often used as a template molecule for the preparation of imprinted materials. In this study, methacrylic acid and 2-vinylpyridine were employed for the first time as dual functional monomers for synthesizing ciprofloxacin imprinted polymers. Methods: The chemical and physicochemical properties of synthesized polymers were characterized using Fourier transform-infrared spectroscopy, thermogravimetric analysis-differential scanning calorimetry, scanning electron microscopy, and nitrogen adsorption-desorption isotherm. The adsorption properties of ciprofloxacin onto synthesized polymers were determined by batch experiments. The extraction performances were studied using the solid phase extraction and HPLC-UV method. Results: The molecularly imprinted polymer synthesized with dual functional monomers showed a higher adsorption capacity and selectivity toward the template molecule. The adsorbed amounts of ciprofloxacin onto the imprinted and non-imprinted polymer were 2.40 and 1.45 mg g^{-1}, respectively. Furthermore, the imprinted polymers were employed as a selective adsorbent for the solid phase extraction of ciprofloxacin in aqueous solutions with the recovery of 105% and relative standard deviation of 7.9%. This work provides an alternative approach for designing a new adsorbent with high adsorption capacity and good extraction performance for highly polar template molecules.

Keywords: ciprofloxacin; imprinted polymer; dual functional monomer; solid phase extraction

1. Introduction

Ciprofloxacin (CIP) is an important antibiotic belonging to the class of fluoroquinolones. It is a broad spectrum of an antibacterial agent widely applied to treat various infectious diseases in animals and humans [1]. It is reported that only 30% of CIP can be metabolized inside the body and a large amount of CIP has been emitted into the environment [1]. Therefore, the intense use of CIP worldwide causes potential threats such as bacterial resistance, allergy, and toxicity [2,3]. For the protection of public health, many countries have established different maximum residue limits for CIP in various food samples. For example, CIP's tolerance is 0.1 mg L^{-1} in milk and 0.5 mg kg^{-1} (total of CIP and enrofloxacin) in porcine liver [4].

Many analytical methods have been developed to determine CIP residue in various food and environmental samples, such as spectrophotometry [5–7], capillary electrophoresis [8], and high-performance liquid chromatography using UV [9], fluorescence [10,11], diode array [12], and mass spectrometry [13]. For the instrumental analysis, sample pretreatment is essential for the purification and enrichment of trace CIP residue in complex matrices. Several separation methods have been employed to extract, purify, and preconcentrate CIP from real samples, including solid phase extraction (SPE) [3], stir bar sorptive extraction [14], magnetic nanoparticles SPE [14], liquid–liquid extraction [15], dispersive liquid–liquid microextraction [14], and immunoaffinity chromatography [14]. Overall, SPE is widely employed as a purification and preconcentration method due to its performance and small organic solvent volumes used [16]. Commercial SPE cartridges are available with different adsorbents, such as C_8, C_{18}, Al_2O_3, silica, and resin. However, these sorbents are non-recyclable, non-selective, and easily extract several compounds, including the analyte and the interferences. Thus, it is important to explore new adsorbents for CIP extraction with excellent selectivity, durability, and reusability.

Molecularly imprinted polymers (MIP) are highly selective adsorbents prepared by the copolymerization of suitable functional monomers and crosslinkers with the presence of a target template molecule [17]. These imprinted polymers have attracted considerable research attention due to their interesting properties such as reproductivity, low cost, ease of preparation, and high selectivity toward target molecules [18]. Due to their unique properties, MIPs have been widely employed in various applications such as chemical sensors [19], biomimetic catalyst [20], drug delivery [21], protein crystallization [22], chromatography [23], bioanalysis [24], and solid phase extraction [1].

A CIP-imprinted polymer (CIP-MIP) was synthesized for the first time in 2006 [25]. Then, many studies have reported CIP-MIP synthesis by using different functional monomers, such as methacrylic acid [1,25–32], 2-vinylpyridine [32], 1-vinyl-3-ethylimidazolium bromide [33], and 4-vinylbenzoic acid [3]. CIP imprinted materials were applied for adsorption [34] or detection of trace CIP residues in various biological and environmental samples using chemosensor and biosensor [35], electrochemical sensor [36,37], and optosensor [38] techniques. Furthermore, Tarly et al. proposed using semi-empirical quantum chemistry stimulation for synthesizing CIP-MIP. Both the computational calculation and the practical experiments showed that MIP synthesized with acrylamide monomer exhibited the highest specific selectivity factor and adsorption capacity [39]. Moreover, MIPs were usually synthesized in the non-polar or weakly polar organic solvent and exhibited good recognition performance for non or weakly polar template molecules [40]. Hydrogen bonds are the driving force for forming specific biding sites for templates containing highly polar functional groups. These interactions could interfere with water molecules, resulting in a reduction in the imprinted polymer selectivity [1,31,41]. The competitive adsorption of water onto specific binding sites via hydrogen bond prevents interaction between the template molecules and imprinting site, which led to the formation of highly non-selective adsorption sites [1]. It is reported that the multifunctional monomer imprinting strategy provides multiple types of intermolecular forces with the template structure [42]. The imprinted polymers synthesized with dual functional monomers showed excellent molecular recognition and a high adsorption capacity, especially for templates containing strong polar functional groups [42–45]. In a similar approach, Wang et al. reported the synthesis of CIP-MIP by using methacrylic acid and 2-hydroxyethyl methacrylate as stimuli-recognition elements [46]. Zhu et al. proposed using 1-ally-3-vinylimidazolium chorine and 2-hydroxyethyl methacrylate as a component of bifunctional monomers [1]. These strategies allowed the obtained polymer to interact strongly with CIP molecules in an aqueous solution via hydrogen bonds, electrostatic, hydrophobic, and π-π stacking interactions. Thus, the multifunctional monomer strategy is an effective method for synthesizing imprinted materials, especially for templates with polar functional groups such as CIP. Various advanced imprinting techniques (on surface, co-precipitation, emulsion, and suspension polymerization) have been developed for preparing imprinted materials [47–50]. The conventional bulk imprinting is essential for preparing imprinted materials because a simple, rapid, and pure MIPs' production can be produced without sophisticated instrumentation. The bulk is an appropriate form for the application in solid phase extraction both in academia [33,41] and industrial manufacture [51].

In this study, methacrylic acid and 2-vinylpyridine, which provide complementary structure, size, and chemical properties toward CIP molecules, were employed for the first time as dual functional monomers in MIP preparation using a bulk imprinting technique. The complementary interactions of functional monomers (MAA and 2-VP) with CIP via hydrogen bond, electrostatic, and π-π stacking are expected to greatly improve the adsorption selectivity of the imprinted polymers. To obtain the best performance of CIP-MIP, several preparation parameters, such as porogenic solvent (methanol/acetic acid, chloroform/methanol, and dichloromethane/methanol), template/monomer molar ratio, and methacrylic acid/2-vinylpyridine molar ratio, were investigated and optimized. The FT-IR, TGA-DSC, SEM, and nitrogen adsorption–desorption analyses were employed for the physicochemical characterization of the obtained polymers. The adsorption properties of CIP onto MIP and non-imprinted polymer (NIP) were determined using batch adsorption experiments. Finally, CIP-MIPs were employed as an adsorbent for the selective extraction of CIP in water.

2. Materials and Methods

2.1. Materials

Ciprofloxacin 98.0%, methacrylic acid (MAA) 99%, 2-vinylpyridine (2-VP) 97%, ethylene glycol dimethacrylate (EDGMA) 98%, and azobisisobutyrontrile (AIBN) 12% wt in acetone were purchased from Sigma-Aldrich, St. Louis, MO, USA. Acetonitrile, methanol, acetic acid, phosphoric acid, and triethylamine with HPLC grade were purchased from Merck, Darmstadt, HL, Germany. The standard stock of CIP (500 mg L^{-1}) was prepared in methanol/acetic acid (9:1, v/v), and the working solutions were diluted from the stock solution with deionized water. The standard solutions were stored at 4 °C to be stable for one week. The phosphoric acid solution 0.05% (pH 3) was prepared by adding 3.0 mL of phosphoric acid 85% to 1000 mL of pure water and adjusting the pH to 3.0 by triethylamine.

2.2. Methods

2.2.1. Preparation of CIP-MIPs

The CIP-MIPs were synthesized using the bulk polymerization method. Briefly, CIP (0.1 mmol), methacrylic acid (0.66 mmol), 2-vinylpyridine (0.33 mmol), and 12 mL of porogen $CHCl_3$/MeOH (9:1, v/v) were added into a screw-capped glass bottle. The mixture was sonicated for 15 min to obtain the homogenous solution and kept overnight at 4 °C for the formation of complex pre-polymerization. Next, ethylene glycol dimethacrylate (5 mmol) and azobisisobutyronitrile (20 mg) were added to the mixture. The oxygen in the bottle was removed by argon for 10 min. The polymerization was activated at 60 °C for 24 h in the thermostatic water bath. After that, the obtained bulk polymer was crushed and ground to obtain polymer particles from 35 to 100 μm. The CIP template was eliminated by washing with methanol/acetic acid (9:1, v/v) in an ultrasonic bath until no CIP was detected using HPLC-UV method. Finally, the obtained polymer particles were dried at 110 °C for 6 h. The corresponding non-imprinted polymer (NIP) was also prepared as a similar protocol but without the CIP template. The detailed MIP preparation parameters were summarized in Table 1.

Table 1. Effect of the imprinted polymer preparation conditions on extraction recovery of CIP in water, (SPE condition: 40 mg of polymer, loading: 5 mL CIP solution in water (0.1 µg L^{-1}), washing: 3 mL of deionized water, eluting: 3 mL of methanol/acetic acid (9:1; v/v), $n = 3$).

MIP	Template (mmol)	Functional Monomer (mmol)		Cross-Linker (mmol)	Porogenic Solvent	Recovery (%)	RSD (%)
		MAA	2-VP				
MIP1	1.0	-	10.0	50	MeOH: AcOH	36.5	2.7
MIP2	1.0	-	10.0	50	$CHCl_3$: MeOH	57.7	5.6
MIP3	1.0	-	10.0	5.0	CH_2Cl_2: MeOH	7.1	3.4
MIP4	1.0	-	15.0	75	$CHCl_3$: MeOH	57.2	7.0
MIP5	1.0	-	20.0	100	$CHCl_3$: MeOH	62.0	3.5
MIP6	1.0	5.0	5.0	50	$CHCl_3$: MeOH	63.2	2.8
MIP7	1.0	6.6	3.3	50	$CHCl_3$: MeOH	104.6	7.9
MIP8	1.0	7.0	3.0	50	$CHCl_3$: MeOH	60.8	6.5
MIP9	1.0	8.0	2.0	50	$CHCl_3$: MeOH	70.0	4.5
MIP10	1.0	10.0	-	50	$CHCl_3$: MeOH	23.1	3.8

2.2.2. Polymer Characterization

The FT-IR spectroscopy analyses were conducted using an ATR-FTIR FT/IR6600A spectrometer, Seri A012761790, JASCO, Tokyo, Japan. Thermogravimetric analysis (TGA) and differential scanning calorimetry (DSC) were carried out using a TG-DSC LabSys Evo 1600, SETARAM, Austin, TX, USA. The samples were heated from 25 to 800 °C at a heat rate of 10 °C min^{-1} under nitrogen atmosphere. The nitrogen adsorption–desorption isotherms were conducted at 77 K using a Micromeritic tristar, Norcross, GA, USA. The polymers were outgassed at 100 °C for 8 h at 0.1 mbar before the analysis. The specific surface area was calculated using the Brunauer–Emmett–Teller (BET) model, using a cross-sectional area of 0.162 nm^2 per nitrogen molecule. Scanning electron microscopy (SEM) was recorded using SEM S-4800, 10 kV, 7.9 mm, Hitachi, Tokyo, Japan.

2.2.3. High Performance Liquid Chromatography Method

Chromatography analysis was conducted using HPLC equipment, Shimazu, Japan, with UV SPD-20A detector and a reversed-phase Water InertSustain AQ-C18 column (4.6 × 150 mm^2, 3 µm particle size i.d.), Waters, USA. The mobile phase at pH 3 was applied in the isocratic mode with 82% of H_3PO_4 0.05% and 18% of acetonitrile. The injection volume was set at 20 µL and 289 nm was employed as the detection wavelength. Retention times and areas of chromatographic peaks were used for qualitative and quantitative analyses, respectively.

2.2.4. Adsorption Study

The individual adsorption isotherms of CIP onto polymer were determined by batch experiments. This study was carried out by equilibrating 10 mg of polymer and 20 mL of CIP solution in 50-milliliter centrifuge tubes. The initial CIP concentrations were varied from 0.14 to 10.0 mg L^{-1}. The centrifuge tubes were slowly shaken for 3 h at 25 °C. The supernatants were filtered out by 0.22-micrometer nylon Millipore filters. The concentrations of CIP in the supernatants were determined using HPLC-UV. The equilibrium adsorption capacity was calculated as follows:

$$Q_{ads} = \frac{V_o(C_i - C_e)}{m_p} \quad (1)$$

where Q_{ads} (mg g^{-1}) is the equilibrium adsorbed amount of CIP; C_i and C_e (mg L^{-1}) are the initial and final concentrations of the CIP solution, respectively; V_o (L) is the volume of the CIP solution; and m_p (g) is the mass of the polymer.

2.2.5. Solid Phase Extraction Study

The obtained MIPs (40 mg) were employed for the preparation of SPE cartridges. A volume of 5 mL of CIP solution in water (0.1 mg L^{-1}) was loaded. The SPE cartridge was washed with 3 mL of deionized water and then eluted by 3 mL of methanol/acetic acid. The final solution volume was adjusted to 1.5 mL. The CIP concentration in eluting solutions were determined using HPLC-UV in order to evaluate the SPE performance through recovery values.

3. Results

3.1. Effect of Polymerization Conditions

3.1.1. Porogenic Solvent

The porogenic solvent is one of the most critical factors in preparing an imprinted polymer and determines the intermolecular interaction between the functional monomer and the template. The influence of the porogenic solvent on the extraction recovery of an imprinted polymer is shown in Table 1 (entries 1–3). The imprinted polymer MIP3, synthesized in dichloromethane:methanol (9:1, v/v), showed a slight extraction recovery for CIP (9.7%), while the higher extraction recovery (57.7%) was obtained when the polymer was synthesized in higher polar aprotic solvent chloroform:methanol (9:1, v/v). However, the polymer synthesized in a polar protic solvent as methanol:acid acetic (9:1, v/v) showed a low extraction recovery (36.5%). These results indicated that the highly polar solvent (chloroform/methanol) was suitable for preparing CIP-MIP and the weakly polar solvent unfavored the interaction between the monomer and the template molecule. In contrast, highly polar protic solvents such as methanol and acetic acid exhibited an intensely competitive interaction with functional monomers. Therefore, the mixture chloroform:methanol (9:1, v/v) was chosen as the porogenic solvent for an imprinted polymer preparation.

3.1.2. Template/Functional Monomer Molar Ratio

Template/functional monomer molar ratio determines the composition, affinity, rigidity, and polymerization rate of the monomer. Three different templates/functional monomer ratios (1:10; 1:15, and 1:20) were evaluated. As shown in Table 1, the recovery was statistically comparable with an increasing number of functional monomers. However, the high template/functional monomer molar ratio increased the non-selective site on polymer and, therefore, caused the loss of the recognition properties [1,33]. For this reason, the template/functional monomer molar ratio 1:10 was used in the imprinted polymer preparation.

3.1.3. Methacrylic Acid/2-Vinylpyridine Molar Ratio

The mixture of methacrylic acid and 2-vinylpyridine monomers with different ratios, which provide a complementary intermolecular interaction toward CIP, was evaluated to adjust the affinity of the obtained polymer. The MAA enhances the affinity and hydrophilicity of MIPs via hydrogen bonds, while 2-VP provides a hydrophobic and π-π stacking interaction with the template molecule. The synergetic effect of these functional monomers is expected to improve the extraction performance of the imprinted polymers. Four MAA/2-VP molar ratios (0.5:0.5; 0.66:0.33; 0.7:0.3 and 0.8:0.2) were evaluated in this study. Higher extraction recoveries (63–105%) were observed when dual functional monomers were used to prepare the imprinted polymer in which the MIP7, synthesized with the MAA/2-VP molar ratios of 0.66:0.33, showed the highest recovery (105%). These results indicated that the extraction performance of MIP was considerably improved by combining MAA and 2-VP as dual functional monomers. The complementary interactions of the hydrogen bond and the π-π stacking of dual-functional monomers favored the access to the recognition cavity and improved the extraction performance of a synthesized imprinted polymer [52].

3.2. Characterization of Polymers

The chemical structure of the obtained polymers was analyzed using FT-IR spectroscopy. As showed in Figure 1, the imprinted polymer (MIP7) and their corresponding non-imprinted polymer (NIP7) demonstrated similar FT-IR spectra. The results suggested that these polymers contained similar chemical compositions. The characteristic of the C-H stretching absorption peak appeared at 2950 cm^{-1}. The strong stretching vibration absorption of C = O in EDGMA and MAA was observed at 1716 cm^{-1} [33]. The characteristics absorption peaks of 2-VP appearing at 1637 and 1450 cm^{-1} were due to the C = N and C = C stretching vibration of the pyridine ring [53]. The strong peak at 1141 cm^{-1} was due to the C–O stretching vibration of EDGMA. The band at 755 cm^{-1} was characteristic for C-H an out-of-plan bending vibration [28]. These results indicated that the 2-VP, MAA, and EDGMA monomers were successfully copolymerized during the imprinted polymer preparation.

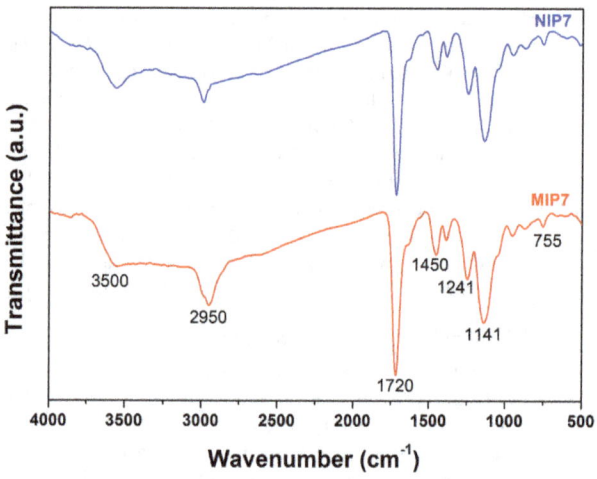

Figure 1. FT-IR spectra of MIP7 and NIP7.

The thermogravimetric analysis-differential scanning calorimetry (TGA-DSC) method was used for analyzing the thermal properties of the polymers. As showed in Figure 2, the first weight loss (9.5%) observed at below 220 °C was due to the loss of the adsorbed water and residue solvent molecules. From 250 to 450 °C, the principal weight loss of 63.1% was characteristic for the thermal decomposition of the polymer. The polymer was completely decomposed at 470 °C. Similar results were observed for the corresponding non-imprinted polymer (NIP7). The weight loss (7.8%) below 220 °C was due to the desorption of water and residue solvent molecules. The thermal decomposition occurred at 250 °C with a major weight loss of 62.6%. The non-imprinted polymer was also completely decomposed at 470 °C. The corresponding DSC thermogram of the imprinted polymer (MIP7) showed two endothermic peaks at 97 and 183 °C, which were due to removing adsorbed water and entrapped solvent molecules. The prominent endothermic peak at 365 °C might be due to the loss of pyridine and CO_2 molecules [54]. The final endothermic peak at 431 °C could be explained by the pyrolysis with carbonization of the backbone. A similar result was observed for the DSC thermogram of the non-imprinted polymer (NIP7).

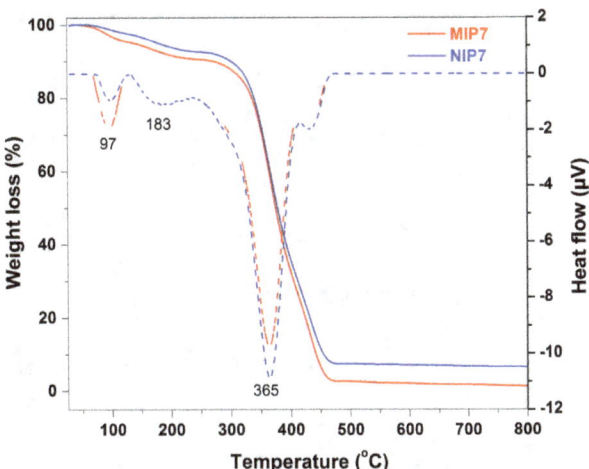

Figure 2. TGA analysis (solid curves) and DSC thermograms (dash curves) of MIP7 and NIP7.

The textural properties of the synthesized polymers were determined using a nitrogen sorption isotherm. The results of the textural properties analyzed using a nitrogen adsorption–desorption isotherm showed that the synthesized polymers were a non-porous material (as shown in Figure 3). The imprinted polymer MIP7 had a specific surface area of 2.5 m^2 g^{-1}, while the non-imprinted NIP7 had a specific surface area of 5.4 m^2 g^{-1}. The morphology properties of polymers were confirmed using scanning electron microscopy (as shown in Figure 4). It is noted that the surface morphologies of the imprinted and non-imprinted polymer were similar. These results indicated that the bulk polymerization process in MIP preparation led to form non-porous and equant polymer particles.

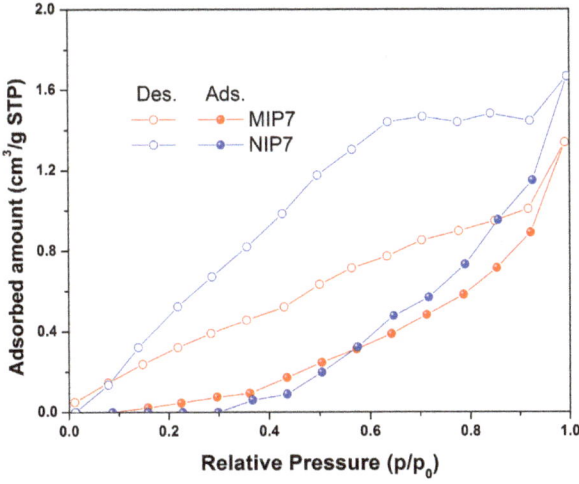

Figure 3. Nitrogen adsorption–desorption isotherms of MIP7 and NIP7 at 77 K.

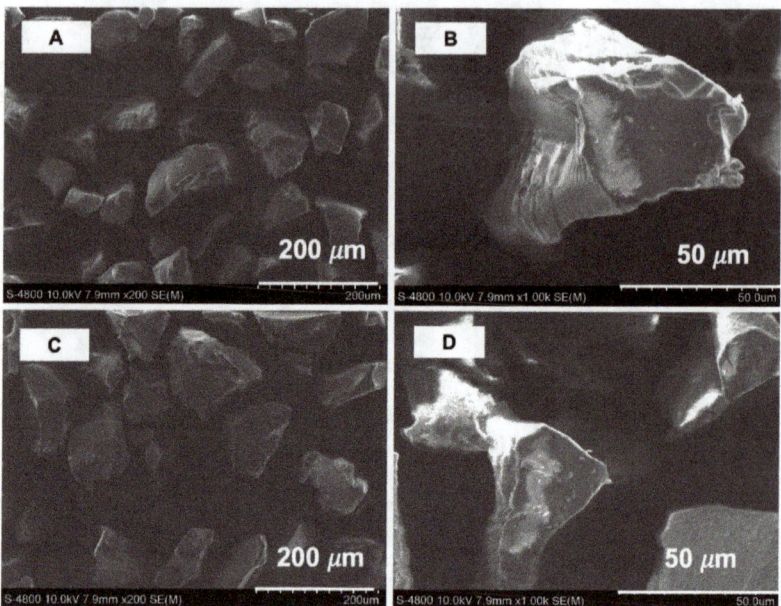

Figure 4. SEM images of MIP7 (**A**,**B**) and NIP7 (**C**,**D**) at different magnifications.

3.3. Adsorption Properties

Batch adsorption experiments were used to determine the adsorption properties of CIP onto the as-prepared polymers. The selectivity of imprinted polymers was evaluated using the imprinting factor (IF = Q_{MIP}/Q_{NIP}) [1]. The adsorption isotherms of MIP2, MIP7, MIP10, and the corresponding non-imprinted polymers are shown in Figure 5.

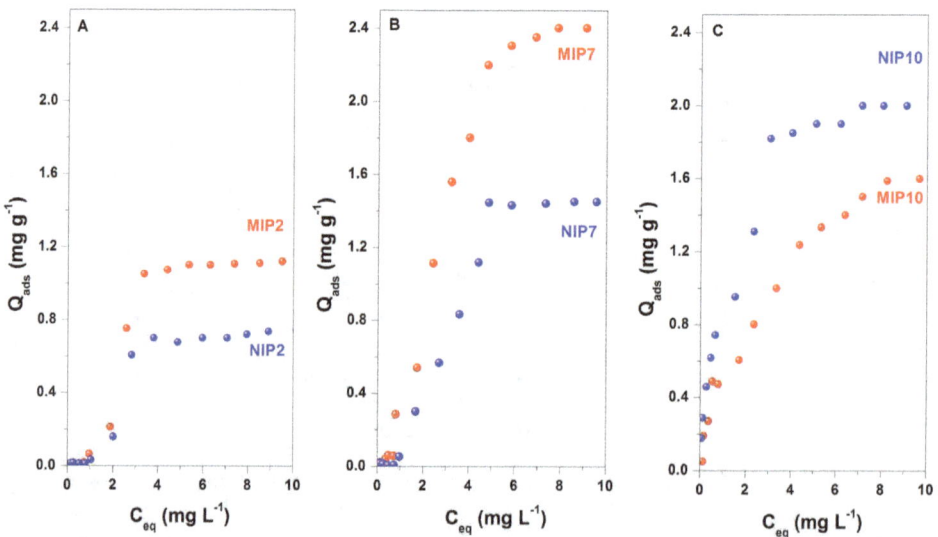

Figure 5. Adsorption isotherms of MIP2 (**A**), MIP7 (**B**), MIP10 (**C**), and their corresponding non-imprinted polymers.

When 2-VP was employed in MIP synthesis, the obtained polymers were non-water-compatible and weakly interacted with CIP molecules. The adsorption isotherms of MIP2 and NIP2 were S-type isotherms, characteristic of a low affinity of adsorbent material [55]. The maximum adsorbed amounts of CIP on MIP2 and NIP2 were 1.12 and 0.74 mg g^{-1}, respectively, with the imprinting factor of 1.51. Thus, MIP2 was a selective adsorbent for CIP. These results indicated that the π-π stacking interaction between 2-VP and CIP could not be disturbed by methanol in the synthesis and water molecules in the rebidding experiments.

It was found that the imprinted polymer obtained from the dual functional monomers had a higher adsorption capacity than the polymers synthesized with one functional monomer. As presented in Figure 5B, the adsorption isotherm of CIP on MIP7 was an L-type isotherm, characteristic of a moderate affinity of adsorbent toward CIP [55]. In contrast, the adsorption isotherm on NIP7 was an S type isotherm. The maximum adsorbed amounts of CIP on MIP7 and NIP7 were 2.40 and 1.45 mg g^{-1}, respectively, with the imprinting factor of 1.66. Moreover, only the imprinted polymer had a good affinity, while the non-imprinted polymer still had a weak interaction toward the target molecule. The results suggested that the hydrogen and π-π stacking interaction between MAA and 2-VP with CIP were responsible for improving the adsorption performances of the polymer.

The MIP10 and NIP10 polymers exhibited a high affinity to the CIP molecule due to the prosperous hydrophilic functional groups of MAA functional monomers. The adsorption isotherms of MIP10 and NIP10 were an H-type isotherm, characteristic of a high affinity of the matrix polymer toward CIP [55]. The maximum adsorbed amounts of CIP on MIP10 and NIP10 were 1.6 and 2.0 mg g^{-1}, respectively, with the imprinting factor of 0.8. Thus, MIP10 was a non-selective adsorbent for CIP. The water-compatible property of polymers is beneficial for the access of target molecules in the adsorption process. However, the hydrogen bond between CIP and functional monomer MAA could be easily disturbed by methanol and water molecules in the synthesis protocol or the rebidding experiments. This phenomenon explained the formation of a non-selective imprinted polymer with MAA functional monomer.

The adsorption data were further analyzed using the Langmuir and Freundlich isotherm models with the non-linear method. The fitted results were summarized in Table 2. Only the adsorption isotherms of CIP on MIP10 and NIP10 were well fitted with Langmuir and Freundlich isotherm models due to their H type isotherms with high correlation coefficients (R^2 = 0.91–0.98). In comparison, the correlation coefficients for the adsorption isotherms for other polymers (MIP2 and MIP7) were low (R^2 = 0.44–0.87). Thus, the Langmuir and Freundlich isotherm models were unsuitable for the adsorption data on MIP2 and MIP7.

Table 2. Fitted results of adsorption data with Langmuir and Freundlich isotherm models using the non-linear method.

Polymer	$Q_{exp.}$ (mg g^{-1})	Langmuir			Freundlich			IF [a]
		Q_{max} (mg g^{-1})	K_L (L mg^{-1})	R^2	K_F	n	R^2	
MIP2	1.12	4.71	0.9353	0.4432	0.0443	0.4789	0.8117	1.51
NIP2	0.74	2.21	0.8952	0.4969	0.0270	0.5073	0.7906	
MIP7	2.40	23.92	0.9487	0.7572	0.1677	0.5873	0.8752	1.66
NIP7	1.45	34.83	0.9962	0.5297	0.0231	0.3704	0.7524	
MIP10	1.60	2.21	0.2664	0.9454	0.3691	1.3408	0.9140	0.80
NIP10	2.00	1.97	0.0754	0.9698	0.8252	2.1160	0.9883	

[a] Imprinting factor = Q_{MIP}/Q_{NIP}.

3.4. Solid Phase Extraction Study

3.4.1. Optimization of Solid Phase Extraction Protocol

Loading condition: The loading solvent provides the appropriate environment for the adsorption of the analytes in the SPE procedure. For this study, 5 mL of CIP solution (0.1 ppm) prepared in deionized water or the mixture of phosphoric acid 0.05% (pH 3)/acetonitrile (8:2, v/v) were loaded on the cartridges. The results showed that CIP was well adsorbed onto the SPE column when deionized water was used as a loading solvent (98–100%), while only 0.1–5% CIP was retained on the cartridges when the CIP solution was prepared in the mixture of phosphoric acid 0.05 % (pH 3)/acetonitrile (8:2, v/v). Thus, ultrapure water was selected as the loading solvent for the SPE experiments.

Washing solvent: A suitable washing solvent is desired to wash out most impurities but not the analyte and improve the sensibility of the analytical method. Different washing solvents, such as dichloromethane/methanol (9:1, v/v), dichloromethane/methanol (5:5, v/v), dichloromethane/methanol (1:9, v/v), chloroform/methanol (9:1, v/v), chloroform/methanol (1:9, v/v), water, and the three steps using water followed by acetonitrile/acetic acid 0.5% (1:9, v/v) and then acetonitrile/ammoniac 0.1% (8:2, v/v) were used to optimize the washing conditions. The mixture of dichloromethane/methanol and chloroform/methanol washed out the most impurities. However, these solvents were also able to elute a large amount of the CIP retained on SPE column. Furthermore, water and the three steps with water, acetonitrile/acetic acid, and acetonitrile/ammoniac showed little influence on CIP recoveries. Therefore, water or the three steps with water, acetonitrile/acetic acid, and acetonitrile/ammoniac can be used as washing solvents, depending on the complexity of analytical samples.

Eluting solvent: The selection of elution is a critical factor to completely desorb the CIP retained on the cartridges. In this study, 5 mL of CIP aqueous solution (0.1 mg L^{-1}) was firstly loaded on the SPE column. Then, 3 mL of different solutions, such as methanol/acetic acid (9:1, v/v), methanol/acetic acid (5:5, v/v), phosphoric acid 0.05% (pH 3)/acetonitrile (8:2, v/v), phosphoric acid 0.05% (pH 3)/acetonitrile (6:4, v/v), acetic acid 0.5% (pH 3)/acetonitrile (8:2, v/v), and acetic acid 0.5% (pH 3)/acetonitrile (6:4, v/v) were employed to elute the analytes. The highest CIP recovery (~100%) was observed when the solution of methanol/acetic acid (9:1, v/v) or phosphoric acid 0.05% (pH 3)/acetonitrile (8:2, v/v) was used as an eluting solvent (Figure 6). Thus, a solution of methanol/acetic acid (9:1, v/v) or phosphoric acid 0.05% (pH 3)/acetonitrile (8:2, v/v) could be used as the eluting solvent.

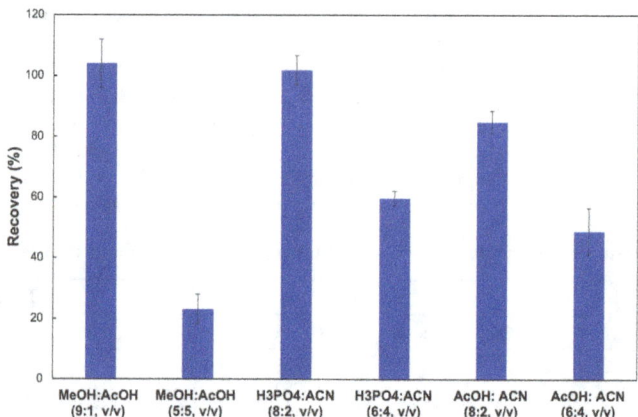

Figure 6. CIP recoveries from MIP-SPE column when using different eluting solvents.

3.4.2. Extraction Performance of Imprinted Polymer

Finally, the synthesized polymers were employed as an adsorbent for the solid phase extraction of CIP in deionized water. As shown in Figure 7, MIP2 and NIP2, which have a weak adsorption affinity, demonstrated low extraction recoveries of CIP (43–58%). The SPE cartridge with MIP7 showed an excellent recovery to CIP (105%), whereas the NIP7 cartridge showed a low recovery (43%). For non-selective polymers MIP10 and NIP10, the CIP extraction recoveries observed were identical. The low extraction recoveries observed for the SPE cartridges with these polymers (23%) could be explained by the high affinity of the MAA monomer toward CIP and that reduced the CIP elimination from the SPE column.

These results indicated that the synergetic effect of dual functional monomers favored the formation of specific adsorption sites on imprinted polymers.

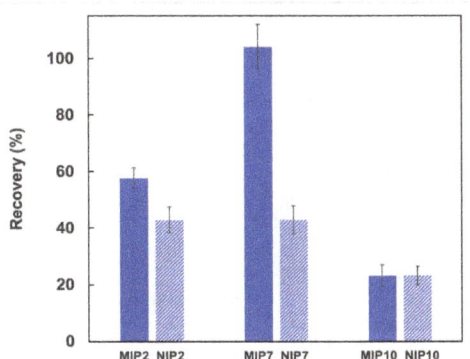

Figure 7. CIP recoveries from different imprinted and non-imprinted polymers (SPE condition: 40 mg of polymer, loading: 5 mL CIP solution in water (0.1 µg L^{-1}), washing: 3 mL of deionized water, eluting: 3 mL of methanol/acetic acid (9:1; v/v), n = 3).

4. Conclusions

In this study, CIP-MIPs have been synthesized using MAA and 2-VP as the dual functional monomers, EDGMA as the cross-linker, and chloroform/methanol (9:1, v/v) as the solvent. The FT-IR, TGA-DSC, SEM, and nitrogen sorption isotherm characterization confirmed the physicochemical properties of the imprinted polymer, and their corresponding non-imprinted polymer was similar. The batch adsorption experiments demonstrated that the MIPs synthesized with the mixture of functional monomers (MAA and 2-VP) showed a higher adsorption capacity and selectivity toward CIP. The CIP-MIP was employed as a SPE adsorbent to extract CIP in pure water with the recovery and the relative standard deviation of 105 and 7.9%, respectively. Further studies are being performing for CIP residue analysis in the wastewater from hospitals and shrimp farms by using this CIP-MIP.

Author Contributions: Funding acquisition, U.D.T.; investigation, H.H.N.T., T.D.P. and H.D.M.; methodology, U.D.T.; project administration, U.D.T.; writing—original draft, U.D.T.; writing—review and editing, T.-T.N.-T. All authors have read and agreed to the published version of the manuscript.

Funding: This research is funded by Graduate University of Science and Technology under grant number GUST.STS.ĐT2018-HH03.

Institutional Review Board Statement: Not applicable.

Informed Consent Statement: Not applicable.

Data Availability Statement: The data presented in this study are available on request from the corresponding author.

Conflicts of Interest: The authors declare no conflict of interest.

References

1. Zhu, G.; Cheng, G.; Wang, P.; Li, W.; Wang, Y.; Fan, J. Water compatible imprinted polymer prepared in water for selective solid phase extraction and determination of ciprofloxacin in real samples. *Talanta* **2019**, *200*, 307–315. [CrossRef]
2. Hanna, N.; Sun, P.; Sun, Q.; Li, X.; Yang, X.; Ji, X.; Zou, H.; Ottoson, J.; Nilsson, L.E.; Berglund, B.; et al. Presence of antibiotic residues in various environmental compartments of Shandong province in eastern China: Its potential for resistance development and ecological and human risk. *Environ. Int.* **2018**, *114*, 131–142. [CrossRef] [PubMed]
3. Fan, Y.; Zeng, G.; Ma, X. Multi-templates surface molecularly imprinted polymer for rapid separation and analysis of quinolones in water. *Environ. Sci. Pollut. Res.* **2020**, *27*, 7177–7187. [CrossRef]

4. Hashemi, S.H.; Ziyaadini, M.; Kaykhaii, M.; Keikha, A.J.; Naruie, N. Separation and determination of ciprofloxacin in seawater human blood plasma and tablet samples using molecularly imprinted polymer pipette-tip solid phase extraction and its optimization by response surface methodology. *J. Sep. Sci.* **2019**, *43*, 505–513. [CrossRef] [PubMed]
5. Neshin, M.V.; Ebrahimi, M. Preconcentration and Determination of Ciprofloxacin with Solid-phase Microextraction and Silica-coated Magnetic Nanoparticles Modified with Salicylic Acid by UV-Vis Spectrophotometry. *Eurasian J. Anal. Chem.* **2018**, *13*, 3 [CrossRef]
6. Kharat, R.; Jadhav, S.; Tamboli, D.; Tamboli, A. Estimation of Ciprofloxacin Hydrochloride in Bulk and Formulation by Derivative UV-Spectrophotometric Methods. *Int. J. Adv. Sci. Res.* **2015**, *1*, 322–328. [CrossRef]
7. Mostafa, S.; El-Sadek, M.; Alla, E.A. Spectrophotometric determination of ciprofloxacin, enrofloxacin and pefloxacin through charge transfer complex formation. *J. Pharm. Biomed. Anal.* **2002**, *27*, 133–142. [CrossRef]
8. Bannefeld, K.-H.; Stass, H.; Blaschke, G. Capillary electrophoresis with laser-induced fluorescence detection, an adequate alternative to high-performance liquid chromatography, for the determination of ciprofloxacin and its metabolite desethyleneciprofloxacin in human plasma. *J. Chromatogr. B Biomed. Sci. Appl.* **1997**, *692*, 453–459. [CrossRef]
9. Wu, S.-S.; Chein, C.-Y.; Wen, Y.-H. Analysis of ciprofloxacin by a simple high-performance liquid chromatography method. *J. Chromatogr. Sci.* **2008**, *46*, 490–495. [CrossRef]
10. Watabe, S.; Yokoyama, Y.; Nakazawa, K.; Shinozaki, K.; Hiraoka, R.; Takeshita, K.; Suzuki, Y. Simultaneous measurement of pazufloxacin, ciprofloxacin, and levofloxacin in human serum by high-performance liquid chromatography with fluorescence detection. *J. Chromatogr. B Anal. Technol. Biomed. Life Sci.* **2010**, *878*, 1555–1561. [CrossRef]
11. Idowu, O.R.; Peggins, J.O. Simple, rapid determination of enrofloxacin and ciprofloxacin in bovine milk and plasma by high-performance liquid chromatography with fluorescence detection. *J. Pharm. Biomed. Anal.* **2004**, *35*, 143–153. [CrossRef]
12. de Oliveira, H.L.; da Silva Anacleto, S.; da Silva, A.T.M.; Pereira, A.; Borges, W.; Figueiredo, E.; de Souza Borges, K.B. Molecularly imprinted pipette-tip solid phase extraction for selective determination of fluoroquinolones in human urine using HPLC-DAD. *J. Chromatogr. B* **2016**, *1033–1034*, 27–39. [CrossRef] [PubMed]
13. Grondin, C.; Zhao, W.; Fakhoury, M.; Jacqz-Aigrain, E. Determination of ciprofloxacin in plasma by micro-liquid chromatography-mass spectrometry: An adapted method for neonates. *Biomed. Chromatogr.* **2011**, *25*, 827–832. [CrossRef] [PubMed]
14. Fan, W.; He, M.; Wu, X.; Chen, B.; Hu, B. Graphene oxide/polyethyleneglycol composite coated stir bar for sorptive extraction of fluoroquinolones from chicken muscle and liver. *J. Chromatogr. A* **2015**, *1418*, 36–44. [CrossRef] [PubMed]
15. Gezahegn, T.; Tegegne, B.; Zewge, F.; Chandravanshi, B.S. Salting-out assisted liquid–liquid extraction for the determination of ciprofloxacin residues in water samples by high performance liquid chromatography–diode array detector. *BMC Chem.* **2019**, *13*, 1–10. [CrossRef] [PubMed]
16. Arabi, M.; Ostovan, A.; Bagheri, A.R.; Guo, X.; Wang, L.; Li, J.; Wang, X.; Li, B.; Chen, L. Strategies of molecular imprinting-based solid-phase extraction prior to chromatographic analysis. *TrAC Trends Anal. Chem.* **2020**, *128*, 115923. [CrossRef]
17. BelBruno, J.J. Molecularly Imprinted Polymers. *Chem. Rev.* **2019**, *119*, 94–119. [CrossRef]
18. Cheong, W.J.; Yang, S.H.; Ali, F. Molecular imprinted polymers for separation science: A review of reviews. *J. Sep. Sci.* **2013**, *36*, 609–628. [CrossRef]
19. Sari, E.; Üzek, R.; Duman, M.; Denizli, A. Detection of ciprofloxacin through surface plasmon resonance nanosensor with specific recognition sites. *J. Biomater. Sci. Polym. Ed.* **2018**, *29*, 1302–1318. [CrossRef]
20. Orozco, J.; Cortés, A.; Cheng, G.; Sattayasamitsathit, S.; Gao, W.; Feng, X.; Shen, Y.; Wang, J. Molecularly Imprinted Polymer-Based Catalytic Micromotors for Selective Protein Transport. *J. Am. Chem. Soc.* **2013**, *135*, 5336–5339. [CrossRef]
21. He, S.; Zhang, L.; Bai, S.; Yang, H.; Cui, Z.; Zhang, X.; Li, Y. Advances of molecularly imprinted polymers (MIP) and the application in drug delivery. *Eur. Polym. J.* **2021**, *143*, 110179. [CrossRef]
22. Huang, Y.; Wang, Y.; Pan, Q.; Wang, Y.; Ding, X.; Xu, K.; Li, N.; Wen, Q. Magnetic graphene oxide modified with choline chloride-based deep eutectic solvent for the solid-phase extraction of protein. *Anal. Chim. Acta* **2015**, *877*, 90–99. [CrossRef] [PubMed]
23. Haginaka, J. Monodispersed, molecularly imprinted polymers as affinity-based chromatography media. *J. Chromatogr. B Anal. Technol. Biomed. Life Sci.* **2008**, *866*, 3–13. [CrossRef] [PubMed]
24. Sobiech, M.; Luliński, P.; Wieczorek, P.P.; Marć, M. Quantum and carbon dots conjugated molecularly imprinted polymers as advanced nanomaterials for selective recognition of analytes in environmental, food and biomedical applications. *TrAC Trends Anal. Chem.* **2021**, *142*, 116306. [CrossRef]
25. Caro, E.; Marcé, R.M.; Cormack, P.A.G.; Sherrington, D.C.; Borrull, F. Direct determination of ciprofloxacin by mass spectrometry after a two-step solid-phase extraction using a molecularly imprinted polymer. *J. Sep. Sci.* **2006**, *29*, 1230–1236. [CrossRef]
26. Lian, Z.; Wang, J. Determination of ciprofloxacin in Jiaozhou Bay using molecularly imprinted solid-phase extraction followed by high-performance liquid chromatography with fluorescence detection. *Mar. Pollut. Bull.* **2016**, *111*, 411–417. [CrossRef]
27. Mirzajani, R.; Kardani, F. Fabrication of ciprofloxacin molecular imprinted polymer coating on a stainless steel wire as a selective solid-phase microextraction fiber for sensitive determination of fluoroquinolones in biological fluids and tablet formulation using HPLC-UV detection. *J. Pharm. Biomed. Anal.* **2016**, *122*, 98–109. [CrossRef]
28. Zhang, X.; Gao, X.; Huo, P.; Yan, Y. Selective adsorption of micro ciprofloxacin by molecularly imprinted functionalized polymers appended onto ZnS. *Environ. Technol.* **2012**, *33*, 2019–2025. [CrossRef] [PubMed]

29. Díaz-Alvarez, M.; Turiel, E.; Martín-Esteban, A. Selective sample preparation for the analysis of (fluoro)quinolones in baby food: Molecularly imprinted polymers versus anion-exchange resins. *Anal. Bioanal. Chem.* **2008**, *393*, 899–905. [CrossRef]
30. Prieto, A.; Schrader, S.; Bauer, C.; Möder, M. Synthesis of a molecularly imprinted polymer and its application for microextraction by packed sorbent for the determination of fluoroquinolone related compounds in water. *Anal. Chim. Acta* **2011**, *685*, 146–152. [CrossRef]
31. Yan, H.; Row, K.H.; Yang, G. Water-compatible molecularly imprinted polymers for selective extraction of ciprofloxacin from human urine. *Talanta* **2008**, *75*, 227–232. [CrossRef]
32. Turiel, E.; Martin-Esteban, A.; Tadeo, J.L. Molecular imprinting-based separation methods for selective analysis of fluoroquinolones in soils. *J. Chromatogr. A* **2007**, *1172*, 97–104. [CrossRef] [PubMed]
33. Ma, W.; Row, K.H. Simultaneous determination of levofloxacin and ciprofloxacin in human urine by ionic-liquid-based, dual-template molecularly imprinted coated graphene oxide monolithic solid-phase extraction. *J. Sep. Sci.* **2019**, *42*, 642–649. [CrossRef] [PubMed]
34. Tegegne, B.; Chimuka, L.; Chandravanshi, B.S.; Zewge, F. Molecularly imprinted polymer for adsorption of venlafaxine, albendazole, ciprofloxacin and norfloxacin in aqueous environment. *Sep. Sci. Technol.* **2021**, *56*, 2217–2231. [CrossRef]
35. Huynh, T.-P.; Sharma, P.S.; Sosnowska, M.; D'Souza, F.; Kutner, W. Functionalized polythiophenes: Recognition materials for chemosensors and biosensors of superior sensitivity, selectivity, and detectability. *Prog. Polym. Sci.* **2015**, *47*, 1–25. [CrossRef]
36. Surya, S.G.; Khatoon, S.; Lahcen, A.A.; Nguyen, A.T.H.; Dzantiev, B.B.; Tarannum, N.; Salama, K.N. A chitosan gold nanoparticles molecularly imprinted polymer based ciprofloxacin sensor. *RSC Adv.* **2020**, *10*, 12823–12832. [CrossRef]
37. Cavalera, S.; Chiarello, M.; Di Nardo, F.; Anfossi, L.; Baggiani, C. Effect of experimental conditions on the binding abilities of ciprofloxacin-imprinted nanoparticles prepared by solid-phase synthesis. *React. Funct. Polym.* **2021**, *163*, 104893. [CrossRef]
38. Naphat, Y.; Nurerk, P.; Chullasat, K.; Kanatharana, P.; Davis, F.; Sooksawat, D.; Bunkoed, O. A nanocomposite optosensor containing carboxylic functionalized multiwall carbon nanotubes and quantum dots incorporated into a molecularly imprinted polymer for highly selective and sensitive detection of ciprofloxacin. *Spectrochim. Acta Part A Mol. Biomol. Spectrosc.* **2018**, *201*, 382–391. [CrossRef]
39. Marestoni, L.D.; Wong, A.; Feliciano, G.T.; Marchi, M.R.R.; Tarley, C.R.T.; Sotomayor, M.D.P.T. Semi-Empirical Quantum Chemistry Method for Pre-Polymerization Rational Design of Ciprofloxacin Imprinted Polymer and Adsorption Studies. *J. Braz. Chem. Soc.* **2015**, *27*, 109–118. [CrossRef]
40. Vasapollo, G.; Del Sole, R.; Mergola, L.; Lazzoi, M.R.; Scardino, A.; Scorrano, S.; Mele, G. Molecularly Imprinted Polymers: Present and Future Prospective. *Int. J. Mol. Sci.* **2011**, *12*, 5908–5945. [CrossRef]
41. De Faria, H.; Abrão, L.C.D.C.; Santos, M.G.; Barbosa, A.F.; Figueiredo, E.C. New advances in restricted access materials for sample preparation: A review. *Anal. Chim. Acta* **2017**, *959*, 43–65. [CrossRef]
42. Zhao, Q.; Zhao, H.; Huang, W.; Yang, X.; Yao, L.; Liu, J.; Li, J.; Wang, J. Dual functional monomer surface molecularly imprinted microspheres for polysaccharide recognition in aqueous solution. *Anal. Methods* **2019**, *11*, 2800–2808. [CrossRef]
43. Cai, X.; Li, J.; Zhang, Z.; Yang, F.; Dong, R.; Chen, L. Novel Pb2+ Ion Imprinted Polymers Based on Ionic Interaction via Synergy of Dual Functional Monomers for Selective Solid-Phase Extraction of Pb2+ in Water Samples. *ACS Appl. Mater. Interfaces* **2014**, *6*, 305–313. [CrossRef]
44. Zhang, Y.; Li, Y.; Hu, Y.; Li, G.; Chen, Y. Preparation of magnetic indole-3-acetic acid imprinted polymer beads with 4-vinylpyridine and β-cyclodextrin as binary monomer via microwave heating initiated polymerization and their application to trace analysis of auxins in plant tissues. *J. Chromatogr. A* **2010**, *1217*, 7337–7344. [CrossRef]
45. Ramstroem, O.; Andersson, L.I.; Mosbach, K. Recognition sites incorporating both pyridinyl and carboxy functionalities prepared by molecular imprinting. *J. Org. Chem.* **1993**, *58*, 7562–7564. [CrossRef]
46. Wang, J.; Dai, J.; Meng, M.; Song, Z.; Pan, J.; Yan, Y.; Li, C. Surface molecularly imprinted polymers based on yeast prepared by atom transfer radical emulsion polymerization for selective recognition of ciprofloxacin from aqueous medium. *J. Appl. Polym. Sci.* **2013**, *131*, 1–10. [CrossRef]
47. Janczura, M.; Luliński, P.; Sobiech, M. Imprinting Technology for Effective Sorbent Fabrication: Current State-of-Art and Future Prospects. *Materials* **2021**, *14*, 1850. [CrossRef] [PubMed]
48. Mayes, A.G.; Mosbach, K. Molecularly Imprinted Polymer Beads: Suspension Polymerization Using a Liquid Perfluorocarbon as the Dispersing Phase. *Anal. Chem.* **1996**, *68*, 3769–3774. [CrossRef] [PubMed]
49. Zhao, G.; Liu, J.; Liu, M.; Han, X.; Peng, Y.; Tian, X.; Liu, J.; Zhang, S. Synthesis of Molecularly Imprinted Polymer via Emulsion Polymerization for Application in Solanesol Separation. *Appl. Sci.* **2020**, *10*, 2868. [CrossRef]
50. Sun, Y.; Zhang, Y.; Ju, Z.; Niu, L.; Gong, Z.; Xu, Z. Molecularly imprinted polymers fabricated by Pickering emulsion polymerization for the selective adsorption and separation of quercetin from Spina Gleditsiae. *New J. Chem.* **2019**, *43*, 14747–14755. [CrossRef]
51. Blasco, C.; Picó, Y. Development of an Improved Method for Trace Analysis of Quinolones in Eggs of Laying Hens and Wildlife Species Using Molecularly Imprinted Polymers. *J. Agric. Food Chem.* **2012**, *60*, 11005–11014. [CrossRef]
52. Duan, F.; Chen, C.; Zhao, X.; Yang, Y.; Liu, X.; Qin, Y. Water-compatible surface molecularly imprinted polymers with synergy of bi-functional monomers for enhanced selective adsorption of bisphenol A from aqueous solution. *Environ. Sci. Nano* **2016**, *3*, 213–222. [CrossRef]

53. Wan, Y.; Wang, M.; Fu, Q.; Wang, L.; Wang, D.; Zhang, K.; Xia, Z.; Gao, D. Novel dual functional monomers based molecularly imprinted polymers for selective extraction of myricetin from herbal medicines. *J. Chromatogr. B Anal. Technol. Biomed. Life Sci.* **2018**, *1097–1098*, 1–9. [CrossRef]
54. Gogoi, A.; Sarma, N.S. Improvement in ionic conductivities of poly-(2-vinylpyridine) by treatment with crotonic acid and vinyl acetic acid. *Bull. Mater. Sci.* **2015**, *38*, 797–803. [CrossRef]
55. Limousin, G.; Gaudet, J.-P.; Charlet, L.; Szenknect, S.; Barthès, V.; Krimissa, M. Sorption isotherms: A review on physical bases, modeling and measurement. *Appl. Geochem.* **2007**, *22*, 249–275. [CrossRef]

Review

Molecularly Imprinted Polymers as State-of-the-Art Drug Carriers in Hydrogel Transdermal Drug Delivery Applications

Aleksandra Lusina * and Michał Cegłowski *

Faculty of Chemistry, Adam Mickiewicz University in Poznań, Uniwersytetu Poznańskiego 8, 61-614 Poznań, Poland
* Correspondence: alelus@amu.edu.pl (A.L.); michal.ceglowski@amu.edu.pl (M.C.)

Abstract: Molecularly Imprinted Polymers (MIPs) are polymeric networks capable of recognizing determined analytes. Among other methods, non-covalent imprinting has become the most popular synthesis strategy for Molecular Imprinting Technology (MIT). While MIPs are widely used in various scientific fields, one of their most challenging applications lies within pharmaceutical chemistry, namely in therapeutics or various medical therapies. Many studies focus on using hydrogel MIPs in transdermal drug delivery, as the most valuable feature of hydrogels in their application in drug delivery systems that allow controlled diffusion and amplification of the microscopic events. Hydrogels have many advantages over other imprinting materials, such as milder synthesis conditions at lower temperatures or the increase in the availability of biological templates like DNA, protein, and nucleic acid. Moreover, one of the most desirable controlled drug delivery applications is the development of stimuli-responsive hydrogels that can modulate the release in response to changes in pH, temperature, ionic strength, or others. The most important feature of these systems is that they can be designed to operate within a particular human body area due to the possibility of adapting to well-known environmental conditions. Therefore, molecularly imprinted hydrogels play an important role in the development of modern drug delivery systems.

Keywords: Molecular Imprinted Polymers (MIP); Molecular Imprinting Technology (MIT); hydrogels; transdermal drug delivery

1. Introduction

Molecular Imprinted Polymers (MIPs) are polymeric systems that possess a unique property to recognize a specific molecule or group of structurally related molecules. MIPs selective recognition's property is determined during the preparation of polymer using a template molecule together with appropriate monomers in a solvent. MIPs are prepared in the presence of template molecules that can be subsequently removed, which determine MIP's cavity selectivity for a specific template or compounds structurally related to this template [1,2]. Created tailor-made sites gain the property to selectively recognize the template molecule's size, shape, and functional groups.

The first reported molecular imprinting concept was proposed in 1931 by Polyakov [3] as "unusual adsorption properties of silica particles prepared using a novel synthesis procedure". The mentioned "unusual adsorption properties" have been reported using numerous polymers, which have been subsequently named as molecularly imprinted polymers—MIPs.

There are three methods to form molecular imprinting. The first one is a covalent method based on reversible covalent bonds, introduced by Wulff in 1995 [4], the second is a method proposed by Mosbach in 1994 [5], which is based on non-covalent interactions between templates-imprinted molecules- and functional monomers and the last one is semi-covalent, reported by Whitcombe et al. [6], in which subsequent rebinding by non-covalent bond can be created after a covalently bounded template is removed (Figure 1) [6,7].

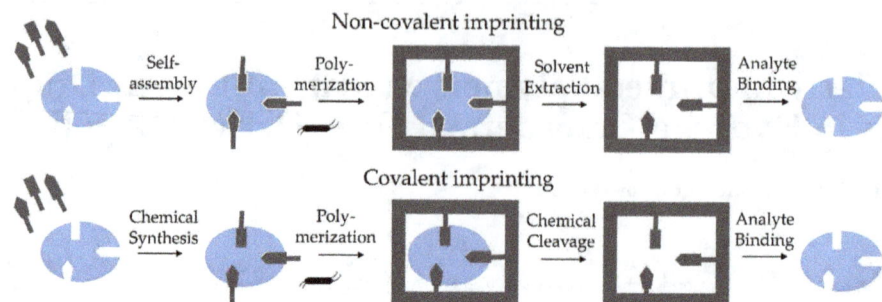

Figure 1. Schematic representation of covalent and non-covalent mechanisms of molecularly imprinted procedures.

The covalent imprinting method is based on creating a covalent bond between the template and the appropriate monomer. Then, during polymerization, the covalent linkage is cleaved, and subsequently, the template is removed from the MIP matrix. Rebinding of the previously removed template causes reappearance of the same covalent linkage. Since the formation of identical rebinding linkages requires rapidly reversible covalent interactions between templates and appropriate monomers, the number of suitable templates for covalent imprinting is limited. Additionally, the robust nature of the covalent interactions and consequent slow dissociation and binding makes it hard to reach thermodynamic equilibrium. The second method, non-covalent imprinting, has no such restrictions and is the most frequently used due to its simplicity [8]. By using an appropriate solvent, the formed various interactions such as hydrogen bonds, π-π and ionic interactions, van der Waals forces, etc., generate template-monomer complexes. After removing the template from the MIP matrix, the interactions can be easily recreated. The removed template can be rebound via the same non-covalent interactions as before polymerization. Therefore, the range of applicative compounds which can be imprinted via non-covalent imprinting is expanded, and non-covalent imprinting has become the most popular and general synthesis strategy for Molecular Imprinting Technology (MIT). The third type of imprinting method is semi-covalent, defined as subsequent rebinding by non-covalent bond after a covalently bounded template is removed. This semi-covalent approach was firstly reported by Whitcombe et al. [6] and offered an intermediate alternative in which the template is bound covalently to functional monomer since the template rebinding is based on non-covalent interactions. Semi-covalent bond can be characterized by the high affinity of covalent binding and mild operation conditions of non-covalent rebinding. The schematic diagram of non-covalent imprinting mechanisms is presented in Figure 2 [7].

In comparison to other well-known recognition systems, MIPs have received considerable attention. Thanks to that, MIPs are widely used in various fields such as purification [9], separation [10], and catalysis [11], and degradation processes [12] but also they have become attractive in drug delivery [13], artificial antibodies [14], or biosensing [15]. The widespread use of MIPs is an aftermath of their favorable characteristics, such as high physical stability to harsh chemical and physical conditions, straightforward preparation, remarkable robustness, excellent reusability, and low-cost synthesis [7,16]. Whereas MIPs present a wide range of advantages, there is some drawback that should be considered. One of them is the design of a new MIP system that will be suitable for a specific template molecule usually requires a lot of work and time to estimate the best synthesis conditions that allow obtaining the intended material. Before finding the optimum conditions, there is a necessity to continually change various experimental parameters [13].

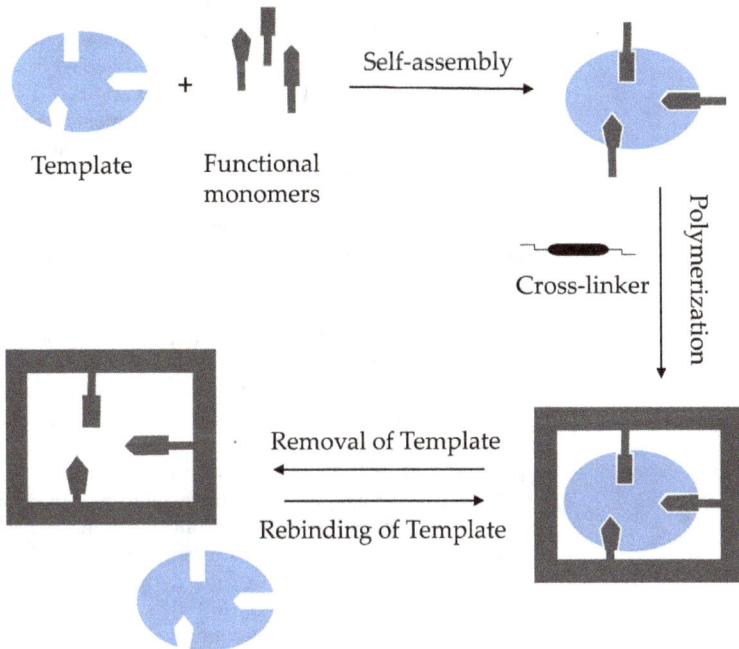

Figure 2. Schematic diagram of the non-covalent method for molecular imprinting process.

2. Fundamentals of MIPs

Molecularly imprinted polymers are polymeric matrices that are moulds for the formation of template complementary binding areas. They can be programmed to recognize a large variety of structures with antibody-like affinities and selectivities. In addition to the already mentioned advantages, these properties have made MIPs attractive in various fields. The main applications of molecularly imprinted polymers are presented in Figure 3 [17].

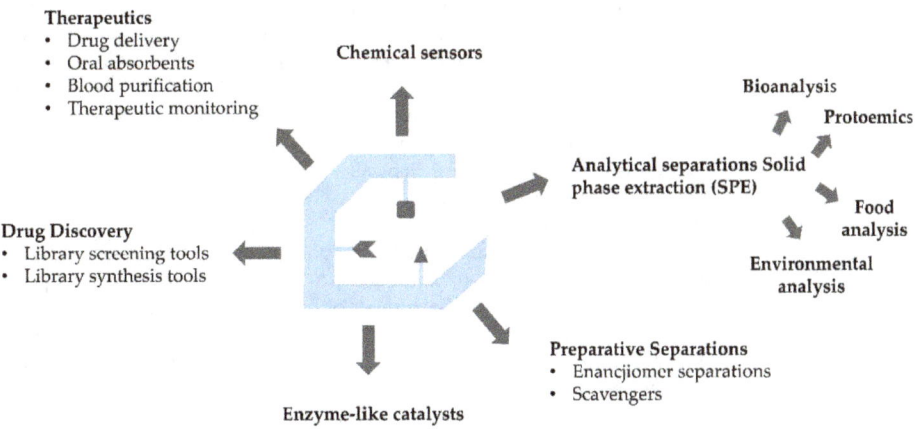

Figure 3. Scheme of the main applications for MIPs.

2.1. Essential Elements of Molecular Imprinting

Generally, MIPs are synthesized using a functional monomer, template, cross-linker, a polymerization initiator, and an appropriate solvent. In short, MIPs are prepared by mixing the mentioned molecules, and then, this pre-polymerization mixture has to be irradiated with UV light or subjected to heat to initiate polymerization [8]. As polymerization is affected by many factors, MIPs can be modified by the appropriate choice of synthesis conditions. There is a possibility to obtain various MIPs with specific properties due to changes of factors such as type and amount of monomer, initiator, cross-linker, and solvent. Additionally, the time and temperature of polymerization reaction also play an important role in creating MIPs with superior targeted properties [16].

The central importance of MIP structure is a template, which can direct the organization of the functional group's pendant to the functional monomers in the molecular imprinting process. Templates should be inert under the polymerization conditions during free radical polymerization [8]. The main goal of molecularly imprinted technology is to create MIPs compared with biological receptors in specificity. Thanks to that, MIPs might replace those entities in real-life applications. Since there is a lot of requirements that should be met, the three mentioned ahead make the template an ideal candidate—it should exhibit excellent chemical stability during the polymerization reaction, it should contain functional groups that do not prevent polymerization, and it should contain functional groups that can form complexes with functional monomers [6]. Additionally, established imprinted small organic molecules such as pharmaceuticals, pesticides, or amino acids are well-known and commonly used. Furthermore, a lot of research proves that not only small molecules are suitable for Molecular Imprinting Technology. Since small molecules have a lot of advantages—like being more rigid to form well-defined binding cavities during the imprinting process—there are some of the protocols that reported using larger organic entities like proteins or even cells. As only a few protocols are reported, imprinting larger organic compounds containing secondary or tertiary structures is still a challenge because these structures may be affected when exposed to the thermal or photolytic treatment involved in the synthesis of MIPs. The rebinding process is also more complicated when using such individuals as large templates, as they do not penetrate the polymeric network easily to reoccupy the binding cavities [8].

It is essential to select a suitable functional monomer that can strongly and selectively react with the template to form specific complexes. Generally, the functional monomers are responsible for the binding interactions present in the imprinted binding sites during the imprinting process. As reported in many protocols, for non-covalent molecular imprinting reaction, the monomer is used in excess of the number of moles of the template to form template-functional monomers assemblies. To maximize expected complex formation, matching the template's functionality with the monomer's functionality plays a crucial role. The imprinting effects increase when the template's functionality is matched with the functionality of the monomer in a complementary fashion, like an H-bond donor with an H-bond acceptor [7]. The amount of monomers that can be used in molecular imprinting is limited. It is imperative to synthesize new functional monomers that form specific interactions with the templates. Typically, monomers include two independent types of units—the recognition unit and the unit, which can be polymerized [16]. Figure 4 presents widely used functional monomers [8].

The amount of cross-linker used in the polymerization process also plays an important role in MIPs properties. Too low cross-linker causes unstable mechanical properties, whereas too high amount will reduce the number of recognition areas per unit mass of MIPs. The primary role of a cross-linker is to form a highly cross-linked polymer. Cross-linker is involved in fixing monomers around template molecules, thus forming a cross-linked polymer. The main aim is to develop a highly cross-linked polymer even after removing templates. Generally, the amount and type of used cross-linker regulate the selectivity and binding capacity of MIPs [8,16]. In MIPs structure, cross-linker fulfills three major functions: controlling the morphology of the polymer matrix, serving to stabilize the imprinted binding site, and giving mechanical stability to the polymer matrix. The structures of commonly used cross-linking agents in molecular imprinting techniques are presented in Figure 5 [8].

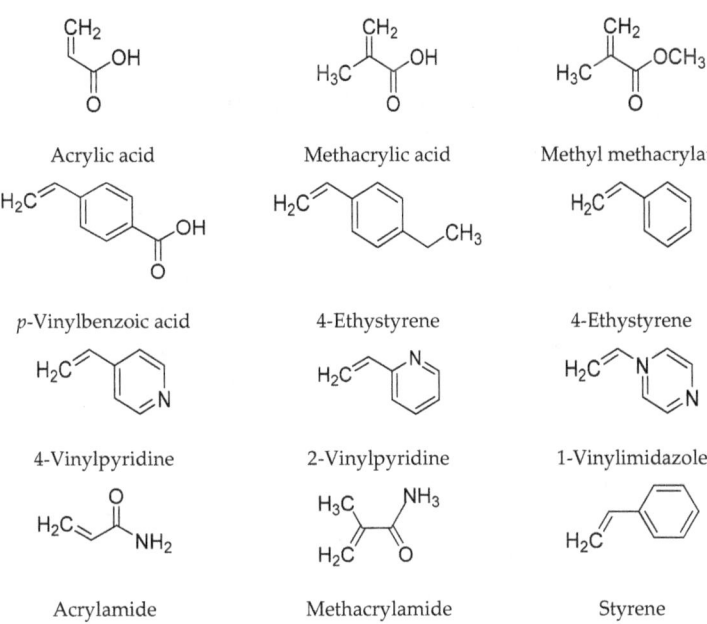

Figure 4. Scheme of commonly used monomers for non-covalent molecularly imprinted technique.

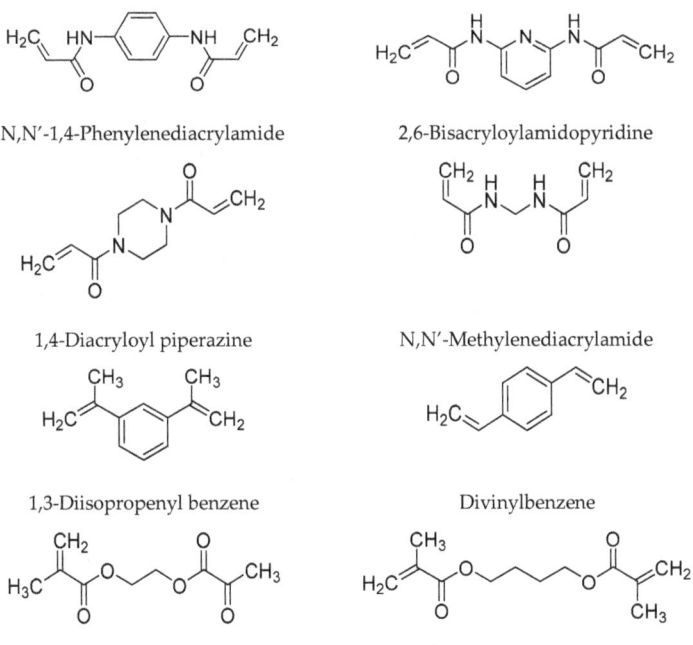

Figure 5. Chemical structures of commonly used cross-linking agents in molecular imprinting technique.

The most useful reactions for preparing MIPs are free radical polymerization (FRP), photopolymerization, and electropolymerization [16]. Plenty of initiators with different chemical properties can be used as the radical source in a free radical polymerization (Figure 6) [8]. In comparison to the monomers, initiators are used at low levels, e.g., 1 wt.%, or 1 mol.% with respect to the total amount of moles of polymerizable double bonds. Moreover, there are several existing ways in which the rate control and mode of decomposition of an initiator to radicals, including heat, light, and by chemical/electrochemical means, depending upon its chemical nature, can be achieved [8].

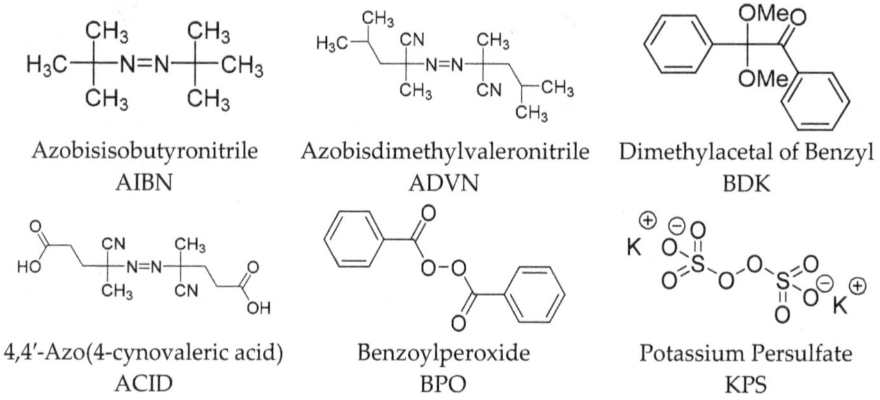

Azobisisobutyronitrile
AIBN

Azobisdimethylvaleronitrile
ADVN

Dimethylacetal of Benzyl
BDK

4,4′-Azo(4-cynovaleric acid)
ACID

Benzoylperoxide
BPO

Potassium Persulfate
KPS

Figure 6. Structures of commonly used initiators in free radical polymerization.

The last chemical individuum that strongly impacts proper MIPs formation is solvent. During polymerization, it generally plays an important role as dispersion media and pore-forming agent. Commonly used solvents include 2-methoxyethanol, methanol, tetrahydrofuran, acetonitrile, dichloromethane, chloroform, N,N-dimethylformamide, and toluene [18]. Porogenic solvent needs to have the following features: all of the used chemical individuals should be well-soluble in the chosen solvent, the solvent should produce large pores to assure flow-through properties of the polymer, and the last on, the solvent should possess low polarity to avoid interferences during complex formation between imprinted molecules and monomers, which is important to obtain high selectivity MIPs [8]. The interactions between templates and monomers depend on the used solvent's polarity. Non-polar or low polar solvents such as chloroform are used for non-covalent imprinting. Due to that, the obtained MIPs gain good imprinting efficiency because the adsorption properties and morphology of polymer depend on the type of used solvent [16].

2.2. Molecular Imprinting in Drug Delivery

Molecular imprinting is one of the most promising ways to recreate biological molecular recognition and mimic properties of antibodies and enzymes in synthetic materials. Many researchers are trying to mimic molecular interactions present in these systems as high recognition characteristics seem to be the fundamental requirement of living systems. Therefore, most approaches focus on creating a binding cavity in which functional chemical groups may be strictly positioned. The mechanism of selective recognition and subsequent drug release from the MIPs structure is presented in Figure 7.

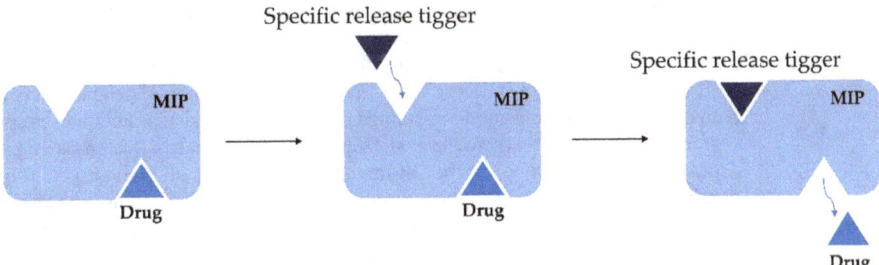

Figure 7. Scheme of intelligent drug release from MIP.

Whereas MIPs are widely used in various areas of science, the area with one of the greatest potential seems to be pharmaceutical chemistry, where MIPs, as synthetic molecularly selective receptors, may be used in therapeutics or medical therapy [17]. Due to properties such as biocompatibility, low toxicity, and biodegradability, MIPs receive extensive attention as drug delivery systems (DDS). MIPs have already been used as selective oral adsorbents for cholesterol [19] and imprinted bile acid sequestrants [20,21]. Commonly, MIPs are widely used as DDS in various diseases such as cancer [22], arrhythmia [23,24], avitaminosis [25], cardiovascular and cerebrovascular disease [26], inflammation [27], addictive disease [28,29] and other [30]. There are also several applications in which MIPs are incorporated into membranes to be used for bio-separation and bio-purification [31]. MIPs are also used in controlled release delivery systems. It is reported that MIPs are widely used for modifying drug release from solid dosage forms, which results in tuned composition release. Whereas many studies are based on a simple modification of non-imprinted polymers, there is also a huge potential in another MIPs application area—intelligent drug release. This release refers to the predictable release of the therapeutic agent in response to specific stimuli like the presence of another specific molecule or change in pH. An example of this intelligent release might be a cell surface epitope. The general mechanism of drug release from the cell surface is illustrated in Figure 8 [17].

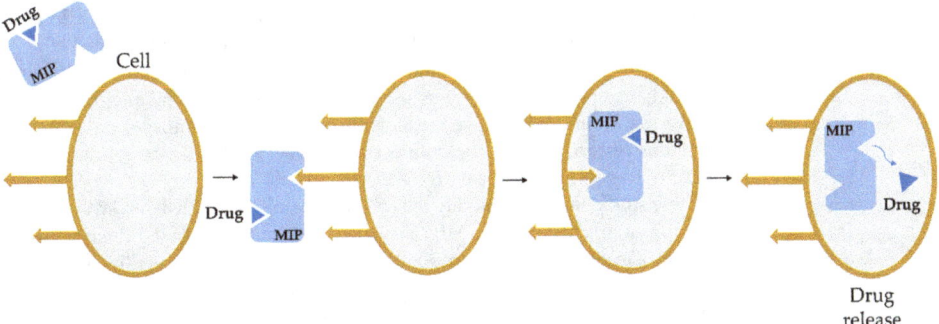

Figure 8. Scheme of targeted drug delivery into a cell with MIP.

3. MIP Challenges in Transdermal Delivery

The oral route is the most common and convenient route of drug administration. Low oral bioavailability and side effects restricted oral administration of most drugs with poor solubility in clinical use [30]. However, there are a lot of medicines that implementation by the oral route is characterized by a low adsorption rate from the gastrointestinal tract or extensive first-pass metabolism [32,33]. In these examples, the skin may be an alternative route for applying drugs, as it has been widely used as a route of administration for local and systemic drugs [34,35]. Transport across the biological barrier creates a problem resulting from the interaction or reaction of the experimental drug with the epithelial barrier [36]. The use of MIT may be integrated to design DDS with properties tuned with pharmaceutical agents produced for dermal or transdermal use [2].

There are a lot of studies that involve MIPs in transdermal drug delivery applications. In one of them, a molecularly imprinted nicotine transdermal system was proposed. In this study reported in 2014 by Ruela et al. [28], an imprinted matrix was prepared by free radical polymerization of a copolymer of methacrylic acid and ethylene glycol dimethacrylate using the bulk technique in the presence of nicotine, which acted as the template. The bulk material was then crushed to obtain particles of 75–106 µm. These particles were dispersed in mineral oil or propylene glycol and formed into a disk with a surface equal to 1.8 cm^2. As results showed, mineral oil was the most promising vehicle due to its hydrophobic characteristics, which improve the molecular recognition of nicotine in MIP particles. Polymeric particles present in the transdermal system differed in polarity. In this study, the pH of the nicotine imprinted polymeric delivery system was similar to the skin's pH. The imprinted polymers were characterized using various techniques to study the morphology of the particles, drug-polymer interactions, and compatibility. The results of controlled release were compared with the commercially available product—Nicopath®. Results obtained after in vitro experiments showed that the amounts of permeated nicotine from the imprinted matrix were similar to commercial patches. The results are 655 and 709 µg cm^{-2} for 24 h, respectively. According to the results obtained in the study made by Ruela, a MIP created with a nicotine template showed promising results. It was demonstrated that non-covalent MIP drug interactions might modify the profile of drug release and skin penetration. Additional studies, such as FT-IR or SEM, also confirmed that prepared MIPs with nicotine as a template have high thermal stability and are resistant to chemical degradation. Although both molecularly imprinted polymer and non-molecular imprinted polymer could bind the templates in their matrixes, MIPs showed better performance during the transdermal release. It is caused by the presence of selective recognition sites in the MIP structure [28,34].

In the further studies made by the same research group, experiments involving the synthesis of several MIPs using precipitation polymerization technique to find optimized materials able to selectively absorb nicotine were performed. As a result, release and skin permeation by nicotine was optimized using MIPs synthesized by precipitation polymerization technique. Obtained polymers showed improved adsorption capacity and selectivity, additionally, MIPs were also able to modulate the transdermal delivery of templated nicotine [29].

Another application of imprinted polymers is shown in one of the studies presented by Bodhibukkana et al. [37]. MIPs were used as a composite material integrated with cellulose to form a membrane to improve the biocompatibility of the transdermal system (Figure 9). As previous research shows, the cellulose membrane is a biocompatible and biodegradable material with good mechanical properties.

These studies aimed to modify the cellulose membrane with a thin layer of R-propranolol or S-propranolol entrapped in MIPs structure. The results showed the potential of molecularly imprinted polymer composite membranes based on cellulose in controlling S-propranolol release into the skin. The degree of stereoselectivity demonstrated a higher therapeutic advantage when considering the two enantiomers of propranolol (R/S). Due to selectivity towards S-propranolol of the MIPs present in the surface of the cellulose membranes, a limited release was achieved [34,37].

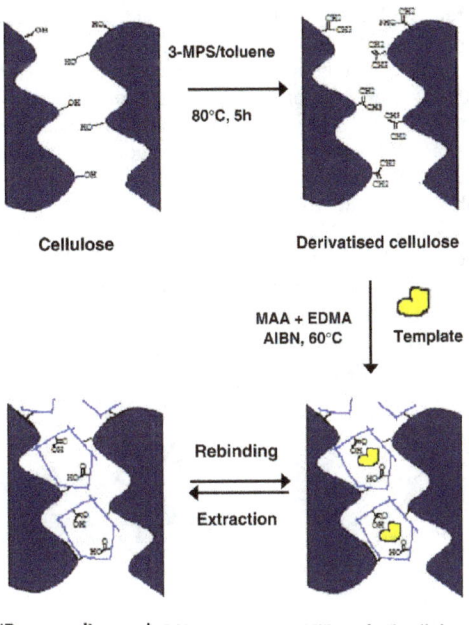

Figure 9. Scheme of cellulose membrane modified with a MIP. Reprinted from [37], Copyright (2022), with permission from Elsevier.

4. Basic Characteristics of Hydrogels

Polymeric hydrogel networks are insoluble, cross-linked, and composed of hydrophilic homo- or hetero-co-polymers, absorbing significant amounts of water and retaining their shape without dissolving. Cross-links within hydrogels may be covalent bonds, permanent entanglements, ionic interactions, or microcrystalline regions incorporating various chains. Loading therapeutics into the hydrogel network takes place using one of two possible scenarios. One of them refers to producing the appropriate gel in the presence of the drug, whereas the second is a path of firstly synthesizing the gel and then loading therapeutic into the gel [38]. Using appropriate monomers with defined properties allows the formed hydrogels to be environmentally responsive, for example, toward changes in pH. Generally, molecular imprinting technologies allow hydrogels to: recognize and selectively bind the specific substrate into the hydrogel. Hydrogels have many advantages compared with other imprinting materials, such as milder synthesis conditions at lower temperatures or relatively high solubility of biological templates like DNA, protein, nucleic acid, etc. Due to that, molecularly imprinted hydrogels play an important role in modern drug delivery systems [39].

The most valuable feature of hydrogels in drug delivery systems is their ability to control diffusion and ability to amplify the microscopic events, which occur at the mesh chain level into macroscopic phenomena [38,40,41]. It is well known that the delivery of certain drugs directly to localized sites beneath the skin is highly desirable in some cases since it would allow local pathology to be treated without significant systemic side effects [42]. Additionally, there are some benefits of transdermal drug delivery within using hydrogels. The drugs dosage can be interrupted on-demand by simply removing the devices, and that drugs can bypass hepatic first-pass metabolism. Using hydrogels is also beneficial because of their dual structure, involving a macroscale three-dimensional macromolecular network with a highly hydrated microscale environment where the former characteristic supplies necessary macroscale rigidity, whereas the latter provides the potential for relatively efficient mass transfer [43]. What is important, swollen hydrogels, due to their high water content, may provide a better feeling for the skin in comparison to conventional ointment and patches [40].

The behavior of hydrogels in a changing environment is presented in Figure 10. Immersing a dry hydrogel in a compatible solvent causes the solvent movement into the hydrogel polymer chain followed by volume expansion and macromolecular rearrangement depending on the extent of crosslinking within the network (presented in Figure 10). Two factors are decisive for the rate at which a polymer expands or swells—the rates of polymer-chain relaxation and solvent penetration into the hydrogel network. Moving from an unperturbed, glassy state to a solvated, rubbery hydrogel state leads to unlimited exchange in transport. This is an important feature for swelling-controlled hydrogels, in which we can obtain a zero-order release or constant release rate. These release rates can be achieved by keeping the constant rate of solvent front penetration, which should be smaller than the drug diffusion.

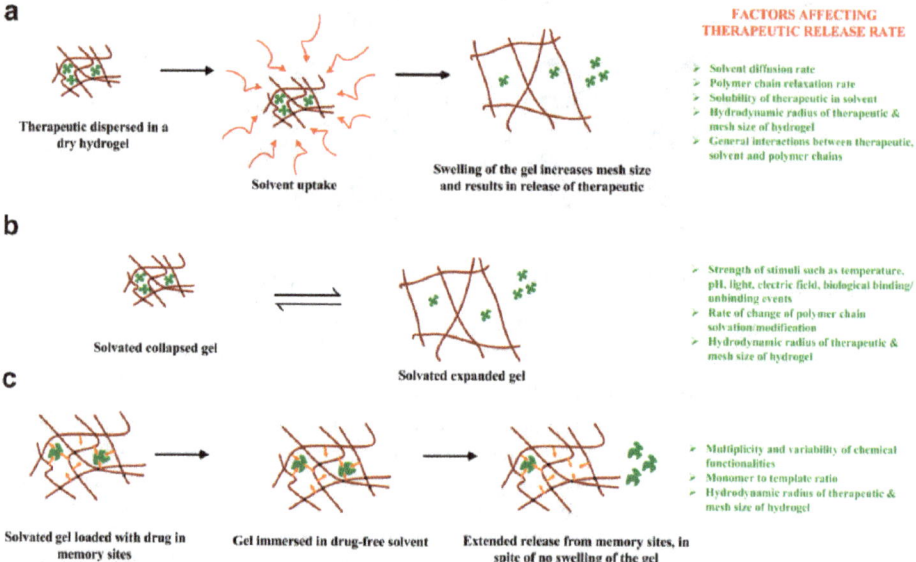

Figure 10. Scheme of (**a**) controlled drug release in hydrogels upon their swelling either by solvent uptake from a dry state or (**b**) thermodynamic compatibility with the solvent. (**c**) Macromolecular memory is obtained by imprinting a multifunctional pre-polymerization complex with the drug as an alternative strategy to release the drug in a controlled fashion when the gels are already solvated and fully swollen. Reprinted from [38], Copyright (2022), with permission from Elsevier.

In Figure 10, a modified hydrogel molecule is shown. The presented hydrogel molecules may contain a specific chemical/biological species along their backbone chain to obtain sensitivity to environmental hydrogels. This feature may be achieved by controlling drug transport by swelling controlled systems (i.e., drug-loaded dry state with water uptake) or swellable systems (e.g., pH, temperature, etc.). In a widely used Fickian model of release kinetics, the relaxation rate is high, resulting in the rate-limiting diffusion process. Thus, the release rate of the drug is proportional to the concentration gradient between the drug source and the environment. The achieved rate is proportional to the concentration gradient between the drug source and the surroundings. The main aim is to find a drug source to achieve zero-order release. Many strategies try to achieve zero-order release, such as biodegradable systems with solvent penetration moving with similar velocities the outer eroding [38,44].

One of the new methods is to obtain hydrogel with macromolecular memory for the drug within the network and delay the transport of drug from the hydrogel matrix by the presence of interactions with various functional groups within the network. This can be achieved by using molecular imprinting methods presented in Figure 10 Interactions between the drug and matrix cavities slow drug release from the hydrogel. This type of hydrogel optimization of slowed release, caused by the amount and strength of functional monomer interactions, crosslinking structure, and mobility of polymer chains, might be a potentially synthetic solid way to gain many hydrogels [38].

5. Mechanism of Controlled Release within Molecular Imprinted Hydrogels

Molecular Imprinted Hydrogels can be classified as anionic, cationic, or neutral, which also determines their behavior. Thermodynamically, the swelling behavior of the hydrogels network is related to the balance between the polymer-water Gibbs free energy of mixing and the Gibbs free energy associated with the elastic nature of the entire polymer [45]. The quantities of the mentioned free energies become equal when achieving the swelling equilibrium [46]. What highlights the hydrogels from others is the advantage of milder synthesis conditions at lower temperatures and in aqueous mediums regarding the fragility and solubility of biological templates, including DNA, protein, or even nucleic acids [39]. Two main solvent-activated systems can be indicated—an osmotic-controlled system and a swelling-controlled system—the rate of water influx controls the overall rate of the drug release. The controlled drug release mechanism is based on water diffusion and polymer chain relaxation [46]. Generally, the time dependence of the drug release rate can be determined by the rate of water diffusion and chain relaxation [47].

It is well-known that the limitations of transdermal drug delivery are controlled by skin anatomy. Generally, the skin permits a painless and compliant network for systemic drug administration [48]. The fact that the skin has evolved and thus impedes the flux of toxins into the body and minimizes water loss means that it naturally has a low permeability to the penetration of foreign molecules. Because the skin provides a barrier to the delivery of many drugs, various chemical additives have been tested to achieve better results in transdermal penetration. Chemical penetration additives offer many advantages, such as design flexibility with formulation chemistry and a more accessible patch application over a large area [49]. The mentioned transdermal patches have been widely helpful in developing new applications for existing therapeutics and reducing first-pass drug-degradation effects. Patches also gain the ability to reduce some side effects. For example, estradiol patches are commonly used and, in contrast to the popular oral formulations, do not cause liver damage [50].

Whereas the mechanism of controlled release of the drug from hydrogel structure is relatively easy to design, implementing imprinted recognition release systems requires consideration of many environmental impacts and the expected properties of the desired hydrogel. Basically, the controlled release mechanism and associated swelling characteristics of polyhydrogels' networks result from cross-links (also known as tie-points or junctions), permanent entanglements, ionic interactions, or microcrystalline regions incorporating various chains. In general, as an analyte replaces pendant analyte groups (attached to the copolymer chains), the polymeric network loses effective cross-links, opening the network's mesh size and regulating the release. Otherwise, as an analyte decreases in concentration within the bulk phase, the molecule rebinds with the analyte groups attached to the copolymer chains, which role is to close the network structure [46].

It is not a surprise that one of the most desirable controlled drug delivery applications is stimuli-responsive hydrogels that can modulate the release in response to pH, temperature, ionic strength, electric field, or specific analyte concentration differences. The most important feature of these systems is that they can be designed to operate within a particular human body area due to the possibility of adapting to well-known environmental conditions [46,47].

5.1. Stimuli-Responsive MIP Hydrogels

The need for creating intelligent materials based on chemical compounds that can mimic the natural receptors inspires the development of imprinting technologies and expand the MIPs synthesis into the synthesis of stimuli-responsive MIPs (SR-MIPs) by stimuli-responsive technology for molecular imprinting. SR-MIPs included thermo-responsive

MIPs, pH-responsive MIPs, dual- or multiple-responsive MIPs, and other-responsive MIPs. Due to their great applications properties, these intelligent polymers play an important role in many fields such as drug delivery, biotechnology, separation science, cell encapsulation in biochemistry, and chemo-biosensing [51]. The combination of stimuli-responsivity and Molecular Imprinting Technology helps to obtain valuable functionalities. Generally, imprinting provides a high loading capacity of specific molecules, whereas the ability to respond to stimuli modulates the affinity to network for the target molecules. The whole process provides a regulatory or switching capability of the release process [52].

Connecting molecularly imprinting technology with the synthesis of stimuli-responsive hydrogels requires conducting the polymerization reaction in the presence of a template in the conformation corresponding to the minimum energy. Imprinted cavities' recognition properties after swelling can be maintained only if the network folds back into the conformation adopted during the synthesis [53]. Generally, when the centers of molecular recognition are present in the stimuli-responsive hydrogel, the conformation of the receptors may be deformed or re-constituted as a function of an external or a physiological signal. There are plenty of functions that stimuli-responsive polymeric hydrogels can perform, such as selectively and effectively load of a particular drug, releasing the drug at a rate modulated by a stimulus, and uptake the released drug again from the environment if the drug remains around the hydrogel when the stimulus stop or diminishes its intensity and the cavities are reformed (Figure 11) [52].

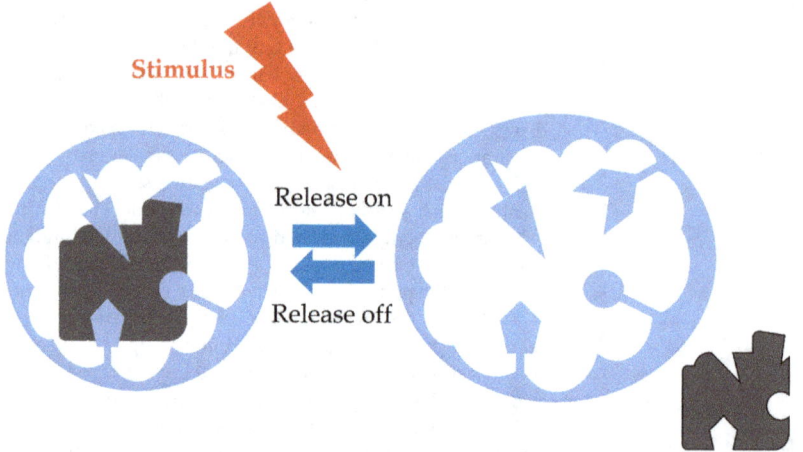

Figure 11. Schematic view of the effect of a stimulus on the conformation of the drug-imprinted cavities in a responsive hydrogel.

Generalizing, stimuli-responsive imprinted hydrogels can be synthesized by combining responsive monomers with functional monomers that interact with the appropriate drug molecules. After polymerization reaction, during hydrogel swelling, the structure of receptors is altered, and the drug is released. The receptors can be reconstituted following stimulus disappear or decrease in its intensity. As a consequence, the release slows down or even stops. Whereas there is a necessity to recognize cavities structure after several swelling/collapse cycles, optimizing stimuli-responsive imprinted hydrogels is still challenging [52,54].

5.1.1. Thermo-Responsive Hydrogels

The thermo-responsive gels have been widely used as smart materials in various fields such as drug delivery systems, tissue engineering, or even cell encapsulation in biochemistry [55–59]. Thermo-responsive MIPs have gained the researchers' curiosity due to the similar recognition mechanisms to the proteins from natural systems and the ability of the hydrogels to swell or deswell thanks to changes in temperature in the surrounding area [40]. Therefore, thermo-responsive polymeric hydrogels can be used in the design of protein-imprinted polymeric materials, which have already been reported in many applications [55,56,59,60]. The important thing is that there is dependence on the availability of binding sites from the MIP structure based on cross-linking. Highly cross-linked MIPs have a more rigid structure thus, the number of binding sites is limited, whereas lightly cross-linked polymer gels can undergo reversible swelling and shrink in response to environmental temperature changes [16]. There are two classes of thermo-responsive hydrogel materials—positive and negative temperature-responsive systems. The main difference is critical solution temperature—the positive temperature-responsive hydrogels have an upper critical solution temperature (UCST), so it means that they contract upon cooling below the UCST. In contrast, the negative-sensitive hydrogels have a lower critical solution temperature (LCST), and they contract upon heating above the mentioned LCST [40]. Generally, thermo-responsive polymers contain both hydrophilic and hydrophobic groups. Due to that, they can form appropriate structures, swelling and shrinking, in response to temperature changes. The mechanism of this response is based on hydrogen bond interactions. It is well known that in lower temperatures, the hydrogen bond interactions are formed between hydrophilic areas in polymer chains and templates, whereas in higher temperatures, higher than the low-critical solution temperature (LCST), the hydrogen bond interactions are destroyed, thus hydrophobic interactions increase. An increase of hydrophobic bond interactions causes the aggregation of polymer chains and then contraction of the gel network [51].

The most known thermo- responsive polymer is poly(N-isopropylacrylamide) (PNI-PAAm). Its low-critical solution temperature is around 32 °C in an aqueous solution, so it means that due to the close to natural body temperature, it may be used widely in smart drug delivery systems [61,62]. Generally, the combination of thermo-responsive properties with Molecular Imprinting Technology can develop networks that provide a promising synthetic strategy ensuring the system responds rapidly to external temperature changes. The schematic mechanism of thermo-responsive hydrogel's action is presented in Figure 12. It is shown that the template can be easily removed from the MIPs network by reducing the external temperature (Figure 12) [51].

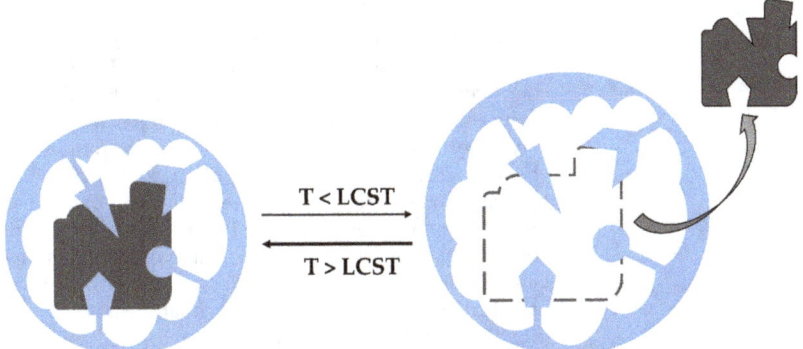

Figure 12. Scheme of the template's removal mechanism from thermo-responsive MIPs.

A lot of studies are reported, showing that the N-isopropylacrylamide (NIPAAm) can be used as a functional monomer for preparing thermo-responsive MIPs, applicable in various fields. The NIPAAm has been used for many target species such as proteins [59,63–65], organic molecules (like 4-aminopyridine) [63], cisplatin [66], or even metal ions (like Cu^{2+} ions) [67].

Wang et al. [68] reported the results of research in which a preparation of pH/thermo-responsive MIPs by frontal polymerization using acrylic acid and N-isopropylacrylamide (NIPAAm) was performed. The proposed MIPs were applied to deliver Gemifloxacin, a fourth-generation fluoroquinolone antibiotic that acts by inhibiting DNA gyrase and topoisomerase IV. The reported data showed that the obtained drug delivery devices based on MIPs possessed higher relative bioavailability of Gemifloxacin than those of the corresponding non-imprinted polymers [68].

One of the recently studied MIP hydrogels with the thermo-responsive feature is a molecularly imprinted polymer based on konjac glucomannan (polysaccharide) imprinted with 5-fluorouracil as a template reported by Ann et al. [69]. 5-Fluorouracil is a compound with a high affinity to a range of tumors such as gastric, intestinal, pancreatic, ovarian, liver, brain, breast, etc. In the reported studies, a novel thermo-responsive MIP was prepared by graft copolymerization using konjac glucomannan (KGM) as a matrix, N-isopropylacrylamide (NIPAAm) as a thermo-responsive monomer, acrylamide (AM) as co-monomer, N,N'-methylenebis(acrylamide) (NBAM) as a cross-linking agent, and 5-fluorouracil (5-Fu) as a template (Figure 13).

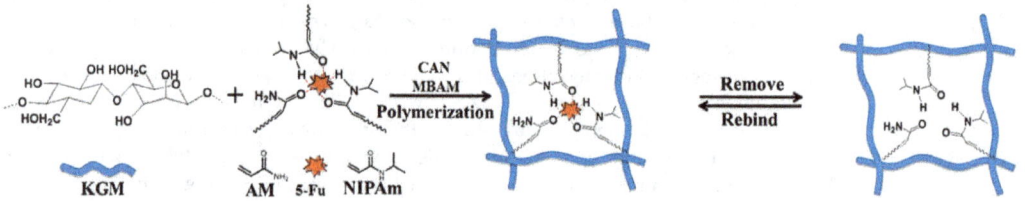

Figure 13. Schematic procedure of synthesis of 5-fluorouracil thermo- responsive MIP reported by Ann et al. Reprinted [69], Copyright (2022), with permission from Elsevier.

5-Fluorouracil selective MIP was characterized by thermo-responsive features. The results showed that the system could quickly respond to an external change in temperature. The swelling or shrinking of the imprinted sites resulted in the adsorption or desorption of 5-fluorouracil. As a result, the prepared MIPs could be used as a sustained-release network controlling the release of 5-Fu by changing the environmental temperature. Obtained data of the release kinetics was fit with the Higuchi release model [69].

5.1.2. pH-Responsive Hydrogels

Hydrogels, which are pH-responsive, must contain many chemical groups that can be easily ionized. Carboxyl or amino groups can be noted as examples of these groups that can, accordingly, donate or accept a proton, which determines the pH-sensitivity feature of the entire MIPs. As in the case of thermo-sensitivity, the mechanism of pH-response is based on hydrogen bonds interaction between the chains and template [51]. When the chemical group is ionized during changes of environmental pH, at the same time, the hydrogen bonds between chains are destroyed, which causes a decrease in the crosslinking points in the hydrogel network. This results in a discontinuous change in the hydrogel volume [70]. Considering the possibilities of ionized groups, pH-responsive polymers can be divided into two types—anionic and cationic ones. Within the anionic types, the most useful group is a carboxyl group, which can be protonated and thus determine hydrophobic interactions at low pH. Additionally, a low pH environment cause leading the volume shrinkage. Opposite to that, at high pH, the behavior of hydrogel is quite different. In these conditions, carboxyl groups dissociate into carboxylate ions, resulting in a high charge density in the polymer network, which causes swelling. Similarly, the pH-responsive feature of the cationic hydrogel network is dependent on the protonation of basic groups in the polymer chains (e.g., amino groups or pyridine groups). At low pH, basic cationic groups are protonated, which leads to internal charge repulsions between neighboring protonated groups. In contrast, at higher pH, the groups become less ionized, resulting in a reduction in the overall hydrodynamic diameter of the polymer [51].

The first reported pH-responsive MIPs were proposed by Tao et al. [71]. In this research, novel pH-responsive MIPs by using amylose as the host matrix, bisphenol A (BPA) as the template, and acrylic acid (AA) as the co-functional monomer prepared. Changing the acidity of the entire solution could reversibly control the rebinding ability towards the template. In that case, the rebinding ability of polymers decreased with the increasing pH of the solution. Comparing two pH conditions, $pH_1 = 4.5$ and $pH_2 = 8.5$, the binding amount was, accordingly, 2.5 $\mu M \cdot g^{-1}$ and 1.0 $\mu M \cdot g^{-1}$ [51,71]. The higher pH caused the loss of the MIP affinity for bisphenol A because of the conformational changes in the amylose chains caused by the electrostatic repulsions among the ionized groups of acrylic acid and the subsequent disruption of the imprinted cavities (Figure 14) [52].

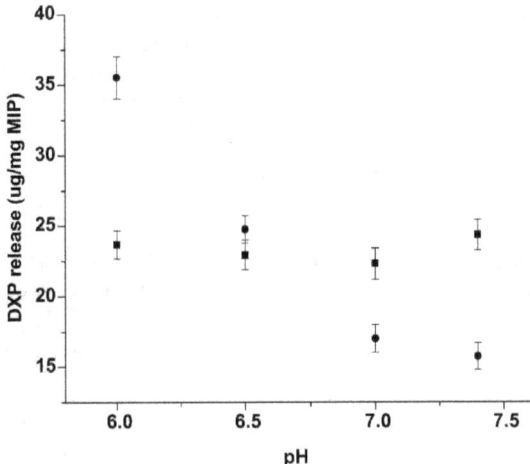

Figure 14. Scheme presenting DXP release profiles at various pH PBS solution at 37 °C [(■) non-imprinted polymer; (●) imprinted polymer]. Reprinted from [72], Copyright (2022), with permission from Elsevier.

The example of pH-responsive MIP prepared for controlled release of dexamethasone-21-phosphate disodium salt (DXP) is nanosphere/hydrogel composite reported by Wang et al. [72]. The chosen DXP is a potential coating for implantable biosensors that should improve their biocompatibility [51]. As results show, the DXP release rate from the MIP structure increased significantly with a decrease of pH value, while the DXP release rate from the non-imprinted polymer structure didn't change with the pH change (Figure 14).

The behavior of DXP in various pH conditions can be explained by interactions between the template and polymer network. When the pH value decreases, some template anions are protonated, which causes weak ionic interactions. It subsequently leads to a faster drug release rate [72]. Obtained hydrogel has been applied in glucose sensors to improve their biocompatibility as well as their lifetime. The results showed that the obtained DXP hydrogel MIPs can potentially suppress the inflammation response, causing an increase in the pH of the imprinted sensors, effectively improving their lifespan [51].

Interesting studies of dual drug release with materials based on poly(L-lactide)-*co*-polyethylene glycol-*co*-poly(L-lactide) dimethacrylate as a degradable polymeric cross-linker were reported by Xu et al. in 2016 [73]. They used acrylic acid and N-isopropylacrylamide as monomers, anti-cancer drug DOX, and the antibiotic tetracycline, which both were used as templates loaded into the hydrogels with dual drug loading efficiency. The drug release at pH 7.4 and 1.2 were examined. The obtained results confirmed the original thesis and showed that the synthesized copolymers are pH-responsive, shrinking at pH = 1.3 and swelling at pH = 7.4. As experiments demonstrated, the dual-drug-loaded hydrogels released drugs in different patterns and successfully killed targeted cells [73].

5.1.3. Dual/Multiple-Responsive MIPs

The dual/multiple-responsive hydrogels are polymeric systems responsive to two or more external stimuli [51]. While the studies on dual stimuli-responsive polymers are widely common, they are relatively less explored than single-responsive polymers. However, dual-responsive MIPs gain the researcher's curiosity due to their feature, such as enhancing the versatility of polymeric materials as they allow tuning of their properties in multiple ways rather than in single-responsive polymers [16]. From the synthetic point of view, by using suitable monomers, the dual-responsive MIPs may be obtained by transforming the single-responsive polymer by replacing the traditional functional monomer with a stimuli-responsive functional monomer [51]. Generally, dual-responsive polymers mainly include: magnetic/photo; magnetic/thermo; thermo/pH; thermo/photo; and thermo/salt dual responsive MIPs. The mentioned types of well-reported dual-responsive MIPs useful in various fields are presented in Table 1.

Table 1. Examples of well-reported dual/multiple-responsive MIPs with their applications [16].

Type of Polymer	Template	Responsive Element	Application	Reference
Thermo/Magnetic	2,4,5-Trichlorophenol	NIPAAm, Fe_3O_4	Selective separation and enrichment fields	[74]
	Sulfamethazine	NIPAAm, γ-Fe_3O_4	Separation, drug release, protein recognition	[75]
	BSA	NIPAAm, Fe_3O_4	Chromatographic separation, solid-phase extraction, drug delivery. Medical diagnosis and biosensors	[76]
Photo/Magnetic	Caffeine	Fe_3O_4, MPABA	Trace caffeine analysis	[77]
pH/Thermo	Ovalbumin	NIPAAm, boronic acid	Chemical sensing and biosensing	[78]
Thermo/Photo	2,4-D	Azobenzene, NIPAAm	Separation, extraction, assays, drug delivery, and bioanalytical analysis	[79]
Thermo/Salt	BSA	NIPAAm, NaCl	Solid-phase extraction, sensors, and protein delivery agents	[80]
Thermo/Salt/Biomolecule	Lysozyme or Cytochrome 4	NIPAAm, NaCl, Bio-molecule	Non-protein acetous receptor	[65]

One of the most interesting uses of dual-responsive MIPs is reported by Zhao et al. [77], a multi-responsive MIP consisting of thermo-responsive and salt-responsive MIPs hydrogel. The research presents dual-responsive MIPs hydrogel for BSA by self-assembly of a basic functional monomer N-[3-(dimethylamino)propyl]methacrylamide (DMAPMA) with bovine serum albumin (BSA) that can be polymerized in the presence of NIPAAm. Obtained dual-responsive polymeric hydrogel proved that it possesses a clear memory of the template protein and can respond to changes in temperature and ionic strength. In the recognition process mechanism, salt ions play an important role in screening the electrostatic interactions between protein molecules and the charged polymer chains. It was observed that the increase of salt concentration caused a screening of electrostatic interactions between polymer chains, while the addition of NaCl to the adjusted volume of polymer caused inhibition. The demonstrated features of obtained dual-responsive MIPs made them attractive for applications such as solid electrolyte membranes, electrode devices, protein delivery agents, and sensors with the controlled release [80].

There are also some studies reporting multiple-responsive MIPs, however, the amount of such systems is much lower than dual-response MIPs. One example proposed by Chen et al. [78] is multi-responsive protein imprinted polymers responsive to temperature, the corresponding template protein, and salt concentration, which results in specific volume shrinking. Cytochrome c or lysozyme were used as templates, NIPAAm as a major functional monomer, MAA and AAm as functional co-monomers, and N,N-methylenebisacrylamide as a crosslinker. As the results showed, the combination of molecular imprinting technique and a stimuli-responsive feature may be useful for preparing protein-responsive polymeric hydrogels that can undergo specific binding and shrinking in the presence of template [65].

Recently, interesting research was reported by Wang et al. [81] based on the fabrication of core-shell imprinted nanospheres with multiple responsive properties. The scheme of the main synthesis is shown in Figure 15.

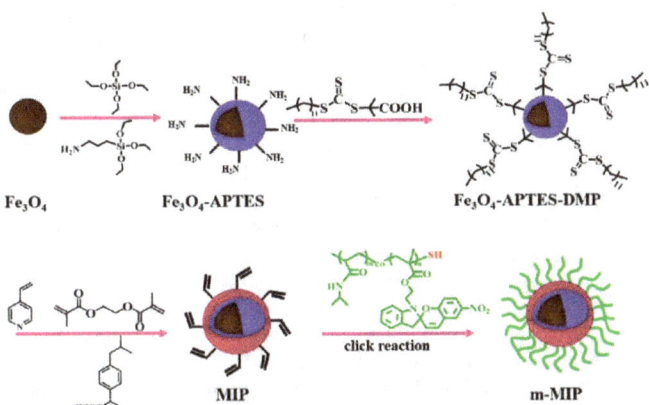

Figure 15. Schematic synthesis of core-shell MIP nanospheres with multiple responsive properties. Reprinted with permission from [81]. Copyright (2022) American Chemical Society.

As results showed, fabricated magnetic core-shell MIP nanospheres, with an imprinted layer with a thickness ranging from 40 to 150 nm, gain fine hydrophilicity, high binding capacity, and favorable selectivity adsorption in aqueous solution. Obtained polymers were also used in the application of drug delivery systems and expressed a sustained release effect triggered by temperature and UV light [81].

5.1.4. Other-Responsive Hydrogels

There is a lot of possibilities to obtain a stimuli-responsive polymeric hydrogel. Hydrogels may be fabricated to be responsive for various, additional to mentioned in previous paragraphs, external stimuli such as light, magnetic field, or other stimuli including salt ions and biomolecules [16]. Whereas many studies are reporting all of the mentioned external stimuli responsive materials, from pharmaceutical point of view, the most interesting thought is to be the last, using biomolecule responsive polymeric networks. Recently, biomolecule-responsive hydrogels have become more important for drug delivery systems and molecular diagnostics because they can sense the target biomolecule, which results in structural changes. The mechanism of biomolecule-responsive MIPs is based on changes in the volume as a function of the concentration of a target biomolecule such as carbohydrates or proteins [82]. The fabricated MIP's hydrogels may be an optimal way to overcome tumors as some studies confirmed that suppositions. A good example is a research presented by Miyata, who prepared tumor-marker-imprinted hydrogel, which can shrink in the presence of a target-tumor marker glycoprotein by using various cross-linkers (low-molecular-weight/high-molecular-weight). The obtained results provided a basis for developing useful biomolecule-responsive hydrogels and permitted the knowledge of critical factors of molecular imprinting. The presented studies were focused not only on the proper preparation of biomolecule-responsive hydrogel but also on the effect of the molecular weight of cross-linkers on the glycoprotein-responsive behavior of imprinted hydrogels. Generally, whereas long chains of the high-molecular-weight cross-linker and network chains undergo conformational changes by complex formation of ligands with a target glycoprotein, short chains of the low-molecular-weight cross-linkers do not undergo these changes [83].

6. Conclusions and Future Work

Molecularly imprinted polymers are promising materials in the synthesis of advanced drug delivery networks due to their ability to increase release times and extend the residency of the entire drug. As the presented studies showed, the main application fields in which MIPs hydrogels can be used are sustained release, controlled release, and targeted delivery system based on its distinct advantages. The future perspectives for transdermal imprinted drug delivery devices are very promising because of noted enormous progress in synthetic and material approaches. The main advantage of using MIPs is their high stability, from which there is a possibility to note: resistance to pressure, high temperatures, extreme pH, and possibility for long-term storage.

In this review, there was also summarized a mechanism and application of various stimuli-responsive MIPs. As presented, stimuli-responsive MIPs can be divided into single-responsive and multi-responsive MIPs. The main conclusion is that the second group has received more attention between non-stimuli and stimuli-responsive MIPs because of their excellent properties. One of the main benefits is their response to external stimuli, which makes it possible to alter their volume and affinity for target molecules by changing the environmental conditions.

Although various achievements have been attained in molecularly imprinted technology and stimuli-responsive MIPs, there are still lots of development challenges and opportunities. For instance, it is still challenging to transfer the imprinting process from organic to aqueous phase, reaching the level of natural molecular recognition, exploring various stimuli-responsive systems to develop stimuli-responsive MIPs, and developing within dual/multiple-responsive MIPs with good biocompatibility with increasing requirements for functional polymer materials.

Funding: This work was supported by the National Science Centre, Poland, under Grant Number 2020/37/B/ST5/01938.

Institutional Review Board Statement: Not applicable.

Informed Consent Statement: Not applicable.

Conflicts of Interest: The authors declare no conflict of interest.

References

1. Haupt, K.; Mosbach, K. Molecularly imprinted polymers and their use in biomimetic sensors. *Chem. Rev.* **2000**, *100*, 2495–2504. [CrossRef] [PubMed]
2. Suedee, R. The Use of Molecularly Imprinted Polymers for Dermal Drug Delivery. *Pharm. Anal. Acta* **2013**, *4*, 1–23. [CrossRef]
3. Polyakov, M.V. Adsorption properties and structure of silica gel. *Zhurnal Fiz. Khimii* **1931**, *2*, 799–805.
4. Wulff, G. REVIEWS Molecular Imprinting in Cross-Linked Materials with the Aid of Molecular Templates—A Way towards Artificial Antibodies. *Angew. Chem. Int. Ed.* **1995**, *34*, 1812–1832. [CrossRef]
5. Mosbach, K. Molecular imprinting. *Trends Biochem. Sci.* **1994**, *19*, 9–14. [CrossRef]
6. Whitcombe, M.J.; Rodriguez, M.E.; Villar, P.; Vulfson, E.N. A New Method for the Introduction of Recognition Site Functionality into Polymers Prepared by Molecular Imprinting: Synthesis and Characterization of Polymeric Receptors for Cholesterol. 1995. Available online: https://pubs.acs.org/doi/pdf/10.1021/ja00132a010 (accessed on 24 January 2022).
7. Chen, L.; Xu, S.; Li, J. Recent advances in Molecular Imprinting Technology: Current status, challenges and highlighted applications. *Chem. Soc. Rev.* **2011**, *40*, 2922–2942. [CrossRef]
8. Yan, H.; Row, K.H. Characteristic and Synthetic Approach of Molecularly Imprinted Polymer. *Int. J. Mol. Sci.* **2006**, *7*, 155–178. [CrossRef]
9. Caro, E.; Marcé, R.M.; Borrull, F.; Cormack, P.A.G.; Sherrington, D.C. Application of molecularly imprinted polymers to solid-phase extraction of compounds from environmental and biological samples. *TrAC—Trends Anal. Chem.* **2006**, *25*, 143–154. [CrossRef]
10. Ramström, O.; Ye, L.; Gustavsson, P.-E. Chiral Recognition by Molecularly Imprinted Polymers in Aqueous Media. *Chromatographia* **1998**, *48*, 197–202. [CrossRef]
11. Martín-Esteban, A. Molecularly-imprinted polymers as a versatile, highly selective tool in sample preparation. *TrAC—Trends Anal. Chem.* **2013**, *45*, 169–181. [CrossRef]
12. Lai, C.; Wang, M.-M.; Zeng, G.-M.; Liu, Y.-G.; Huang, D.-L.; Zhang, C.; Wang, R.-Z.; Xu, P.; Cheng, M.; Huang, C.; et al. Synthesis of surface molecular imprinted TiO 2/graphene photocatalyst and its highly efficient photocatalytic degradation of target pollutant under visible light irradiation. *Appl. Surf. Sci.* **2016**, *390*, 368–376. [CrossRef]

13. Vasapollo, G.; del Sole, R.; Mergola, L.; Lazzoi, M.R.; Scardino, A.; Scorrano, S.; Mele, G. Molecularly imprinted polymers: Present and future prospective. *Int. J. Mol. Sci.* **2011**, *12*, 5908–5945. [CrossRef] [PubMed]
14. Matsui, J.; Nicholls, I.A.; Karube, I.; Mosbach, K. Carbon-Carbon Bond Formation Using Substrate Selective Catalytic Polymers Prepared by Molecular Imprinting: An Artificial Class II Aldolase. *J. Org. Chem.* **1996**, *61*, 5414–5417. [CrossRef]
15. Whitcombe, M.J.; Chianella, I.; Larcombe, L.; Piletsky, S.A.; Noble, J.; Porter, R.; Horgan, A. The rational development of molecularly imprinted polymer-based sensors for protein detection. *Chem. Soc. Rev.* **2011**, *40*, 1547–1571. [CrossRef] [PubMed]
16. Chen, L.; Wang, X.; Lu, W.; Wu, X.; Li, J. Molecular imprinting: Perspectives and applications. *Chem. Soc. Rev.* **2016**, *45*, 2137–2211. [CrossRef] [PubMed]
17. Sellergren, B.; Allender, C.J. Molecularly imprinted polymers: A bridge to advanced drug delivery. *Adv. Drug Deliv. Rev.* **2005**, *57*, 1733–1741. [CrossRef]
18. Gladis, J.M.; Rao, T.P. Effect of porogen type on the synthesis of uranium ion imprinted polymer materials for the preconcentration/separation of traces of uranium. *Microchim. Acta* **2004**, *146*, 251–258. [CrossRef]
19. Sellergren, B.; Wieschemeyer, J.; Boos, K.S.; Seidel, D. Imprinted polymers for selective adsorption of cholesterol from gastrointestinal fluids. *Chem. Mater.* **1998**, *10*, 4037–4046. [CrossRef]
20. Huval, C.C.; Chen, X.; Holmes-Farley, S.R.; Mandeville, W.H.; Polomoscanik, S.C.; Sacchiero, R.J.; Dhal, P.K. Molecularly Imprinted Bile Acid Sequestrants: Synthesis and Biological Studies. *MRS Proc.* **2003**, *787*, 63. [CrossRef]
21. Huval, C.C.; Bailey, M.J.; Braunlin, W.H.; Holmes-Farley, S.R.; Mandeville, W.H.; Petersen, J.S.; Polomoscanik, S.C.; Sacchiro, R.J.; Chen, A.X.; Dhal, P.K. Novel cholesterol lowering polymeric drugs obtained by molecular imprinting. *Macromolecules* **2001**, *34*, 1548–1550. [CrossRef]
22. Hashemi-Moghaddam, H.; Zavareh, S.; Karimpour, S.; Madanchi, H. Evaluation of molecularly imprinted polymer based on HER2 epitope for targeted drug delivery in ovarian cancer mouse model. *React. Funct. Polym.* **2017**, *121*, 82–90. [CrossRef]
23. Mohebali, A.; Abdouss, M.; Mazinani, S.; Zahedi, P. Synthesis and characterization of poly(methacrylic acid)-based molecularly imprinted polymer nanoparticles for controlled release of trinitroglycerin. *Polym. Adv. Technol.* **2016**, *27*, 1164–11716. [CrossRef]
24. Suedee, R.; Bodhibukkana, C.; Tangthong, N.; Amnuaikit, C.; Kaewnopparat, S.; Srichana, T. Development of a reservoir-type transdermal enantioselective-controlled delivery system for racemic propranolol using a molecularly imprinted polymer composite membrane. *J. Control. Release* **2008**, *129*, 170–178. [CrossRef] [PubMed]
25. Mokhtari, P.; Ghaedi, M. Water compatible molecularly imprinted polymer for controlled release of riboflavin as drug delivery system. *Eur. Polym. J.* **2019**, *118*, 614–618. [CrossRef]
26. Hemmati, K.; Masoumi, A.; Ghaemy, M. Tragacanth gum-based nanogel as a superparamagnetic molecularly imprinted polymer for quercetin recognition and controlled release. *Carbohydr. Polym.* **2016**, *136*, 630–640. [CrossRef]
27. Parisi, O.I.; Ruffo, M.; Scrivano, L.; Malivindi, R.; Vassallo, A.; Puoci, F. Smart bandage based on molecularly imprinted polymers (Mips) for diclofenac controlled release. *Pharmaceuticals* **2018**, *11*, 92. [CrossRef]
28. Ruela, A.L.M.; Figueiredo, E.C.; Pereira, G.R. Molecularly imprinted polymers as nicotine transdermal delivery systems. *Chem. Eng. J.* **2014**, *248*, 1–8. [CrossRef]
29. Ruela, A.L.M.; de Figueiredo, E.C.; Carvalho, F.C.; de Araújo, M.B.; Pereira, G.R. Adsorption and release of nicotine from imprinted particles synthesised by precipitation polymerisation: Optimising transdermal formulations. *Eur. Polym. J.* **2018**, *100*, 67–76. [CrossRef]
30. He, S.; Zhang, L.; Bai, S.; Yang, H.; Cui, Z.; Zhang, X.; Li, Y. Advances of molecularly imprinted polymers (MIP) and the application in drug delivery. *Eur. Polym. J.* **2021**, *143*, 110179. [CrossRef]
31. Ulbricht, M. Membrane separations using molecularly imprinted polymers. *J. Chromatogr. B Anal. Technol. Biomed. Life Sci.* **2004**, *804*, 113–125. [CrossRef]
32. Sabbagh, F.; Kim, B.S. Recent advances in polymeric transdermal drug delivery systems. *J. Control. Release* **2022**, *341*, 132–146. [CrossRef] [PubMed]
33. Sanjay, S.T.; Zhou, W.; Dou, M.; Tavakoli, H.; Ma, L.; Xu, F.; Li, X. Recent advances of controlled drug delivery using microfluidic platforms. *Adv. Drug Deliv. Rev.* **2018**, *128*, 3–28. [CrossRef] [PubMed]
34. Luliński, P. Molecularly imprinted polymers based drug delivery devices: A way to application in modern pharmacotherapy. A review. *Mater. Sci. Eng. C* **2017**, *76*, 1344–1353. [CrossRef] [PubMed]
35. Gato, K.; Fujii, M.Y.; Hisada, H.; Carriere, J.; Koide, T.; Fukami, T. Molecular state evaluation of active pharmaceutical ingredients in adhesive patches for transdermal drug delivery. *J. Drug Deliv. Sci. Technol.* **2020**, *58*, 101800. [CrossRef]
36. Jeong, W.Y.; Kwon, M.; Choi, H.E.; Kim, K.S. Recent advances in transdermal drug delivery systems: A review. *Biomater. Res.* **2021**, *25*, 1–15. [CrossRef] [PubMed]
37. Bodhibukkana, C.; Srichana, T.; Kaewnopparat, S.; Tangthong, N.; Bouking, P.; Martin, G.P.; Suedee, R. Composite membrane of bacterially-derived cellulose and molecularly imprinted polymer for use as a transdermal enantioselective controlled-release system of racemic propranolol. *J. Control. Release* **2006**, *113*, 43–56. [CrossRef] [PubMed]
38. Venkatesh, S.; Saha, J.; Pass, S.; Byrne, M.E. Transport and structural analysis of molecular imprinted hydrogels for controlled drug delivery. *Eur. J. Pharm. Biopharm.* **2008**, *69*, 852–860. [CrossRef] [PubMed]
39. Zhang, N.; Xu, Y.; Li, Z.; Yan, C.; Mei, K.; Ding, M.; Ding, S.; Guan, P.; Qian, L.; Du, C.; et al. Molecularly Imprinted Materials for Selective Biological Recognition. *Macromol. Rapid Commun.* **2019**, *40*, e1900096. [CrossRef] [PubMed]
40. Peppas, N.A.; Bures, P.; Leobandung, W.; Ichikawa, H. Hydrogels in pharmaceutical formulations. [CrossRef]

41. Peppas, N.A.; Korsmeyer, R.W. Dynamically Swelling Hydrogel in Controlled Release Application Chapter 6 Dynamically Swelling Hydrogels in Controlled Release Applications. 2016. Available online: https://www.researchgate.net/publication/279542219 (accessed on 24 January 2022).
42. Kim, M.-K.; Chung, S.-J.; Lee, M.-H.; Cho, A.-R.; Shim, C.-K. Targeted and sustained delivery of hydrocortisone to normal and stratum corneum-removed skin without enhanced skin absorption using a liposome gel. *J. Control. Release* 1997, *46*, 243–251. [CrossRef]
43. Zhang, I.; Shung, K.K.; Edwards, D.A. Hydrogels with Enhanced Mass Transfer for Transdermal Drug Delivery. *J. Pharm. Sci.* 1996, *85*, 1312–1316. [CrossRef] [PubMed]
44. Tahara, K.; Yamamoto, K.; Nishihata, T. Overall mechanism behind matrix sustained release (SR) tablets prepared with hydroxypropyl methylcellulose 2910. *J. Control. Release* 1995, *35*, 59–66. [CrossRef]
45. Peppas, N.A. Preparation methods and structure of hydrogels. In *Hydrogels in Medicine and Pharmacy*, 1st ed.; Taylor & Francis Group, LLC: Oxfordshire, UK, 1986; Volume 1, pp. 1–26.
46. Byrne, M.E.; Park, K.; Peppas, N.A. Molecular imprinting within hydrogels. *Adv. Drug Deliv. Rev.* 2002, *54*, 146–161. [CrossRef]
47. Peppas, N.A.; Colombo, P. Analysis of drug release behavior from swellable polymer carriers using the dimensionality index. *J. Control. Release* 1997, *45*, 35–40. [CrossRef]
48. Zaffaroni, A. Overview and Evolution of Therapeutic Systems. *Ann. N. Y. Acad. Sci.* 1991, *618*, 405–421. [CrossRef] [PubMed]
49. Prausnitz, M.R.; Mitragotri, S.; Langer, R. Current status and future potential of transdermal drug delivery. *Nat. Rev. Drug Discov.* 2004, *3*, 115–124. [CrossRef] [PubMed]
50. Cramer, M.P.; Saks, S.R. Translating Safety, Efficacy and Compliance into Economic Value for Controlled Release Dosage Forms. *PharmacoEconomics* 1994, *5*, 482–504. [CrossRef]
51. Xu, S.; Lu, H.; Zheng, X.; Chen, L. Stimuli-responsive molecularly imprinted polymers: Versatile functional materials. *J. Mater. Chem. C* 2013, *1*, 4406–4422. [CrossRef]
52. Alvarez-Lorenzo, C.; González-Chomón, C.; Concheiro, A. Molecularly Imprinted Hydrogels for Affinity-controlled and Stimuli-responsive Drug Delivery. In *Smart Materials for Drug Delivery*; Alvarez-Lorenzo, C., Concheiro, A., Eds.; RSC Publishing: London, UK, 2013; Chapter 21; pp. 228–260. [CrossRef]
53. Alvarez-Lorenzo, C.; Guney, O.; Oya, T.; Sakai, Y.; Kobayashi, M.; Enoki, T.; Takeoka, Y.; Ishibashi, T.; Kuroda, K.; Tanaka, K.; et al. Polymer gels that memorize elements of molecular conformation. *Macromolecules* 2000, *33*, 8693–8697. [CrossRef]
54. Ito, K.; Chuang, J.; Alvarez-Lorenzo, C.; Watanabe, T.; Ando, N.; Grosberg, A.Y. Multiple point adsorption in a heteropolymer gel and the Tanaka approach to imprinting: Experiment and theory. *Prog. Polym. Sci.* 2003, *28*, 1489–1515. [CrossRef]
55. Pan, G.; Guo, Q.; Ma, Y.; Yang, H.; Li, B. Thermo-responsive hydrogel layers imprinted with RGDS peptide: A system for harvesting cell sheets. *Angew. Chem. Int. Ed.* 2013, *52*, 6907–6911. [CrossRef]
56. Pan, G.; Guo, Q.; Cao, C.; Yang, H.; Li, B. Thermo-responsive molecularly imprinted nanogels for specific recognition and controlled release of proteins. *Soft Matter* 2013, *9*, 3840–3850. [CrossRef]
57. Turan, E.; Özçetin, G.; Caykara, T. Dependence of protein recognition of temperature-sensitive imprinted hydrogels on preparation temperature. *Macromol. Biosci.* 2009, *9*, 421–428. [CrossRef] [PubMed]
58. Zhang, C.; Jia, X.; Wang, Y.; Zhang, M.; Yang, S.; Guo, J. Thermosensitive molecularly imprinted hydrogel cross-linked with N-malely chitosan for the recognition and separation of BSA. *J. Sep. Sci.* 2014, *37*, 419–426. [CrossRef] [PubMed]
59. Qin, L.; He, X.-W.; Zhang, W.; Li, W.-Y.; Zhang, Y.-K. Macroporous thermosensitive imprinted hydrogel for recognition of protein by metal coordinate interaction. *Anal. Chem.* 2009, *81*, 7206–7216. [CrossRef]
60. Adrus, N.; Ulbricht, M. Molecularly imprinted stimuli-responsive hydrogels for protein recognition. *Polymer* 2012, *53*, 4359–4366. [CrossRef]
61. Temtem, M.; Pompeu, D.; Barroso, T.; Fernandes, J.; Simões, P.C.; Casimiro, T.; Rego, A.M.B.D.; Aguiar-Ricardo, A. Development and characterization of a thermoresponsive polysulfone membrane using an environmental friendly technology. *Green Chem.* 2009, *11*, 638–645. [CrossRef]
62. Zhao, Q.; Sun, J.; Ling, Q.; Zhou, Q. Synthesis of macroporous thermosensitive hydrogels: A novel method of controlling pore size. *Langmuir* 2009, *25*, 3249–3254. [CrossRef]
63. Liu, X.; Zhou, T.; Du, Z.; Wei, Z.; Zhang, J. Recognition ability of temperature responsive molecularly imprinted polymer hydrogels. *Soft Matter* 2011, *7*, 1986–1993. [CrossRef]
64. Reddy, S.M.; Phan, Q.T.; El-Sharif, H.F.; Govada, L.; Stevenson, D.; Chayen, N.E. Protein crystallization and biosensor applications of hydrogel-based molecularly imprinted polymers. *Biomacromolecules* 2012, *13*, 3959–3965. [CrossRef] [PubMed]
65. Chen, Z.; Hua, Z.; Xu, L.; Huang, Y.; Zhao, M.; Li, Y. Protein-responsive imprinted polymers with specific shrinking and rebinding. *J. Mol. Recognit.* 2008, *21*, 71–77. [CrossRef] [PubMed]
66. Singh, B.; Chauhan, N.; Sharma, V. Design of molecular imprinted hydrogels for controlled release of cisplatin: Evaluation of network density of hydrogels. *Ind. Eng. Chem. Res.* 2011, *50*, 13742–13751. [CrossRef]
67. Tokuyama, H.; Kanazawa, R.; Sakohara, S. Equilibrium and kinetics for temperature swing adsorption of a target metal on molecular imprinted thermosensitive gel adsorbents. *Sep. Purif. Technol.* 2005, *44*, 152–159. [CrossRef]
68. Wang, X.-L.; Yao, H.-F.; Li, X.-Y.; Wang, X.; Huang, Y.-P.; Liu, Z.-S. PH/temperature-sensitive hydrogel-based molecularly imprinted polymers (hydroMIPs) for drug delivery by frontal polymerization. *RSC Adv.* 2016, *6*, 94038–94047. [CrossRef]

69. An, K.; Kang, H.; Zhang, L.; Guan, L.; Tian, D. Preparation and properties of thermosensitive molecularly imprinted polymer based on konjac glucomannan and its controlled recognition and delivery of 5-fluorouracil. *J. Drug Deliv. Sci. Technol.* **2020**, *60*, 101977. [CrossRef]
70. Suedee, R.; Jantarat, C.; Lindner, W.; Viernstein, H.; Songkro, S.; Srichana, T. Development of a pH-responsive drug delivery system for enantioselective-controlled delivery of racemic drugs. *J. Control. Release* **2010**, *142*, 122–131. [CrossRef] [PubMed]
71. Kanekiyo, Y.; Naganawa, R.; Tao, H. pH-Responsive Molecularly Imprinted Polymers. *Angew. Chem.* **2003**, *115*, 3122–3124. [CrossRef]
72. Wang, C.; Javadi, A.; Ghaffari, M.; Gong, S. A pH-sensitive molecularly imprinted nanospheres/hydrogel composite as a coating for implantable biosensors. *Biomaterials* **2010**, *31*, 4944–4951. [CrossRef] [PubMed]
73. Xu, L.; Qiu, L.; Sheng, Y.; Sun, Y.; Deng, L.; Li, X.; Bradley, M.; Zhang, R. Biodegradable pH-responsive hydrogels for controlled dual-drug release. *J. Mater. Chem. B* **2018**, *6*, 510–517. [CrossRef] [PubMed]
74. Pan, J.; Wang, B.; Dai, J.; Dai, X.; Hang, H.; Ou, H.; Yan, Y. Selective recognition of 2,4,5-trichlorophenol by temperature responsive and magnetic molecularly imprinted polymers based on halloysite nanotubes. *J. Mater. Chem.* **2012**, *22*, 3360–3369. [CrossRef]
75. Xu, L.; Pan, J.; Dai, J.; Li, X.; Hang, H.; Cao, Z.; Yan, Y. Preparation of thermal-responsive magnetic molecularly imprinted polymers for selective removal of antibiotics from aqueous solution. *J. Hazard. Mater.* **2012**, *233–234*, 48–56. [CrossRef]
76. Li, X.; Zhang, B.; Li, W.; Lei, X.; Fan, X.; Tian, L.; Zhang, H.; Zhang, Q. Preparation and characterization of bovine serum albumin surface-imprinted thermosensitive magnetic polymer microsphere and its application for protein recognition. *Biosens. Bioelectron.* **2014**, *51*, 261–267. [CrossRef] [PubMed]
77. Xu, S.; Li, J.; Song, X.; Liu, J.; Lu, H.; Chen, L. Photonic and magnetic dual responsive molecularly imprinted polymers: Preparation, recognition characteristics and properties as a novel sorbent for caffeine in complicated samples. *Anal. Methods* **2013**, *5*, 124–133. [CrossRef]
78. Gao, F.-X.; Ma, X.-T.; He, X.-W.; Li, W.-Y.; Zhang, Y.-K. Smart surface imprinting polymer nanospheres for selective recognition and separation of glycoprotein. *Colloids Surf. A Physicochem. Eng. Asp.* **2013**, *433*, 191–199. [CrossRef]
79. Fang, L.; Chen, S.; Guo, X.; Zhang, Y.; Zhang, H. Azobenzene-containing molecularly imprinted polymer microspheres with photo- and thermoresponsive template binding properties in pure aqueous media by atom transfer radical polymerization. *Langmuir* **2012**, *28*, 9767–9777. [CrossRef]
80. Hua, Z.; Chen, Z.; Li, Y.; Zhao, M. Thermosensitive and salt-sensitive molecularly imprinted hydrogel for bovine serum albumin. *Langmuir* **2008**, *24*, 5773–5780. [CrossRef]
81. Lin, F.; Chen, J.; Lee, M.; Lin, B.; Wang, J. Multi-Responsive Ibuprofen-Imprinted Core-Shell Nanocarriers for Specific Drug Recognition and Controlled Release. *ACS Appl. Nano Mater.* **2020**, *3*, 1147–1152. [CrossRef]
82. Miyata, T. Preparation of smart soft materials using molecular complexes. *Polym. J.* **2010**, *42*, 277–289. [CrossRef]
83. Miyata, T.; Hayashi, T.; Kuriu, Y.; Uragami, T. Responsive behavior of tumor-marker-imprinted hydrogels using macromolecular cross-linkers. *J. Mol. Recognit.* **2012**, *25*, 336–343. [CrossRef]

Review

A Review on Molecularly Imprinted Polymers Preparation by Computational Simulation-Aided Methods

Zhimin Liu [1,2], Zhigang Xu [2,*], Dan Wang [2,*], Yuming Yang [2], Yunli Duan [2], Liping Ma [1,*], Tao Lin [3] and Hongcheng Liu [3]

1. Faculty of Environmental Science and Engineering, Kunming University of Science and Technology, Kunming 650500, China; lab_chem@126.com
2. Faculty of Science, Kunming University of Science and Technology, Kunming 650500, China; yangym1205@163.com (Y.Y.); chemdyl@163.com (Y.D.)
3. Institute of Quality Standard and Testing Technology, Yunnan Academy of Agriculture Science, Kunming 650223, China; lintaonj@126.com (T.L.); liuorg@163.com (H.L.)
* Correspondence: chemxuzg@kust.edu.cn (Z.X.); wang_dan_l@163.com (D.W.); lpma2522@hotmail.com (L.M.)

Abstract: Molecularly imprinted polymers (MIPs) are obtained by initiating the polymerization of functional monomers surrounding a template molecule in the presence of crosslinkers and porogens. The best adsorption performance can be achieved by optimizing the polymerization conditions, but this process is time consuming and labor-intensive. Theoretical calculation based on calculation simulations and intermolecular forces is an effective method to solve this problem because it is convenient, versatile, environmentally friendly, and inexpensive. In this article, computational simulation modeling methods are introduced, and the theoretical optimization methods of various molecular simulation calculation software for preparing molecularly imprinted polymers are proposed. The progress in research on and application of molecularly imprinted polymers prepared by computational simulations and computational software in the past two decades are reviewed. Computer molecular simulation methods, including molecular mechanics, molecular dynamics and quantum mechanics, are universally applicable for the MIP-based materials. Furthermore, the new role of computational simulation in the future development of molecular imprinting technology is explored.

Keywords: computational simulation; molecularly imprinted polymers; intermolecular interaction

1. Introduction

Molecularly imprinted polymers (MIPs) are porous materials with specific recognition capacity towards the template molecule, which are obtained by self-assembly of template molecules and functional monomers in a porogen, and then polymerization is initiated in the presence of a cross-linking agent. The process of preparing MIPs is outlined in Figure 1. When the template molecule interacts with the functional monomer, the imprinting site is memorized through multiple action effects and fixed through the polymerization process. After the template is removed, the adsorption cavity complementary in shape and structure to the template molecule is left in the polymer matrix, which can selectively recognize the target molecule. Molecular imprinting technology originated from antibody immunology, that is, the specific combination of "lock and key" between antibody and antigen [1]. In 1973, Wulff [2] prepared organic MIPs for the first time. Since then, MIPs have attracted widespread attention. At present, MIPs, as a kind of intelligent adsorption material, are widely used in various fields, such as chromatographic separation [3], solid phase extraction [4–6], sensors [7–9], and biomedicine [10,11]. In the past two decades, great progress in MIPs has been achieved (Figure 2). A variety of novel and interesting imprinted polymers, including supramolecular imprinted polymers [12,13], multitemplate imprinted polymers [14,15], multifunctional monomer imprinted polymers [16,17], dummy template imprinted polymers [18,19], and chiral recognition polymers [20,21], have been developed.

In fact, synthesis parameters have been obtained through experimental optimization in most cases. Finding complex and cumbersome conditions is time consuming and laborious. Moreover, numerous organic reagents are used. These factors severely restrict the application and promotion of molecular imprinting technology.

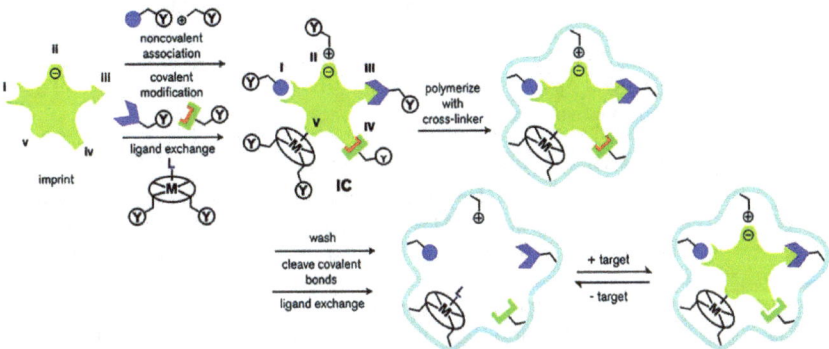

Figure 1. Schematic diagram of the molecular imprinting process: (**I**) non-covalent, (**II**) electrostatic/ionic, (**III**) covalent, (**IV**) semi-covalent, and (**V**) coordination to a metal center (Reprinted with permission from [22]. Copyright 2014 Royal Society of Chemistry).

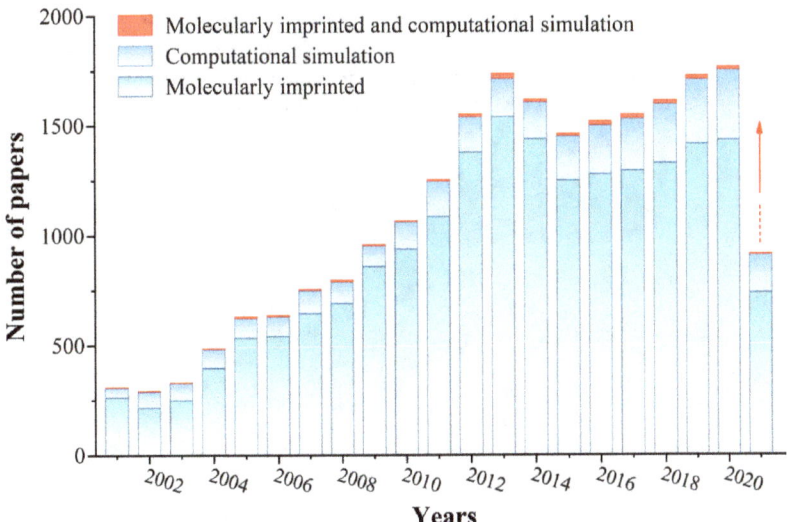

Figure 2. The literature statistics of MIPs and computational simulation. (Database: Scifinder; Search keywords: molecularly imprinted, computational simulation, molecularly imprinted and computational simulation, respectively. Search time: 13 June 2021).

Computational simulation has rapidly developed in recent years. It uses computer technology as a carrier and combines the theoretical basis of quantum mechanics and statistical mechanics as a tool-based cross-discipline. Molecular simulation calculation employs computer technology to simulate changes in the static structure and dynamic motion of molecules by calculating and comparing the relationship between the form and energy of the interaction between molecules to effectively explain the mechanism of action at the molecular level. The method is simple to operate and not restricted by the space environment, and the calculation is accurate and efficient. At present, many reports on the application of computational simulation in molecular imprinting technology have been published [23–26]. Computational simulation greatly reduces the cost of condition optimization during the polymerization of MIPs. Furthermore, it can effectively predict the more stable conformational composition between the template and the monomer. It can even simulate and calculate the types of porogens, crosslinkers, and initiators [27,28]. In this paper, the theoretical methods for simulating MIPs are briefly summarized, and the progress in the application of MIP simulation in molecular imprinting technology over the past 20 years has been reviewed. This provides insights into the cross application of molecular imprinting technology and computational simulation and the development of green chemistry.

2. Theoretical Methods of Computational Simulation for MIPs

The methods used in the theoretical calculation and simulation of various MIP designs are molecular mechanics (MM), molecular dynamics (MD), and quantum mechanics (QM). The computational cost of MM optimization is considerably lower than that of QM, and thus it is orders of magnitude faster than the latter. However, the accuracy of MM results is limited by simplified calculation models, which allow the reduction in calculation costs. The QM approach can better solve the problem of choosing the appropriate initial direction of interacting molecules because it is more accurate than the other methods. However, the computational complexity of the QM approach exponentially increases as the number of molecules involved in the calculation system increases. The MD method can effectively address this problem. When simulating the dynamic process of the interaction between molecules, changes in the molecule itself are often not considered, thereby making the calculation of the simulation method more efficient. Therefore, the MD method is most widely used when numerous molecules are involved in designing MIPs, such as in optimizing the ratio of template, monomer, and cross-linking agent. The application of MM, MD, and QM methods in MIP simulation is given in Table 1.

Table 1. Theoretical simulation calculation methods for the design of MIPs.

Simulation Method	Template	Force Field/Method	Software	MIPs Design
Molecular mechanics (MM)	Myoglobin [29]	OPLS3	Prime	Screening functional monomers
	Morphine [30]	CHARMM and MMFF94	Discovery Studio	Template-monomer ratio
	Metolachlor deschloro [31], metsulfuron-methyl [32]	AMBER MM	SPSS Statistics	Screening functional monomers/template-monomer ratio
	Norfloxacin [33]	MMFF94X	Discovery Studio	Screening functional monomers/template-monomer ratio
Molecular dynamics (MD)	Curcumin [34], fenthion [35], N-3-oxo-dodecanoyl-L-homoserine lactone [36], methidathion [37], endotoxins [38], phosmet insecticide [39], cocaine [40], methyl parathion [41], aflatoxin B1 [42]	Tripos	SYBYL	Screening functional monomers/template-monomer ratio
	Bisphenol A [43], carbamazepine [44], phthalates [45], norfloxacin [46], sulfamethoxazole [47]	COMPASS	Materials Studio/accelrys.com	Screening functional monomers/template-monomer ratio
	Thiamethoxam [48]	AMBER	Gaussian	Template-monomer ratio and solvent
	Rhodamine B [49]	GROMOS	GROMACS	Template-monomer ratio and solvent
Quantum mechanics (QM)	Vancomycin [50], primaquine [51], tramadol [52], thiamethoxam [48], clenbuterol [53], sulfadimidine [54], bilobalide [55], chloramphenicol [56], paclitaxel [57], acetamiprid [58], acetazolamide [59], lamotrigine [60], cyanazine [61], 3-methylindole [62], polybrominated diphenyl ethers [63], pirimicarb [64], metoprolol [65], ciprofloxacin or norfloxacin [66]	DFT	Gaussian	Screening functional monomers/template-monomer ratio/solvent
	Aspartame [67], pinacolyl methylphosphonate [68], metolachlor deschloro [31], metsulfuron-methyl [32], thiocarbohydrazide [69]	Semiempirical method	Spartan/SPSS Statistics	Screening functional monomers/template-monomer ratio
	Benzo[a]pyrene [70], tryptophan [71], furosemide [72], buprenorphine [73], hydroxyzine and cetirizine [74], atenolol [75], diazepam [76], metolachlor deschloro [31], metsulfuron-methyl [32], allopurinol [77], methadone [78], clonazepam [79], theophylline [80], ametryn [81], mosapride citrate [82], baicalein [83],	Ab initio	HyperChem/Gaussian/AutoDockTools/SPSS Statistics	Screening functional monomers/template-monomer ratio

2.1. MM Method

The MM method treats molecules as a collection of atoms held together by elasticity or resonance force. It uses energy functions, such as internal energy terms, including bond length, bond angle, and dihedral angle changes, to calculate changes in the molecular internal energy caused by changes in molecular structure. Combined with nonbonding energy (electrostatic interaction), these potential energy functions are called potential functions, and their parameters can be obtained by fitting quantum chemistry calculation results or experimental data. The MM method can optimize the molecular static structure of thousands of atomic systems; perform molecular structure optimization, system dynamics, and thermodynamic calculations; and select the smallest energy and the most stable molecular conformation in the space structure. However, this method ignores the movement of electrons and cannot simulate the state of electron movement in chemical reactions, thereby lowering the accuracy of the calculation [84–86].

Common force fields used in MIP calculation simulation by MM method are OPLS3, CHARMM, MMFF94, AMBER MM and MMFF94X. Compared with other commonly used small molecule force fields, the OPLS3 supplies reference data and related parameter types which exceed one order of magnitude. Therefore, this force field achieves a high level of accuracy in the performance benchmark for assessing conformational propensity and solvation of small molecules. It is mainly suitable for liquid systems, such as peptides, proteins, nucleic acids, and organic solvents. In addition, it employs lots of reference data and related parameter types; this also limits its use in the simulation of small molecule systems. The CHARMM force field is mainly used for the simulation of biological macromolecules, including energy minimization, molecular dynamics and Monte Carlo simulation. The simulation process provides information about molecular structure, interaction, energy, etc. The MMFF94 force field provides good accuracy in a series of organic and pharmaceutical molecular simulation calculations. The core parameterization is provided by high-quality quantum computing without a large amount of experimental data for testing molecular systems. It performs well in optimizing geometry, bond length, angle, as well electrostatic and hydrogen bonding effects. The AMBER force field also has significant and extensive applications in the field of simulation and calculation of biological macromolecules. Its advantage lies in the calculation of biological macromolecules, but the calculation results of small molecule systems are often unsatisfactory.

2.2. MD Method

MD methods can clarify the macroscopic properties of particles in dynamic motion. In 1957, Alder and Wainwright developed the MD technology. They computationally simulated the behavior of hard balls in boxes at different temperatures and densities. This method aims to establish a particle system via Newtonian mechanics and statistical mechanics, calculate the speed and position of molecules, obtain the state of motion of molecules, and integrate the dynamics and thermodynamic properties of systems. It is based on molecular mechanics and considers the influence of external environment, such as temperature and pressure, to calculate the molecular structures (crystallization, expansion and compression, vitrification, and deformation) and thermodynamic parameters of molecules in motion. The calculation result is close to the real state. The application of MD in molecular imprinting is illustrated in Figure 3. An MIP pre-polymerization system can possibly be established, the components and concentrations used in synthesizing the corresponding polymer copied, and the interactions and conformational changes between molecules observed.

The Tripos force field in the MD method has been proven to produce molecular geometry close to the crystal structure for different molecular selections. This force field has good calculation results in both protein and organic molecular simulations. The COMPASS force field is the first molecular force field based on ab initio calculations, which can accurately predict the molecular structure, conformation, vibration, and thermodynamic properties of isolated and condensed molecules. The COMPASS force field is also the first molecular force field that unifies the organic molecular system and the inorganic molecular system that were previously treated separately. It can simulate organic and inorganic small molecules, macromolecules, some metal ions, metal oxides and metals. The GROMOS force field guarantees the accuracy of the parameters through a series of quantum chemical calculations and existing databases. Most importantly, this force field fully considers the symmetry of the molecular structure to make the simulation more perfect.

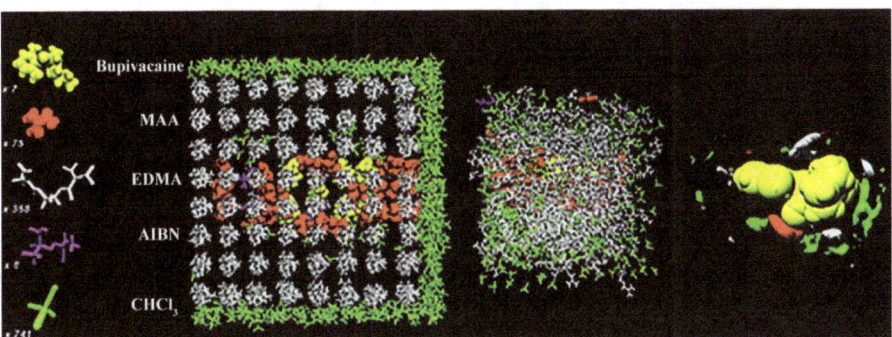

Figure 3. Schematic diagram of molecular dynamics calculation simulation molecular imprinting pre-assembly (Reprinted with permission from [85]. Copyright 2009 American Chemical Society).

2.3. QM Method

Many reports on the application of computational simulation in molecular imprinting technology have been published. Most of them used QM-based calculation methods. The primary QM simulation calculation methods are ab initio calculation methods, semiempirical calculation methods, and density functional theory (DFT) methods.

The ab initio calculation method is based on the Hartree–Fock method. This method uses some of the most basic physical constants, such as the speed of light and Planck's constant, as known parameters, and it adopts mathematical methods to calculate molecular physics and chemistry without introducing empirical parameters. In this method, the SDCI method is applied to both the single excited and double excited states of molecules. Other methods, such as CISD(T) and MCSCF, consider different variables. The ab initio calculation method is not limited to the structure of small molecules as it can also calculate the static and dynamic properties of macromolecular systems, including intramolecular and intermolecular interactions [87,88]. The ab initio calculation method is a quantum chemical calculation method that directly solves the Schrodinger equation based on the basic principles of QM method. Compared with the semiempirical method, the ab initio calculation method is more accurate, but time-consuming.

The semiempirical method introduces some experimentally measured parameters on the basis of the ab initio calculation method, simplifies the Hartree–Fock method, and reduces the amount of experimental calculation. This method can calculate and simulate the electronic structure and properties of real biological systems, such as enzymes and proteins. Other calculation methods are available: AM1, PM3, EHMO, CNDO, and NDDO. Among these methods, AM1 can predict the existence of hydrogen bonds between molecules by calculating the activation energy of particles. However, when it is used to calculate thermodynamic properties, such as the enthalpy of particles, large errors are committed. PM3 usually has a small error, and it is used in molecular simulations and theoretical calculations [89]. However, if the target analyte is a large molecule, only semiempirical methods have practical calculation meaning. This method is often used as the first step of high-precision calculations to obtain the initial structure of subsequent calculations. Some errors are easily reported due to the use of large reference parameters from the beginning. Thus, the optimization of the basis set results in a very slow optimization speed, or there can even be failure to optimize. In general, the semiempirical method is only suitable for simple organic molecules, and qualitative information, such as molecular orbital, electric charge and normal mode, could be obtained.

The DFT method is based on the Hohenberg–Kohn principle. It uses the electron density function in molecular simulation. The computational complexity of this method is small, and it can calculate molecular bond energy [90], predict compound structures [91], and predict reaction mechanisms [92]. Various molecular modeling and theoretical calculations in molecular imprinting technology apply the DFT method. It mainly calculates the binding energy (ΔE) between the template molecule and the functional monomer. In general, the lower the energy value is, the stronger the intermolecular interaction will be, indicating that the composite system of the template molecule and the functional monomer is more stable, which means that the prepared MIP has superior performance [93–96]. This method expresses the kinetic energy of an atom as a functional of electron density, which adds the classical expressions of the nucleus–electron and electron–electron interaction to calculate the energy of the atom. However, it is still difficult to describe the intermolecular forces, especially van der Waals forces, or the energy gap calculated by DFT method.

The calculation methods of theoretical molecular simulation used in molecular imprinting technology largely adopt the QM method. However, the QM method also has shortcomings. This method can only qualitatively predict the type of interaction between molecules but cannot accurately describe the energy changes of complex mixtures. Therefore, the previous research using this method for theoretical simulation has mostly applied to the qualitative description of molecular systems [97,98]. The amount of calculation involved in a quantitative system exponentially increases with the increase in the number of molecules, a condition that seriously affects calculation efficiency and accuracy.

3. Computational Simulation and Design of New MIPs

The application of theoretical calculations in designing MIPs is primarily achieved by theoretical simulations and selection of appropriate functional monomers, template molecules, crosslinkers, and their ratios. The binding energy (i.e., electronic interaction energy) between the template molecule and the functional monomer can be simulated and calculated provided that the binding energy between the template molecule and the functional monomer is high, indicating that the corresponding MIPs have excellent selectivity and adsorption performance. In addition, the ratio of the molecular and monomer system is closely related to the imprint factor of MIPs. In general, this ratio is calculated and optimized by performing the computational simulation in a vacuum environment to obtain the Eqaution (1) for the binding energy between the template molecule and the functional monomer.

$$\Delta E = E_{(Template\text{-}monomer\ complex)} - E_{Template} - E_{monomer} \tag{1}$$

In most cases, vacuum simulation calculations often differ from the actual situation as they consider the effects of spatial media, including the addition of solvents, to make the simulation calculation highly consistent with experimental results. The solvent (i.e., porogen) affects the energy of the system during the synthesis of MIPs. The results of molecular modeling can be made closer to real situation and the reliability of the results can be increased by conducting the simulation of a molecular fingerprint polymer in a solvent medium. The binding energy is calculated by Equation (2):

$$\Delta E_{Solvent} = E_{(Template\text{-}monomer\ complex\ in\ solvent\ (pore\text{-}forming\ agent))} - E_{(template\text{-}functional\ complex\ in\ the\ gas\ phase)} \tag{2}$$

where $\Delta E_{Solvent}$ is the energy difference between a template molecule and a functional monomer in solution and in a vacuum environment. A weak influence of the solution on noncovalent interactions during molecular fingerprint polymerization results in a small energy difference value, suggesting that the solvent is the best polymerization solvent for obtaining molecular fingerprint polymers [99,100].

The primary factor in MIP imprinting polymerization is the strong bonding force between the template and the functional monomer. Therefore, choosing the right functional monomer is a key factor in designing MIPs. An MIP can be reasonably designed by applying the DFT method in selecting the monomer with the best interaction with 2-isopropoxyphenol; it can be combined with the PM3.5 method to optimize the template-to-monomer ratio [101]. Quantum calculations were performed using the Spartan software, and the complexes' binding energy can be obtained to evaluate their stability. Pyrrole had been selected as the best functional monomer for designing 2-isopropoxyphenol MIPs. PM3 and DFT calculation methods were also used to simulate and calculate the monomers with the strongest interaction with disulfoton [102], chlorogenic acid [103], and amoxicillin [104], as well as the best ratio between the two. This method can be further used to calculate the solution energies of baicalein and acrylamide complexes in different solvents to screen the best polymerization solvent [105].

The strongest interaction site can be further located by obtaining the electrostatic potential map on the surface of the template molecule via the DFT method [106]. Figure 4 shows the electrostatic charge distribution of carvedilol after the geometry was optimized. The hydrogen bonding sites between carvedilol and functional monomer evidently appear in the red, yellow, and blue regions, which were O1, O2, O3, and H1. According to the quantitative information of the electrostatic map, each functional monomer undergoes hydrogen bonding at the four interaction sites in sequence to form hydrogen bonds; thus, the ratio of template and monomer complexes were 1:1 and 1:2, and 1:3 and 1:4. When the functional monomer is methacrylic acid and the template is combined with the monomer at a ratio of 1:4, a stable complex can be formed. The DFT method had also been adopted to study the interaction between p-nitrophenol and β-cyclodextrin [12].

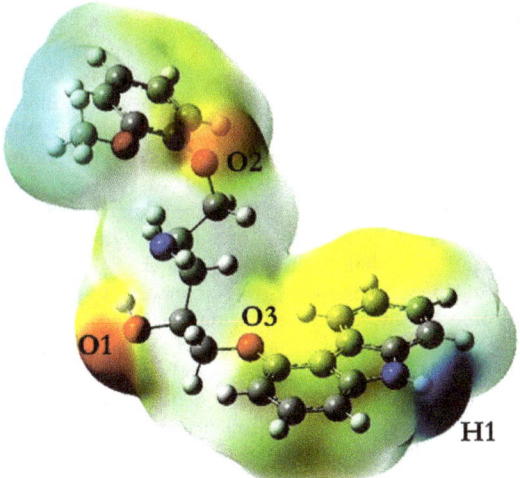

Figure 4. Electrostatic potential energy diagram of endotoxin in template (Reprinted with permission from [106]. Copyright 2019 Elsevier).

The key to the selectivity and enrichment ability of MIPs lies in the formation of a stable complex between the template and the functional monomer. Therefore, choosing the right functional monomer is an important factor in designing MIPs. The DFT method had been employed to study intermolecular interactions between harmane and functional monomer [107]. Firstly, MD simulation was performed at constant energy by combining it with the PM3 method. Subsequently, quenching kinetics and simulated annealing were combined to perform geometric optimization calculations on the structure of the template and monomer complex. The calculated optimized energy was then compared to finding the lowest energy conformation. The DFT method calculates the frequency of the harmane–monomer (1: n) complex system with the smallest energy value. It obtained the theoretical parameters of intermolecular interactions and provided a reliable theoretical basis for the interaction between the template and the monomer. Similar molecular simulations of combining DFT and MD methods had been used in furazolidone [108] and kojic acid [109] imprinted polymer designs. Another work had applied the COMPASS force field in MD simulation to calculate the intermolecular interactions of ibuprofen, naproxen, and diclofenac with 2-vinylpyridine [110]. Before MD simulation, the energy of the geometric optimization of the composite system was minimized in the Materials Studio software, and the intermolecular binding energies of each system were simultaneously calculated. The influence of the solvent on the composite system was also considered, which can be seen from the optimized conformation of the calculation. During synthesis, toluene did not participate in the monomer–template interaction, indicating that the effect of toluene on the selective adsorption of such MIPs was negligible.

The DFT calculation method had also been used to explore the influence of different porogens on the binding energy of nicotinamide to monomer methacrylic acid [111]. Template molecules, functional monomers, and template-monomer complexes are modeled in a vacuum and then optimized for conformation to calculate the single-point energy of the PM3 level. When toluene is used as a porogen, a small dielectric constant and an aprotic solvent may result in a large interaction force between the template and the monomer. The MIP prepared under this condition exhibits the ideal affinity and selectivity for the target molecule. Farhad et al. [112] calculated and synthesized a new MIP of ephedrine. They used the restricted Hartree–Fock method in DFT and then applied the Podient continuum model to simulate and calculate the optimal polymerization solvent. They found that methacrylic acid and methanol were the best monomers and porogens for pseudoephedrine MIP. The calculated simulation results were consistent with the experimental results. The effect of imprinting was further maximized using the DFT method in designing and preparing clenbuterol MIP [53]. The simulation results showed that the best functional monomer was acrylic acid, and the best ratio of the template molecule clenbuterol and its norepinephrine to monomer was 1:3. Given that MeOH has the smallest solvent energy, it was selected as the best porogen. In the simultaneous calculation and simulation of various types of crosslinking agents, when ethylene glycol dimethacrylate was used as the crosslinking agent, the crosslinking agent exerts the least interference to the imprinting process. The same method has been used to study the thermodynamic properties of the polymerization process [113]. Preliminary conformational optimization of the intermolecular electrostatic potential showed that atrazine had five interaction sites. A strong complex interaction was achieved when atrazine and 2-(trifluoromethyl) acrylic acid were combined at a ratio of 1:4. As a porogen, toluene had the least interference to the self-assembly system. The calculation temperature of thermodynamic properties was within the range of 278.15–308.15 K, and the enthalpy (ΔH), free energy (ΔG), and entropy (ΔS) of the best imprinting combination system in vacuum and the toluene medium were calculated. Thermodynamic analysis revealed that atrazine MIP was beneficial to the formation of imprinting sites in a medium with a low polarity and temperature.

Baggiani [114] applied the semiempirical quantum method (AM1) to screen the best combination of six functional monomers and carbamate interactions. They used a simulated annealing algorithm to optimize the structural arrangement of supramolecules, and they calculated the heat of the formation of six composite systems. Because both acrylamide and methacrylic acid can form two different hydrogen bond interactions with the template molecule, they were considered to be the most suitable functional monomers for preparing urethane imprinted polymers. In recent years, the simulated annealing algorithm has been independently used to attain a stable conformation of the template and the monomer complex [43,44,46]. During the annealing process, the temperature of the local environment is always in a dynamic process, ensuring that the templates and the monomers located at different positions and directions are bonded, thereby expanding the available conformations. To screen suitable functional monomers from monomer library quickly, Elena [115] employed the Tripos force field and combined it with the Leapfrog algorithm to simulate and calculate the possible interactions between simazine and 20 functional monomers.

Simazine and methacrylic acid achieved a high binding energy mainly through electrostatic interactions. The combination of the Tripos force field and the Leapfrog algorithm is widely used in screening functional monomers [39–41,116]. Owing to the ability of this combination to simulate imprinting sites visually, this method can quickly determine the monomer library. The monomer with the strongest interaction with the target molecule substantially shortens the time required to screen the best monomer from numerous monomers [117,118].

An imprinted polymer membrane that can selectively adsorb atrazine in an aqueous medium can be developed [119] using the Hyperchem software to find a functional monomer suitable for atrazine. When methacrylic acid was used as a functional monomer, two ionic bonds and three hydrogen bonds can form between atrazine and methacrylic acid. Liang et al. [120] designed and prepared a porous core–shell MIP for the selective separation and enrichment of ursodeoxycholic acid through computational simulation. They selected the molecular force field AMBER for computational simulation. The conformation of the template, monomer, and porogen are optimized, and then the MD optimization method was used to calculate the binding energy of the template and the monomer complex in the three solvents. MIP with toluene as the porogen had the strongest affinity and selectivity for ursodeoxycholic acid. Viveiros et al. [121] applied the supercritical carbon dioxide technology for the molecular imprinting of acetamide for the first time. They selected the best functional monomer, crosslinking agent, and their molar ratio for each template molecule. They then introduced CO_2 as a solvent into the model. The theoretically designed itaconic acid MIP exhibited a high affinity and selectivity to acetamide.

Molecular simulation methods are also used in designing and synthesizing dummy template MIPs. Feng [122] adopted MM combined with AM1 to select the best alternative template for clenbuterol and its metabolites. Dummy template MIP for sulfonylurea herbicides can also be obtained by theoretical calculations by using the Hyperchem software for molecular modeling [32]. The lowest energy conformation of the template and monomer is optimized by the MM and PM3 methods, and the binding energy of the complex was obtained by AMBER MM force field. The difference between mesulfuron-methyl and nine other sulfonylurea herbicide molecules was the smallest. Thus, it was considered to be the best dummy template molecule.

Molecular simulation calculation can be also performed in designing chiral enantioselective imprinted polymers. Sobiech [123] prepared an octopamine chiral selective MIP via computational simulations. Discovery Studio can visualize the interface to build molecular models. The DFT method can be used to optimize the geometry of all compounds, whereas its combination with the Breneman model can reproduce the atomic part electrostatic potential of each molecule. During modeling, the monomers were randomly distributed around the template, and intermolecular interactions occurred during the energy minimization process. The simulation results showed that the binding energy of octopamine and 4-vinyl benzoic acid was the smallest. Furthermore, a two-step method of MM and QM were used to simulate the design of enantioselective tBOC-tyrosine imprinted polymers [124]. The geometric structure of the molecule was greatly optimized after the two-step simulation calculations from the MM method to the MD method. This multiscale "coarse to fine" technology can address the shortcomings of using the MM method or the QM method alone and combine their advantages. The molecular structure can be roughly optimized under the MMFF94 force field. The QM method was then used to further refine its geometry. After optimizing the system, the imprinted target molecule was deleted from the QM-optimized geometry, leaving behind the "imprint" binding site. Further analysis of single monomer–target interactions in the binding sites suggested that the hydrogen atom in the chiral center would be more conducive to bonding with the functional group than the one in the chiral center. The enantioselectivity factor was obtained by determining the ratio of the binding energy of the target to the binding energy of its enantiomers. If the value is greater than 1, then the imprinted polymer can be considered to have enantioselective recognition ability for the chiral target molecule.

4. Computational Simulation and MIP Identification Mechanism

Theoretical simulation can also provide a theoretical basis for the identification mechanism of MIPs. The formation process of experimentally proposed magnetic molecularly imprinted polypyrrole at the molecular level can be understood via the DFT method to obtain the thermodynamic properties of the prepolymerized template and the monomer complex in the presence of water. On the basis of the negative values of ΔG and ΔH, this results in the complexation of the monomer with praziquantel in aqueous solution spontaneously forming stable complexes. Moreover, the results of molecular geometric conformation simulation showed that four hydrogen bonds and one π–π stacking interaction are established between praziquantel and pyrrole, which explains the formation of praziquantel and pyrrole prepolymer complex at the molecular level [125]. Through PM3 and DFT theoretical simulation methods, the Muliken charge on each atom of the fluazuron optimized geometric structure can be obtained, which can quantitatively reveal the existence of six regions with a high electron charge density. These local regions can interact with methacrylic acid molecules and build hydrogen bonds. If the value of enthalpy and Gibbs free energy is less than zero, then the prepolymer complex of flusulfuron–methyl and methacrylic acid can be considered to have spontaneously and stably formed. These simulation results explain the polymerization mechanism of fluazuron MIP [126].

The selective mechanism of ciprofloxacin-imprinted membrane was also further explained through molecular simulations [127]. The binding energy of the interaction between the functional monomer and ciprofloxacin and its structural analogs, including norfloxacin hydrochloride, enrofloxacin hydrochloride and ofloxacin hydrochloride, was obtained through molecular simulation calculations. Kinetic simulations had also been performed using GROMACS software. The parameters of bond, angle, dihedral angle, and Lennard–Jones interaction had been directly taken from the GAFF force field. Part of the charge was obtained using the restricted electrostatic potential method at the theoretical level of B3LYP/6-31+G (d, p). The recombination ability of the imprinted site of ciprofloxacin was dominated by hydrogen bond interactions, whereas its structural analogs were dominated by van der Waals interaction. Thus, strong hydrogen bond interactions led to a high tendency for the imprinted site of ciprofloxacin to recombine with the template molecule. Theoretical simulations of the recombination mechanism and selective permeation experiments mutually confirmed the superior selectivity of ciprofloxacin-imprinted membranes. Zhang [128] further explained the specific selective recognition mechanism of molecularly imprinted nanocomposite membranes for artemisinin by dynamic simulations. A comparison of the binding energy of the imprinted membrane with artemisinin and its structural analogs shows that the strong interaction between artemisinin and the imprinted polymer matrix contributes to its large adsorption capacity and high selectivity. A similar DFT method has been used to explain the selective recognition mechanism of the alternative template N-(4-isopropylphenyl)-N'-butyleneurea MIP to phenylurea herbicides [129]. This method also explained the mechanism of experimentally preferred dummy template imprinted polymer [130] and the strength of the bonding force of chiral naproxen MIP [21] at the molecular level. These observations provide a theoretical basis to explain the experimental results from the perspective of intermolecular interactions.

Yang [61] performed molecular simulation to reveal the essential reason for the difference between single-template and double-template MIP stirring bars in their ability to recognize target analytes by using the YASARA software to study the recognition mechanism. The 3D shape and size of the imprinted cavity in the MIPs are the corresponding template molecules. Given that the dual-template MIP contains imprinted cavities of the two template molecules, it had a fairly high recovery for nine fluoroquinolones, and the simulation results are consistent with the experimental findings. However, the influence of template–template interactions on the performance of multitemplate MIPs has been further verified via the DFT method [62]. The results of both theoretical simulations and experiments indicated that the interaction between more template molecules affects the formation of specific recognition sites and even reduces the formation of effective imprinting sites.

5. Conclusions and Outlook

Computer molecular modeling technology has been applied to the screening and optimization of molecules in many materials, and it is also a feasible method for preliminary exploration of MIP. Computer simulation reduces the time and reagent-related costs required to obtain the appropriate MIP adsorbent, and significantly reduces the consumption of organic solvents. In addition, it can explain the specific recognition mechanism of imprinted materials at the molecular level. For all the above reasons, the use of computer molecular simulation to design MIP adsorbents in analytical practice not only conforms to the principles of green analytical chemistry, but also explains the nature of MIPs binding to target molecules from the intermolecular forces. The QM method, compared with other methods, can ensure more accurate simulation results in the MIP system dominated by non-bonding interaction, because the smallest structural unit electron was studied and the quantum effect was considered in the method. Therefore, the QM method is also the most widely used in MIP simulation operations. However, in the simulation of macromolecules and polyatomic systems, this method is very time-consuming and even prone to errors. MM and MD are classical mechanics methods. Their smallest structural unit is no longer an electron but an atom. Therefore, the simulation operation complexity of the imprinting system is greatly reduced, and the operation speed is faster. MM method directly utilizes the potential function to study the problem, without considering the kinetic energy and the corresponding structure of the atom. However, the MD method focuses on the movement of atoms in the MIP system and establishes the relationship between temperature and time, which can simulate the imprinting system more realistically, and the simulation results are more representative. In general, the DFT procedure in the QM method was recommended in the MIPs design and mechanism interpretation simulation calculation. However, this also means that the computational complexity of this method increases dramatically for large molecules and systems with a large number of molecules. MD method may be the best solution at this situation, simulated annealing process in particular, which can complete the lowest energy conformation search in a very short time. At present, an increasing number of research have been using multiple calculation methods to achieve complementary advantages when designing and optimizing the experimental parameters of MIPs preparation, so as to ensure more efficient and accurate simulation results. In addition, simulation is also the direction of current efforts. A more realistic simulation environment can make the calculation results accurate and reliable.

Author Contributions: Conceptualization, Z.L.; editing, supervision, funding acquisition, Z.X.; writing—Original draft preparation, D.W., Y.Y. and Y.D.; investigation, reviewed references. L.M.; writing—Review and editing, T.L. and H.L. All authors have read and agreed to the published version of the manuscript.

Funding: This work was funded by the National Natural Science Foundation of China (No. 21565018), the Scientific Research Fund Project of Yunnan Provincial Department of Education (No. 2021Y114) and the Yunnan Key Research and Development Program (No. 202002AE320005).

Institutional Review Board Statement: Not applicable.

Informed Consent Statement: Not applicable.

Data Availability Statement: There are no data associated with this publication.

Conflicts of Interest: The authors declare no conflict of interest.

Abbreviations

MIP	Molecularly imprinted polymers
MM	Molecular mechanics
MD	Molecular dynamics
QM	Quantum mechanics
DFT	Density functional theory
OPLS	Optimized potentials for liquid simulations
CHARMM	Chemistry at Harvard macromolecular mechanics
MMFF	Merck molecular force field
Tripos	Tripos force field
COMPASS	Computer plan appraisal system for ship
GROMOS	Gromos force field
Ab initio	Latin term meaning "from the beginning"
B3LYP	Becke, three-parameter
GAFF	Generation amber force field
SDCI	Single and double excitation configuration interaction
CISD(T)	Configuration interactions with single and double substitutions
MCSCF	Multiconfiguration self-consistent field
AM1	AM1 semiempirical method
PM3	Semiempirical PM3 method
EHMO	Extended huckel molecular orbital theory
CNDO	Complete neglect of differential overlap method
NDDO	Neglect of diatomic differential overlap method

References

1. Pauling, L. A Theory of the structure and process of formation of antibodies. *J. Am. Chem. Soc.* **1940**, *62*, 2643–2657. [CrossRef]
2. Wulff, G.; Sarhan, A.; Zabrocki, K. Enzyme-analogue built polymers and their use for the resolution of racemates. *Tetrahedron Lett.* **1973**, *14*, 4329–4332. [CrossRef]
3. Kubo, T.; Nomachi, M.; Nemoto, K.; Sano, T.; Hosoya, K.; Tanaka, N.; Kaya, K. Chromatographic separation for domoic acid using a fragment imprinted polymer. *Anal. Chim. Acta* **2006**, *577*, 1–7. [CrossRef] [PubMed]
4. Sarafraz-Yazdi, A.; Razavi, N. Application of molecularly-imprinted polymers in solid-phase microextraction techniques. *TrAC Trends Anal. Chem.* **2015**, *73*, 81–90. [CrossRef]
5. Song, Y.P.; Zhang, L.; Wang, G.N.; Liu, J.X.; Liu, J.; Wang, J.P. Dual-dummy-template molecularly imprinted polymer combining ultra performance liquid chromatography for determination of fluoroquinolones and sulfonamides in pork and chicken muscle. *Food Control* **2017**, *82*, 233–242. [CrossRef]
6. Lian, Z.; Li, H.-B.; Wang, J. Experimental and computational studies on molecularly imprinted solid-phase extraction for gonyautoxins 2,3 from dinoflagellate Alexandrium minutum. *Anal. Bioanal. Chem.* **2016**, *408*, 5527–5535. [CrossRef] [PubMed]
7. Ahmad, O.S.; Bedwell, T.S.; Esen, C.; Garcia-Cruz, A.; Piletsky, S.A. Molecularly imprinted polymers in electrochemical and optical sensors. *Trends Biotechnol.* **2019**, *37*, 294–309. [CrossRef]
8. Su, C.; Li, Z.; Zhang, D.; Wang, Z.; Zhou, X.; Liao, L.; Xiao, X. A highly sensitive sensor based on a computer-designed magnetic molecularly imprinted membrane for the determination of acetaminophen. *Biosens. Bioelectron.* **2020**, *148*, 111819. [CrossRef]
9. Goud, K.Y.; Reddy, K.K.; Gobi, K.V. Development of highly selective electrochemical impedance sensor for detection of sub-micromolar concentrations of 5-Chloro-2,4-dinitrotoluene. *J. Chem. Sci.* **2016**, *128*, 763–770. [CrossRef]
10. Gu, X.; Huang, J.; Zhang, L.; Zhang, Y.; Wang, C.-Z.; Sun, C.; Yao, D.; Li, F.; Chen, L.; Yuan, C.-S. Efficient discovery and capture of new neuronal nitric oxide synthase-postsynaptic density protein-95 uncouplers from herbal medicines using magnetic molecularly imprinted polymers as artificial antibodies. *J. Sep. Sci.* **2017**, *40*, 3522–3534. [CrossRef]
11. Dong, Y.; Li, W.; Gu, Z.; Xing, R.; Ma, Y.; Zhang, Q.; Liu, Z. Inhibition of HER2-positive breast cancer growth by blocking the HER2 signaling pathway with HER2-glycan-imprinted nanoparticles. *Angew. Chem. Int. Ed.* **2019**, *58*, 10621–10625. [CrossRef] [PubMed]
12. Liu, Y.; Liu, Y.; Liu, Z.; Du, F.; Qin, G.; Li, G.; Hu, X.; Xu, Z.; Cai, Z. Supramolecularly imprinted polymeric solid phase microextraction coatings for synergetic recognition nitrophenols and bisphenol A. *J. Hazard. Mater.* **2019**, *368*, 358–364. [CrossRef]
13. Li, S.; Yin, G.; Wu, X.; Liu, C.; Luo, J. Supramolecular imprinted sensor for carbofuran detection based on a functionalized multiwalled carbon nanotube-supported Pd-Ir composite and methylene blue as catalyst. *Electrochim. Acta* **2016**, *188*, 294–300. [CrossRef]
14. Wang, S.; She, Y.; Hong, S.; Du, X.; Yan, M.; Wang, Y.; Qi, Y.; Wang, M.; Jiang, W.; Wang, J. Dual-template imprinted polymers for class-selective solid-phase extraction of seventeen triazine herbicides and metabolites in agro-products. *J. Hazard. Mater.* **2019**, *367*, 686–693. [CrossRef] [PubMed]

15. Liu, Y.; Liu, Y.; Liu, Z.; Hill, J.P.; Alowasheeir, A.; Xu, Z.; Xu, X.; Yamauchi, Y. Ultra-durable, multi-template molecularly imprinted polymers for ultrasensitive monitoring and multicomponent quantification of trace sulfa antibiotics. *J. Mater. Chem. B* **2021**, *9*, 3192–3199. [CrossRef] [PubMed]
16. Gao, R.; Hao, Y.; Zhao, S.; Zhang, L.; Cui, X.; Liu, D.; Tang, Y.; Zheng, Y. Novel magnetic multi-template molecularly imprinted polymers for specific separation and determination of three endocrine disrupting compounds simultaneously in environmental water samples. *RSC Adv.* **2014**, *4*, 56798–56808. [CrossRef]
17. Li, Y.; Zhang, L.; Dang, Y.; Chen, Z.; Zhang, R.; Li, Y.; Ye, B. A robust electrochemical sensing of molecularly imprinted polymer prepared by using bifunctional monomer and its application in detection of cypermethrin. *Biosens. Bioelectron.* **2019**, *127*, 207–214. [CrossRef]
18. Marća, M.; Wieczorek, P.P. The preparation and evaluation of core-shell magnetic dummy-template molecularly imprinted polymers for preliminary recognition of the low-mass polybrominated diphenyl ethers from aqueous solutions. *Sci. Total. Environ.* **2020**, *724*, 138151. [CrossRef]
19. Yuan, X.; Yuan, Y.; Gao, X.; Xiong, Z.; Zhao, L. Magnetic dummy-template molecularly imprinted polymers based on multi-walled carbon nanotubes for simultaneous selective extraction and analysis of phenoxy carboxylic acid herbicides in cereals. *Food Chem.* **2020**, *333*, 127540. [CrossRef]
20. Zhang, Y.; Wang, H.-Y.; He, X.-W.; Li, W.-Y.; Zhang, Y.-K. Homochiral fluorescence responsive molecularly imprinted polymer: Highly chiral enantiomer resolution and quantitative detection of L-penicillamine. *J. Hazard. Mater.* **2021**, *412*, 125249. [CrossRef]
21. Liu, Y.; Liu, Y.; Liu, Z.; Zhao, X.; Wei, J.; Liu, H.; Si, X.; Xu, Z.; Cai, Z. Chiral molecularly imprinted polymeric stir bar sorptive extraction for naproxen enantiomer detection in PPCPs. *J. Hazard. Mater.* **2020**, *392*, 122251. [CrossRef]
22. Lofgreen, J.E.; Ozin, G.A. Controlling morphology and porosity to improve performance of molecularly imprinted sol–gel silica. *Chem. Soc. Rev.* **2014**, *43*, 911–933. [CrossRef] [PubMed]
23. Marć, M.; Kupka, T.; Wieczorek, P.P.; Namieśnik, J. Computational modeling of molecularly imprinted polymers as a green approach to the development of novel analytical sorbents. *TrAC Trends Anal. Chem.* **2018**, *98*, 64–78. [CrossRef]
24. Paredes-Ramos, M.; Bates, F.; Rodríguez-González, I.; López-Vilariño, J.M. Computational approximations of molecularly imprinted polymers with sulphur based monomers for biological purposes. *Mater. Today Commun.* **2019**, *20*, 100526. [CrossRef]
25. Cowen, T.; Karim, K.; Piletsky, S. Computational approaches in the design of synthetic receptors—A review. *Anal. Chim. Acta* **2016**, *936*, 62–74. [CrossRef] [PubMed]
26. Khan, M.S.; Pal, S.; Krupadam, R.J. Computational strategies for understanding the nature of interaction in dioxin imprinted nanoporous trappers. *J. Mol. Recognit.* **2015**, *28*, 427–437. [CrossRef] [PubMed]
27. Paredes-Ramos, M.; Sabín-López, A.; Peña-García, J.; Pérez-Sánchez, H.; López-Vilariño, J.M.; Sastre De Vicente, M.E. Computational aided acetaminophen—Phthalic acid molecularly imprinted polymer design for analytical determination of known and new developed recreational drugs. *J. Mol. Graph. Model.* **2020**, *100*, 107627. [CrossRef] [PubMed]
28. Lai, W.; Zhang, K.; Shao, P.; Yang, L.; Ding, L.; Pavlostathis, S.G.; Shi, H.; Zou, L.; Liang, D.; Luo, X. Optimization of adsorption configuration by DFT calculation for design of adsorbent: A case study of palladium ion-imprinted polymers. *J. Hazard. Mater.* **2019**, *379*, 120791. [CrossRef] [PubMed]
29. Sullivan, M.V.; Dennison, S.R.; Archontis, G.; Reddy, S.M.; Hayes, J.M. Toward rational design of selective molecularly imprinted polymers (MIPs) for proteins: Computational and experimental studies of acrylamide based polymers for myoglobin. *J. Phys. Chem. B* **2019**, *123*, 5432–5443. [CrossRef]
30. Xi, S.; Zhang, K.; Xiao, D.; He, H. Computational-aided design of magnetic ultra-thin dummy molecularly imprinted polymer for selective extraction and determination of morphine from urine by high-performance liquid chromatography. *J. Chromatogr. A* **2016**, *1473*, 1–9. [CrossRef] [PubMed]
31. Zhang, L.; Han, F.; Hu, Y.; Zheng, P.; Sheng, X.; Sun, H.; Song, W.; Lv, Y. Selective trace analysis of chloroacetamide herbicides in food samples using dummy molecularly imprinted solid phase extraction based on chemometrics and quantum chemistry. *Anal. Chim. Acta* **2012**, *729*, 36–44. [CrossRef] [PubMed]
32. Han, F.; Zhou, D.B.; Song, W.; Hu, Y.Y.; Lv, Y.N.; Ding, L.; Zheng, P.; Jia, X.Y.; Zhang, L.; Deng, X.J. Computational design and synthesis of molecular imprinted polymers for selective solid phase extraction of sulfonylurea herbicides. *J. Chromatogr. A* **2021**, *1651*, 462321. [CrossRef]
33. Fizir, M.; Wei, L.; Muchuan, N.; Itatahine, A.; Mehdi, Y.A.; He, H.; Dramou, P. QbD approach by computer aided design and response surface methodology for molecularly imprinted polymer based on magnetic halloysite nanotubes for extraction of norfloxacin from real samples. *Talanta* **2018**, *184*, 266–276. [CrossRef] [PubMed]
34. Piletska, E.V.; Abd, B.H.; Krakowiak, A.S.; Parmar, A.; Pink, D.L.; Wall, K.S.; Wharton, L.; Moczko, E.; Whitcombe, M.J.; Karim, K.; et al. Magnetic high throughput screening system for the development of nano-sized molecularly imprinted polymers for controlled delivery of curcumin. *Analyst* **2015**, *140*, 3113–3120. [CrossRef] [PubMed]
35. Bakas, I.; Ben Oujji, N.; Istamboulié, G.; Piletsky, S.; Piletska, E.; Ait-Addi, E.; Ait-Ichou, I.; Noguer, T.; Rouillon, R. Molecularly imprinted polymer cartridges coupled to high performance liquid chromatography (HPLC-UV) for simple and rapid analysis of fenthion in olive oil. *Talanta* **2014**, *125*, 313–318. [CrossRef]
36. Karim, K.; Cowen, T.; Guerreiro, A.; Piletska, E.; Whitcombe, M. A protocol for the computational design of high affinity molecularly imprinted polymer synthetic receptors. *Glob. J. Biotechnol. Biomater. Sci.* **2017**, *3*, 1–7. [CrossRef]

37. Bakas, I.; Hayat, A.; Piletsky, S.; Piletska, E.; Chehimi, M.M.; Noguer, T.; Rouillon, R. Electrochemical impedimetric sensor based on molecularly imprinted polymers/sol–gel chemistry for methidathion organophosphorous insecticide recognition. *Talanta* **2014**, *130*, 294–298. [CrossRef]
38. Abdin, M.J.; Altintas, Z.; Tothill, I.E. In silico designed nanoMIP based optical sensor for endotoxins monitoring. *Biosens. Bioelectron.* **2015**, *67*, 177–183. [CrossRef]
39. Aftim, N.; Istamboulié, G.; Piletska, E.; Piletsky, S.; Calas-Blanchard, C.; Noguer, T. Biosensor-assisted selection of optimal parameters for designing molecularly imprinted polymers selective to phosmet insecticide. *Talanta* **2017**, *174*, 414–419. [CrossRef]
40. Smolinska-Kempisty, K.; Ahmad, O.S.; Guerreiro, A.; Karim, K.; Piletska, E.; Piletsky, S. New potentiometric sensor based on molecularly imprinted nanoparticles for cocaine detection. *Biosens. Bioelectron.* **2017**, *96*, 49–54. [CrossRef]
41. Esen, C.; Czulak, J.; Cowen, T.; Piletska, E.; Piletsky, S.A. Highly efficient abiotic assay formats for methyl parathion: Molecularly imprinted polymer nanoparticle assay as an alternative to enzyme-linked immunosorbent assay. *Anal. Chem.* **2019**, *91*, 958–964. [CrossRef] [PubMed]
42. Sergeyeva, T.; Yarynka, D.; Piletska, E.; Lynnik, R.; Zaporozhets, O.; Brovko, O.; Piletsky, S.; El'Skaya, A. Fluorescent sensor systems based on nanostructured polymeric membranes for selective recognition of Aflatoxin B1. *Talanta* **2017**, *175*, 101–107. [CrossRef] [PubMed]
43. Qiu, C.; Xing, Y.; Yang, W.; Zhou, Z.; Wang, Y.; Liu, H.; Xu, W. Surface molecular imprinting on hybrid SiO_2-coated CdTe nanocrystals for selective optosensing of bisphenol A and its optimal design. *Appl. Surf. Sci.* **2015**, *345*, 405–417. [CrossRef]
44. He, Q.; Liang, J.-J.; Chen, L.-X.; Chen, S.-L.; Zheng, H.-L.; Liu, H.-X.; Zhang, H.-J. Removal of the environmental pollutant carbamazepine using molecular imprinted adsorbents: Molecular simulation, adsorption properties, and mechanisms. *Water Res.* **2020**, *168*, 115164. [CrossRef]
45. Li, X.; Wan, J.; Wang, Y.; Ding, S.; Sun, J. Improvement of selective catalytic oxidation capacity of phthalates from surface molecular-imprinted catalysis materials: Design, mechanism, and application. *Chem. Eng. J.* **2021**, *413*, 127406. [CrossRef]
46. Kong, Y.; Wang, N.; Ni, X.; Yu, Q.; Liu, H.; Huang, W.; Xu, W. Molecular dynamics simulations of molecularly imprinted polymer approaches to the preparation of selective materials to remove norfloxacin. *J. Appl. Polym. Sci.* **2016**, *133*. [CrossRef]
47. Xu, W.; Wang, Y.; Huang, W.; Yu, L.; Yang, Y.; Liu, H.; Yang, W. Computer-aided design and synthesis of CdTe@SiO_2 core-shell molecularly imprinted polymers as a fluorescent sensor for the selective determination of sulfamethoxazole in milk and lake water. *J. Sep. Sci.* **2017**, *40*, 1091–1098. [CrossRef]
48. Silva, C.F.; Menezes, L.F.; Pereira, A.C.; Nascimento, C.S. Molecularly Imprinted Polymer (MIP) for thiamethoxam: A theoretical and experimental study. *J. Mol. Struct.* **2021**, *1231*, 129980. [CrossRef]
49. Liu, R.; Li, X.; Li, Y.; Jin, P.; Qin, W.; Qi, J. Effective removal of rhodamine B from contaminated water using non-covalent imprinted microspheres designed by computational approach. *Biosens. Bioelectron.* **2009**, *25*, 629–634. [CrossRef]
50. Yu, H.; Yao, R.; Shen, S. Development of a novel assay of molecularly imprinted membrane by design-based gaussian pattern for vancomycin determination. *J. Pharm. Biomed. Anal.* **2019**, *175*, 112789. [CrossRef] [PubMed]
51. Prasad, B.B.; Kumar, A.; Singh, R. Molecularly imprinted polymer-based electrochemical sensor using functionalized fullerene as a nanomediator for ultratrace analysis of primaquine. *Carbon* **2016**, *109*, 196–207. [CrossRef]
52. Fonseca, M.C.; Nascimento, C.S.; Borges, K.B. Theoretical investigation on functional monomer and solvent selection for molecular imprinting of tramadol. *Chem. Phys. Lett.* **2016**, *645*, 174–179. [CrossRef]
53. Zhang, B.; Fan, X.; Zhao, D. Computer-aided design of molecularly imprinted polymers for simultaneous detection of clenbuterol and its metabolites. *Polymers* **2018**, *11*, 17. [CrossRef]
54. Zhang, L.; He, L.; Wang, Q.; Tang, Q.; Liu, F. Theoretical and experimental studies of a novel electrochemical sensor based on molecularly imprinted polymer and GQDs-PtNPs nanocomposite. *Microchem. J.* **2020**, *158*, 105196. [CrossRef]
55. Huang, X.; Zhang, W.; Wu, Z.; Li, H.; Yang, C.; Ma, W.; Hui, A.; Zeng, Q.; Xiong, B.; Xian, Z. Computer simulation aided preparation of molecularly imprinted polymers for separation of bilobalide. *J. Mol. Model.* **2020**, *26*, 198. [CrossRef] [PubMed]
56. Xie, L.; Xiao, N.; Li, L.; Xie, X.; Li, Y. Theoretical insight into the interaction between chloramphenicol and functional monomer (methacrylic acid) in molecularly imprinted polymers. *Int. J. Mol. Sci.* **2020**, *21*, 4139. [CrossRef] [PubMed]
57. Wang, L.; Yang, F.; Zhao, X.; Li, Y. Screening of functional monomers and solvents for the molecular imprinting of paclitaxel separation: A theoretical study. *J. Mol. Model.* **2020**, *26*, 26. [CrossRef]
58. Silva, C.F.; Borges, K.B.; Nascimento, C.S. Computational study on acetamiprid-molecular imprinted polymer. *J. Mol. Model.* **2019**, *25*, 1–5. [CrossRef]
59. Khodadadian, M.; Ahmadi, F. Computer-assisted design and synthesis of molecularly imprinted polymers for selective extraction of acetazolamide from human plasma prior to its voltammetric determination. *Talanta* **2010**, *81*, 1446–1453. [CrossRef]
60. Wang, W.; Qian, D.; Xiao, X.; Gao, S.; Cheng, J.; He, B.; Liao, L.; Deng, J. A highly sensitive and selective sensor based on a graphene-coated carbon paste electrode modified with a computationally designed boron-embedded duplex molecularly imprinted hybrid membrane for the sensing of lamotrigine. *Biosens. Bioelectron.* **2017**, *94*, 663–670. [CrossRef]
61. Gholivand, M.B.; Torkashvand, M.; Malekzadeh, G. Fabrication of an electrochemical sensor based on computationally designed molecularly imprinted polymers for determination of cyanazine in food samples. *Anal. Chim. Acta* **2012**, *713*, 36–44. [CrossRef]
62. Yu, R.; Zhou, H.; Li, M.; Song, Q. Rational selection of the monomer for molecularly imprinted polymer preparation for selective and sensitive detection of 3-methylindole in water. *J. Electroanal. Chem.* **2019**, *832*, 129–136. [CrossRef]

63. Marc, M.; Panuszko, A.; Namiesnik, J.; Wieczorek, P.P. Preparation and characterization of dummy-template molecularly imprinted polymers as potential sorbents for the recognition of selected polybrominated diphenyl ethers. *Anal. Chim. Acta* **2018**, *1030*, 77–95. [CrossRef] [PubMed]
64. He, C.; Lay, S.; Yu, H.; Shen, S. Synthesis and application of selective adsorbent for pirimicarb pesticides in aqueous media using allyl-β -cyclodextrin based binary functional monomers. *J. Sci. Food Agric.* **2018**, *98*, 2089–2097. [CrossRef]
65. Nezhadali, A.; Mojarrab, M. Computational design and multivariate optimization of an electrochemical metoprolol sensor based on molecular imprinting in combination with carbon nanotubes. *Anal. Chim. Acta* **2016**, *924*, 86–98. [CrossRef]
66. Gao, B.; He, X.-P.; Jiang, Y.; Wei, J.-T.; Suo, H.; Zhao, C. Computational simulation and preparation of fluorescent magnetic molecularly imprinted silica nanospheres for ciprofloxacin or norfloxacin sensing. *J. Sep. Sci.* **2014**, *37*, 3753–3759. [CrossRef] [PubMed]
67. Tiu, B.D.B.; Pernites, R.B.; Tiu, S.B.; Advincula, R.C. Detection of aspartame via microsphere-patterned and molecularly imprinted polymer arrays. *Colloids Surf. A Physicochem. Eng. Asp.* **2016**, *495*, 149–158. [CrossRef]
68. Vergara, A.V.; Pernites, R.B.; Pascua, S.; Binag, C.A.; Advincula, R.C. QCM sensing of a chemical nerve agent analog via electropolymerized molecularly imprinted polythiophene films. *J. Polym. Sci. Part A Polym. Chem.* **2012**, *50*, 675–685. [CrossRef]
69. Nezhadali, A.; Shadmehri, R. Computer-aided sensor design and analysis of thiocarbohydrazide in biological matrices using electropolymerized-molecularly imprinted polypyrrole modified pencil graphite electrode. *Sens. Actuators B Chem.* **2013**, *177*, 871–878. [CrossRef]
70. Khan, M.S.; Wate, P.S.; Krupadam, R.J. Combinatorial screening of polymer precursors for preparation of benzo[α] pyrene imprinted polymer: An ab initio computational approach. *J. Mol. Model.* **2012**, *18*, 1969–1981. [CrossRef] [PubMed]
71. Prasad, B.B.; Rai, G. Study on monomer suitability toward the template in molecularly imprinted polymer: An ab initio approach. *Spectrochim. Acta Part A Mol. Biomol. Spectrosc.* **2012**, *88*, 82–89. [CrossRef]
72. Gholivand, M.B.; Khodadadian, M.; Ahmadi, F. Computer aided-molecular design and synthesis of a high selective molecularly imprinted polymer for solid-phase extraction of furosemide from human plasma. *Anal. Chim. Acta* **2010**, *658*, 225–232. [CrossRef]
73. Ganjavi, F.; Ansari, M.; Kazemipour, M.; Zeidabadinejad, L. Computer-aided design and synthesis of a highly selective molecularly imprinted polymer for the extraction and determination of buprenorphine in biological fluids. *J. Sep. Sci.* **2017**, *40*, 3175–3182. [CrossRef]
74. Azimi, A.; Javanbakht, M. Computational prediction and experimental selectivity coefficients for hydroxyzine and cetirizine molecularly imprinted polymer based potentiometric sensors. *Anal. Chim. Acta* **2014**, *812*, 184–190. [CrossRef]
75. Hasanah, A.N.; Rahayu, D.; Pratiwi, R.; Rostinawati, T.; Megantara, S.; Saputri, F.A.; Puspanegara, K.H. Extraction of atenolol from spiked blood serum using a molecularly imprinted polymer sorbent obtained by precipitation polymerization. *Heliyon* **2019**, *5*, e01533. [CrossRef]
76. Hasanah, A.N.; Soni, D.; Pratiwi, R.; Rahayu, D.; Megantara, S.; Mutakin. Synthesis of diazepam-imprinted polymers with two functional monomers in chloroform using a bulk polymerization method. *J. Chem.* **2020**, *2020*, 7282415. [CrossRef]
77. Tabandeh, M.; Ghassamipour, S.; Aqababa, H.; Tabatabaei, M.; Hasheminejad, M. Computational design and synthesis of molecular imprinted polymers for selective extraction of allopurinol from human plasma. *J. Chromatogr. B* **2012**, *898*, 24–31. [CrossRef] [PubMed]
78. Ahmadi, F.; Rezaei, H.; Tahvilian, R. Computational-aided design of molecularly imprinted polymer for selective extraction of methadone from plasma and saliva and determination by gas chromatography. *J. Chromatogr. A* **2012**, *1270*, 9–19. [CrossRef] [PubMed]
79. Aqababa, H.; Tabandeh, M.; Tabatabaei, M.; Hasheminejad, M.; Emadi, M. Computer-assisted design and synthesis of a highly selective smart adsorbent for extraction of clonazepam from human serum. *Mater. Sci. Eng. C* **2013**, *33*, 189–195. [CrossRef] [PubMed]
80. Salajegheh, M.; Ansari, M.; Foroghi, M.M.; Kazemipour, M. Computational design as a green approach for facile preparation of molecularly imprinted polyarginine-sodium alginate-multiwalled carbon nanotubes composite film on glassy carbon electrode for theophylline sensing. *J. Pharm. Biomed. Anal.* **2019**, *162*, 215–224. [CrossRef] [PubMed]
81. Khan, S.; Hussain, S.; Wong, A.; Foguel, M.V.; Moreira Gonçalves, L.; Pividori Gurgo, M.I.; Taboada Sotomayor, M.D.P. Synthesis and characterization of magnetic-molecularly imprinted polymers for the HPLC-UV analysis of ametryn. *React. Funct. Polym.* **2018**, *122*, 175–182. [CrossRef]
82. Nashar, R.M.E.; Ghani, N.T.A.; Gohary, N.A.E.; Barhoum, A.; Madbouly, A. Molecularly imprinted polymers based biomimetic sensors for mosapride citrate detection in biological fluids. *Mater. Sci. Eng. C* **2017**, *76*, 123–129. [CrossRef] [PubMed]
83. He, H.; Gu, X.; Shi, L.; Hong, J.; Zhang, H.; Gao, Y.; Du, S.; Chen, L. Molecularly imprinted polymers based on SBA-15 for selective solid-phase extraction of baicalein from plasma samples. *Anal. Bioanal. Chem.* **2015**, *407*, 509–519. [CrossRef] [PubMed]
84. Lipkowitz, K.B.; Peterson, M.A. Molecular mechanics in organic synthesis. *Chem. Rev.* **1993**, *93*, 2463–2486. [CrossRef]
85. Karlsson, B.C.G.; O'Mahony, J.; Karlsson, J.G.; Bengtsson, H.; Eriksson, L.A.; Nicholls, I.A. Structure and dynamics of monomer–template Complexation: An Explanation for Molecularly Imprinted Polymer Recognition Site Heterogeneity. *J. Am. Chem. Soc.* **2009**, *131*, 13297–13304. [CrossRef] [PubMed]
86. Alder, B.J.; Wainwright, T.E. Phase transition for a hard sphere system. *J. Chem. Phys.* **1957**, *27*, 1208–1209. [CrossRef]
87. Wormer, P.E.S.; Avoird, A.V.D. Chapter 37—Forty years of ab initio calculations on intermolecular forces. *Theory Appl. Comput. Chem.* **2005**, 1047–1077.

88. Seminario, J.M. Chapter 6—Ab initio and DFT for the strength of classical molecular dynamics. *Theor. Comput. Chem.* **1999**, *7*, 187–229.
89. Boeyens, J.C.A.; Comba, P. ChemInform abstract: Molecular mechanics: Theoretical basis, rules, scope and limits. *ChemInform* **2001**, *32*, 3–10. [CrossRef]
90. Kazemi, S.; Daryani, A.S.; Abdouss, M.; Shariatinia, Z. DFT computations on the hydrogen bonding interactions between methacrylic acid-trimethylolpropane trimethacrylate copolymers and letrozole as drug delivery systems. *J. Theor. Comput. Chem.* **2016**, *15*, 1–20. [CrossRef]
91. Huang, Y.; Zhu, Q. Computational modeling and theoretical calculations on the interactions between spermidine and functional monomer (methacrylic acid) in a molecularly imprinted polymer. *J. Chem.* **2015**, *2015*, 216983. [CrossRef]
92. Li, L.; Chen, L.; Zhang, H.; Yang, Y.; Liu, X.; Chen, Y. Temperature and magnetism bi-responsive molecularly imprinted polymers: Preparation, adsorption mechanism and properties as drug delivery system for sustained release of 5-fluorouracil. *Mater. Sci. Eng. C* **2016**, *61*, 158–168. [CrossRef]
93. Cohen, A.J.; Mori-Sánchez, P.; Yang, W. Challenges for density functional theory. *Chem. Rev.* **2012**, *112*, 289–320. [CrossRef] [PubMed]
94. Kohn, W.; Sham, L.J. Self-Consistent Equations Including Exchange and Correlation Effects. *Phys. Rev.* **1965**, *140*, A1133–A1138. [CrossRef]
95. Dong, W.; Yan, M.; Zhang, M.; Liu, Z.; Li, Y. A computational and experimental investigation of the interaction between the template molecule and the functional monomer used in the molecularly imprinted polymer. *Anal. Chim. Acta* **2005**, *542*, 186–192. [CrossRef]
96. Wu, L.; Sun, B.; Li, Y.; Chang, W. Study properties of molecular imprinting polymer using a computational approach. *Analyst* **2003**, *128*, 944. [CrossRef]
97. Nicholls, I.A.; Andersson, H.S.; Charlton, C.; Henschel, H.; Karlsson, B.C.G.; Karlsson, J.G.; O'Mahony, J.; Rosengren, A.M.; Rosengren, K.J.; Wikman, S. Theoretical and computational strategies for rational molecularly imprinted polymer design. *Biosens. Bioelectron.* **2009**, *25*, 543–552. [CrossRef] [PubMed]
98. Zhang, L.; Chen, L.; Zhang, H.; Yang, Y.; Liu, X. Recognition of 5-fluorouracil by thermosensitive magnetic surface molecularly imprinted microspheres designed using a computational approach. *J. Appl. Polym. Sci.* **2017**, *134*, 45468. [CrossRef]
99. Dong, W.; Yan, M.; Liu, Z.; Wu, G.; Li, Y. Effects of solvents on the adsorption selectivity of molecularly imprinted polymers: Molecular simulation and experimental validation. *Sep. Purif. Technol.* **2007**, *53*, 183–188. [CrossRef]
100. Douhaya, Y.V.; Barkaline, V.V.; Tsakalof, A. Computer-simulation-based selection of optimal monomer for imprinting of tri-O-acetyl adenosine in a polymer matrix: Calculations for benzene solution. *J. Mol. Model.* **2016**, *22*, 157. [CrossRef]
101. Qader, B.; Baron, M.; Hussain, I.; Gonzalez-Rodriguez, J. Electrochemical determination of 2-isopropoxyphenol in glassy carbon and molecularly imprinted poly-pyrrole electrodes. *J. Electroanal. Chem.* **2018**, *821*, 16–21. [CrossRef]
102. Qader, B.; Baron, M.; Hussain, I.; Sevilla, J.M.; Johnson, R.P.; Gonzalez-Rodriguez, J. Electrochemical determination of disulfoton using a molecularly imprinted poly-phenol polymer. *Electrochim. Acta* **2019**, *295*, 333–339. [CrossRef]
103. Peng, M.; Li, H.; Long, R.; Shi, S.; Zhou, H.; Yang, S. Magnetic porous molecularly imprinted polymers based on surface precipitation polymerization and mesoporous SiO_2 layer as sacrificial support for efficient and selective extraction and determination of chlorogenic acid in duzhong brick Tea. *Molecules* **2018**, *23*, 1554.
104. Ayankojo, A.G.; Reut, J.; Boroznjak, R.; Öpik, A.; Syritski, V. Molecularly imprinted poly(meta-phenylenediamine) based QCM sensor for detecting Amoxicillin. *Sens. Actuators B Chem.* **2018**, *258*, 766–774. [CrossRef]
105. Li, H.; He, H.; Huang, J.; Wang, C.-Z.; Gu, X.; Gao, Y.; Zhang, H.; Du, S.; Chen, L.; Yuan, C.-S. A novel molecularly imprinted method with computational simulation for the affinity isolation and knockout of baicalein from Scutellaria baicalensis. *Biomed. Chromatogr.* **2016**, *30*, 117–125. [CrossRef]
106. Pereira, T.F.D.; Da Silva, A.T.M.; Borges, K.B.; Nascimento, C.S. Carvedilol-imprinted polymer: Rational design and selectivity studies. *J. Mol. Struct.* **2019**, *1177*, 101–106. [CrossRef]
107. Kowalska, A.; Stobiecka, A.; Wysocki, S. A computational investigation of the interactions between harmane and the functional monomers commonly used in molecular imprinting. *J. Mol. Struct.* **2009**, *901*, 88–95. [CrossRef]
108. Rebelo, P.; Pacheco, J.G.; Voroshylova, I.V.; Melo, A.; Cordeiro, M.N.D.S.; Delerue-Matos, C. Rational development of molecular imprinted carbon paste electrode for Furazolidone detection: Theoretical and experimental approach. *Sens. Actuators B Chem.* **2021**, *329*, 129112. [CrossRef]
109. Wang, D.; Yang, Y.; Xu, Z.; Liu, Y.; Liu, Z.; Lin, T.; Chen, X.; Liu, H. Molecular simulation-aided preparation of molecularly imprinted polymeric solid-phase microextraction coatings for kojic acid detection in wheat starch and flour samples. *Food Anal. Methods* **2021**, 1–12. [CrossRef]
110. Madikizela, L.M.; Mdluli, P.S.; Chimuka, L. Experimental and theoretical study of molecular interactions between 2-vinyl pyridine and acidic pharmaceuticals used as multi-template molecules in molecularly imprinted polymer. *React. Funct. Polym.* **2016**, *103*, 33–43. [CrossRef]
111. Wu, L.; Zhu, K.; Zhao, M.; Li, Y. Theoretical and experimental study of nicotinamide molecularly imprinted polymers with different porogens. *Anal. Chim. Acta* **2005**, *549*, 39–44. [CrossRef]

112. Ahmadi, F.; Sadeghi, T.; Ataie, Z.; Rahimi-Nasrabadi, M.; Eslami, N. Computational design of a selective molecular imprinted polymer for extraction of pseudoephedrine from plasma and determination by HPLC. *Anal. Chem. Lett.* **2017**, *7*, 295–310. [CrossRef]
113. Han, Y.; Gu, L.; Zhang, M.; Li, Z.; Yang, W.; Tang, X.; Xie, G. Computer-aided design of molecularly imprinted polymers for recognition of atrazine. *Comput. Theor. Chem.* **2017**, *1121*, 29–34. [CrossRef]
114. Baggiani, C.; Anfossi, L.; Baravalle, P.; Giovannoli, C.; Tozzi, C. Selectivity features of molecularly imprinted polymers recognising the carbamate group. *Anal. Chim. Acta* **2005**, *531*, 199–207. [CrossRef]
115. Piletska, E.V.; Turner, N.W.; Turner, A.P.F.; Piletsky, S.A. Controlled release of the herbicide simazine from computationally designed molecularly imprinted polymers. *J. Control. Release* **2005**, *108*, 132–139. [CrossRef] [PubMed]
116. Sergeyeva, T.A.; Piletska, O.V.; Piletsky, S.A.; Sergeeva, L.M.; Brovko, O.O.; El'Ska, G.V. Data on the structure and recognition properties of the template-selective binding sites in semi-IPN-based molecularly imprinted polymer membranes. *Mater. Sci. Eng. C* **2008**, *28*, 1472–1479. [CrossRef]
117. Altintas, Z.; France, B.; Ortiz, J.O.; Tothill, I.E. Computationally modelled receptors for drug monitoring using an optical based biomimetic SPR sensor. *Sens. Actuators B Chem.* **2016**, *224*, 726–737. [CrossRef]
118. Altintas, Z.; Abdin, M.J.; Tothill, A.M.; Karim, K.; Tothill, I.E. Ultrasensitive detection of endotoxins using computationally designed nanoMIPs. *Anal. Chim. Acta* **2016**, *935*, 239–248. [CrossRef] [PubMed]
119. Sergeyeva, T.A.; Piletsky, S.A.; Brovko, A.A.; Slinchenko, E.A.; Sergeeva, L.M.; El'Skaya, A.V. Selective recognition of atrazine by molecularly imprinted polymer membranes. Development of conductometric sensor for herbicides detection. *Anal. Chim. Acta* **1999**, *392*, 105–111. [CrossRef]
120. Liang, S.; Wan, J.; Zhu, J.; Cao, X. Effects of porogens on the morphology and enantioselectivity of core–shell molecularly imprinted polymers with ursodeoxycholic acid. *Sep. Purif. Technol.* **2010**, *72*, 208–216. [CrossRef]
121. Viveiros, R.; Karim, K.; Piletsky, S.A.; Heggie, W.; Casimiro, T. Development of a molecularly imprinted polymer for a pharmaceutical impurity in supercritical CO_2: Rational design using computational approach. *J. Clean. Prod.* **2017**, *168*, 1025–1031. [CrossRef]
122. Feng, F.; Zheng, J.W.; Qin, P.; Han, T.; Zhao, D.Y. A novel quartz crystal microbalance sensor array based on molecular imprinted polymers for simultaneous detection of clenbuterol and its metabolites. *Talanta* **2017**, *167*, 94–102. [CrossRef] [PubMed]
123. Sobiech, M.; Żołek, T.; Luliński, P.; Maciejewska, D. Separation of octopamine racemate on (R, S)-2-amino-1-phenylethanol imprinted polymer–Experimental and computational studies. *Talanta* **2016**, *146*, 556–567. [CrossRef]
124. Erracina, J.J.; Sharfstein, S.T.; Bergkvist, M. In silico characterization of enantioselective molecularly imprinted binding sites. *J. Mol. Recognit.* **2018**, *31*, e2612. [CrossRef]
125. Nascimento, T.A.; Silva, C.F.; Oliveira, H.L.D.; Da Silva, R.C.S.; Nascimento, C.S.; Borges, K.B. Magnetic molecularly imprinted conducting polymer for determination of praziquantel enantiomers in milk. *Analyst* **2020**, *145*, 4245–4253. [CrossRef] [PubMed]
126. Teixeira, R.A.; Dinali, L.A.F.; Silva, C.F.; De Oliveira, H.L.; Da Silva, A.T.M.; Nascimento, C.S.; Borges, K.B. Microextraction by packed molecularly imprinted polymer followed by ultra-high performance liquid chromatography for determination of fipronil and fluazuron residues in drinking water and veterinary clinic wastewater. *Microchem. J.* **2021**, *168*, 106405. [CrossRef]
127. Lu, J.; Qin, Y.; Wu, Y.; Chen, M.; Sun, C.; Han, Z.; Yan, Y.; Li, C.; Yan, Y. Mimetic-core-shell design on molecularly imprinted membranes providing an antifouling and high-selective surface. *Chem. Eng. J.* **2021**, *417*, 128085. [CrossRef]
128. Zhang, Y.; Tan, X.; Liu, X.; Li, C.; Zeng, S.; Wang, H.; Zhang, S. Fabrication of multilayered molecularly imprinted membrane for selective recognition and separation of artemisinin. *ACS Sustain. Chem. Eng.* **2019**, *7*, 3127–3137. [CrossRef]
129. Wang, J.; Guo, R.; Chen, J.; Zhang, Q.; Liang, X. Phenylurea herbicides-selective polymer prepared by molecular imprinting using N-(4-isopropylphenyl)-N'-butyleneurea as dummy template. *Anal. Chim. Acta* **2005**, *540*, 307–315. [CrossRef]
130. Liu, Y.; Wang, D.; Du, F.; Zheng, W.; Liu, Z.; Xu, Z.; Hu, X.; Liu, H. Dummy-template molecularly imprinted micro-solid-phase extraction coupled with high-performance liquid chromatography for bisphenol A determination in environmental water samples. *Microchem. J.* **2019**, *145*, 337–344. [CrossRef]

MDPI
St. Alban-Anlage 66
4052 Basel
Switzerland
www.mdpi.com

Polymers Editorial Office
E-mail: polymers@mdpi.com
www.mdpi.com/journal/polymers

Disclaimer/Publisher's Note: The statements, opinions and data contained in all publications are solely those of the individual author(s) and contributor(s) and not of MDPI and/or the editor(s). MDPI and/or the editor(s) disclaim responsibility for any injury to people or property resulting from any ideas, methods, instructions or products referred to in the content.

www.ingramcontent.com/pod-product-compliance
Lightning Source LLC
LaVergne TN
LVHW082009090526
838202LV00006B/263